# *Wetland Restoration, Flood Pulsing, and Disturbance Dynamics*

**Beth Middleton**
Department of Plant Biology
Southern Illinois University

***John Wiley & Sons, Inc.***
NEW YORK / CHICHESTER / WEINHEIM / BRISBANE / SINGAPORE / TORONTO

This book is printed on acid-free paper. ♾



Published simultaneously in Canada.



***Library of Congress Cataloging-in-Publication Data:***

Middleton, Beth.
Wetland restoration, flood pulsing, and disturbance dynamics / Beth Middleton.
p. cm.
Includes bibliographical references (p. ) and index.
ISBN 0-471-29263-X (cloth)
1. Wetland ecology. 2. Restoration ecology. I. Title.
QH541.5.M3M54 1998
333.91'8153—dc21 98-4512

Printed in the United States of America.

10 9 8 7 6 5 4 3 2 1

*To my father, who gave to me his love of the land*

# *Preface*

Wetland restoration has rapidly emerged as a technology since the movement began to protect wetlands in the early 1970s. To fill the demand for on-site restoration, consulting firms and government agencies have moved into the arena of wetland restoration. At the same time, university researchers have been working on ecological theories of importance to the success of these projects, including the role of flood pulsing and disturbance dynamics in succession and invasion. This book presents an argument for the incorporation of flood pulsing and disturbance dynamics in the planning of wetland restoration projects.

The features presented in this book will be of great value to wetland restorationists, university researchers, and students. The book is divided into three parts with additional invaluable appendices. Part I explores natural disturbances (Chapter 1) by humans in natural wetlands, for example, flood pulsing, debris in rivers, fire, and herbivory to provide a background for discussing the importance of these disturbances in restored wetlands. To build an argument for the importance of water fluctuation in restored wetlands, Part II links the ecological theory related to restoration (Chapter 2) to life history requirements in wetland plant species (Chapter 3). Part III explores the incorporation of disturbance dynamics in restored landscapes via reengineering wetlands and reestablishing vegetation. The information gathered for the reader in this book includes:

- *Flood pulsing* and *disturbance dynamics* in wetlands (Chapter 1), which looks at the emerging ideas of researchers relevant to restoration technology

- Relevant theory related to restoration (Chapter 2), including the overlapping concepts of *self-design, succession*, and *invasion theory* in wetlands
- The all-important and often overlooked *life history requirements* (Chapter 3) of various life stages of wetland plants in relation to disturbance dynamics, particularly water level fluctuations
- Approaches to the *reengineering* (Chapter 4) of disturbance dynamics in restored wetlands
- *Revegetation technology* (Chapter 5), including *natural restoration and flood pulsing*, reseeding, and transplanting techniques in a variety of world wetland types
- Detailed *case histories* of wetland restoration projects (Chapter 6), including the Kissimmee River project (Florida), the Murray-Darling River (Australia), the Ecological Society of India projects (Pune, India), and the Rhine River (the Netherlands/Germany)
- A compendium of water level tolerances and other traits for world plant species including seed germination, seedling recruitment, and adult survivorship
- An extensive *bibliography* of over 1200 works pertaining to wetland restoration, an invaluable tool for world wetland researchers and practitioners
- An appendix on dispersal of wetland species
- An appendix on seed germination requirements
- *Internet* addresses for access to the fast-changing world of wetland technology and information, especially for students, wetland researchers, and restorationists
- An extensive glossary
- Scores of illustrations and photos

This book is the culmination of my years of work and training as a wetland ecologist. As such, there are many people to thank for their efforts in bringing me to this point in my thinking, particularly my thesis advisors, including Virginia Kline, Dave Schimpf, Craig Davis and Arnold van der Valk. Many individuals have supported me in the preparation of the book. My thanks to the staff at Wiley including Neil Levine, Donna Conte, Rose Leo Kish. Also thanks to Mary Grace Luke-Stefanchik and Jane Kinney for help in the initial stages of the book. Perry Rossa of Mead and Hunt, Inc., and Doris Rusch of the Wisconsin Department of Natural Resources were key persons in assembling the restoration planning information for the Lodi Wildlife Area. Special thanks to Kathy Fahey and the librarians and students at Morris Library of Southern Illinois University, who tirelessly helped me locate the considerable volume of literature required to construct this book. I'd like to thank those who reviewed various portions of the book, including Nancy Rorick and Joy Marburgher. I would also like to thank my graduate students and others who filled in for me at times during my sabbatical to allow me time to write this book.

Thanks to Katie Clark for her help in locating Internet sources.

BETH MIDDLETON

# Contents

# *Part I*

# *Restoration and Disturbance: Background*

# 1

# Disturbance Dynamics in Wetlands

*Wetland restoration has come of age since the 1970s, when wetland conservation became important around the world. Since then, many wetlands have been created or restored, typically without attention to the importance of natural disturbance in a landscape setting. With a few notable exceptions, restorationists attempt to produce an unchanging habitat at an equilibrium, with no allowances for the natural disturbance processes that have guided and shaped species across the millennia.*

Wetlands are besieged by disturbances from which they recover over time. While disturbances are important in resetting the wetland cycle, management of restored wetlands often attempts to maintain an unchanging system at an equilibrium, reflecting a critical misunderstanding of the dynamic nature of wetland ecosystems. Disturbances provide opportunities for regeneration, which is important in the maintenance of the regional biota (Main, 1993). Even though water level fluctuation is a normal and regulating process in most wetland types, managers often attempt to maintain static water levels in wetlands, preventing most plant species from regenerating. Throughout the world, this interference has had a profound effect on wetlands (Keddy, 1990).

The perception of natural systems as stable entities may be rooted in human memory and cultural background. A single person's observational lifetime spans only a short period in the disturbance history of an ecosystem (Steedman et al., 1996). Also, culture, religion, and science are often rooted in ideas of steady-state dynamics; such attitudes can be linked to the human desire to sustain stability in ecosystems (Prigogine, 1980). As such, retrospective historical studies can be invaluable in understanding the importance of disturbance dynamics in ecosystems (Steedman et al., 1996).

While greatly disparate definitions exist in the literature, **disturbance** can be defined as "any relatively discrete event in time that disrupts ecosystem, community or population structure and changes resources, substrate availability or the physical environment," including **perturbations,** whether natural or human induced (White and Pickett, 1985).

The incorporation of natural disturbance regimes into restoration management will ultimately improve the craft. Some restored wetland types may be doomed to failure without natural disturbance because these cannot be maintained without disturbance. Disturbance-dependent palm thickets in southern California depend on heavy rains to clear thickets of debris and dead plant material so that the buried seeds of the California fan palm (*Washingtonia filifera*) can reestablish in newly deposited beds of saturated sand (Vogl and McHargue, 1966). In riparian systems, floods allow the renewal of vegetation across floodplains on surfaces exposed by the meandering of watercourses; accretion or subsidence of deltas; and the formation of point bars, islands, and oxbows. The vegetation in the Everglades has been molded by a multitude of recurring disturbances or pulse events including hurricanes, fires, fluctuating water levels, droughts, and frosts. For mangroves, one of the many vegetation types of the Everglades, regeneration is facilitated on mud flats near wattle (debris) dams deposited by storms (Vogl, 1980).

Disturbance-dependent systems often decline in growth and productivity without a repeating cycle of pulse events to help maintain them (Odum, 1969; Loucks, 1970; Fredrickson and Reid, 1990). Without disturbance, some wetland types become dominated by *Typha* or *Phragmites* (Kantrud, 1986). The effects of specific disturbances vary widely (e.g., fire versus windstorm) and can be distinguished from one another quantitatively by their **severity**, as described by their **frequency, duration, intensity**, and **return interval** within particular regional landscapes (White and Pickett 1985; Ehrenfeld and Toth 1997)

Most human disturbances are unlike natural ones because they threaten **stability**, or the ability of the system both to resist change and to return to its preexisting conditions after disturbance that is, its **recoverability** (Rykiel, 1979). After human disturbance, the sudden, vast changes in the environment may be beyond the ability of species to adapt, a trait acquired through long evolutionary exposure to specific natural disturbances (Odum, 1969; Vogl, 1980). Human disturbances can have profound effects on the ecosystem and can persist for long periods of time. Supporters of development may claim that there is no difference between human and natural disturbance. However, consider the differences between beaver and human dams in their spatial and temporal landscape dynamics (Regier et al., 1989).

Another common shortcoming in restoration perspective is viewing a project in isolation rather than as part of a landscape (Risser, 1992; Bell et al., 1997). As will become clear shortly, landscape function (e.g., overbank flow, nutrient cycling, fire dynamics, animal/plant/nutrient flow patterns across boundaries) is often disrupted by fragmentation (Main, 1993). **Landscape reintegration**, or the restoration of functional aspects of the landscape, is an essential aspect of restoration (Risser, 1992; Main, 1993).

Landscapes are interconnected in critical ways that affect the outcome of restoration projects. For example, in situations where a disturbance removes all of the vegetation and the seed bank at a site, intact nearby patches can play a role in the reestablishment of vegetation. In such cases, resilience is possible only via asynchronous patch change in an interconnected landscape (Willard and Hiller, 1990). Also, considering that natural wetlands change regularly, as long as the function of a restored (or natural) wetland continues unimpaired, some change in species composition should not be of concern to managers (Larson et al., 1980; Willard and Hiller, 1990; Mitsch and Gosselink, 1993).

The importance of incorporating ecosystem change into restoration plans is slowly becoming recognized (Box 1-1; Davis & Ogden, 1997). Yet, in many restoration projects, appropriate natural pulses of disturbance such as fire or overbank flow are not part of the long-range management plan. Indeed, managers should proceed cautiously because a disturbance on a restoration site before the vegetation has had an opportunity to become established may endanger the success of the project (Hammer, 1997). Ultimately, failure to use a dynamic outlook regarding the spatial and temporal variability of communities can lead to the long-term failure of restoration projects (Willard and Hiller, 1990; Hobbs and Norton, 1996).

This chapter will consider the importance of specific disturbances (pulse events) in wetlands and their potential for the improvement of restoration projects. The disturbances considered will include flood pulsing, wood in rivers, beaver, fire, grazing, windstorms, and hurricanes, as well as the ways in which humans interfere with these processes.

## DISTURBANCES IN WETLANDS

### Flood Pulsing

**Flood pulsing**, or the idea that the physical and biotic functions of the floodplain wetland (see Box 1-2) are dependent on the dynamics of water

*BOX 1-1*

**RESTORATION WANNA-BEES**

Not every activity done in the name of restoration is restoration in its strictest sense. The literature is rife with papers on "restoration" that would be better referred to as creation, enhancement, or reclamation.

**Restoration** is returning a site to a condition similar to the one that existed before it was altered, along with its predisturbance functions and related physical, chemical, and biological characteristics. The goal of restoration is to establish a site that is self-regulating and integrated within its landscape, rather than to reestablish an aboriginal condition that can be impossible to define and/or restore within the context of current land use or global climatic change. Wetland restoration requires hydrologic and morphological rejuvenation so that the restored site mimics its original environment, which then mirrors the adaptive abilities of the target array of plants and animals. Sometimes restoration requires the chemical cleanup of toxic materials.

**Creation** is establishing an ecosystem that did not originally occupy the site. A prairie established on a forested floodplain is not a restoration.

**Enhancement** is improving the structure or function of an already existing wetland. Generally, enhancements are performed to increase the value of the site for wildlife.

**Reclamation** is altering an ecosystem, which creates another type with some human utility. This term is used both for processes that destroy natural ecosystems, changing them into agricultural or urban uses, and for the amelioration of severely damaged ecosystems (e.g., mining reclamation). (See Magnuson et al., 1980; Cairns, 1988; Lewis, 1989; National Research Council, 1992; Jackson et al., 1995.)

discharged from the river channel, has recently become an important consideration in the restoration of riverine plant communities (Stromberg and Patten, 1988; Scott et al., 1993a; Middleton, 1995a). While the idea was conceived primarily for the purpose of explaining floodplain dynamics, particularly with regard to fish ecology along large tropical rivers (Welcomme, 1974; Junk, 1982; National Research Council, 1982; Junk et al., 1989; Welcomme, 1992), recently the idea has been expanded to include both temperate regions and other aspects of floodplain dynamics (Fenner et al.,

*BOX 1-2*

## WHAT IS A WETLAND?

An often held misconception among the public is that wetlands are difficult to define, a notion fostered by the bickering surrounding proposed development. Wetlands include polyglot terms such as *beng, billabong, bog, fen, ghiol, jheel, hochmoor, khandar, marsh, pokhar, seep, swamp, ti, toiche*, and *varzea* (Maltby, 1986; Gopal and Sah, 1995). The term **wetland** is defined by either of two definitions as

1. "lands transitional between terrestrial and aquatic systems where the water table is usually at or near the surface or the land is covered by shallow water" (Cowardin et al., 1979), or,
2. "areas of marsh, fen, peatland or water, whether natural or artificial, permanent or temporary, with water that is static or flowing, fresh, brackish, or salt, including areas or marine water the depth of which at low tide does not exceed six metres" (Ramsar Convention, 1971, in Roggeri, 1995).

Particular wetlands may become the subject of debate because of their location (e.g., the ideal shopping mall location), unusual characteristics, or regulatory status. The types of wetlands that are the subject of greatest controversy include permafrost wetlands, riparian ecosystems, isolated and headwater wetlands, shallow wetlands, agricultural wetlands, nonagricultural altered sites, and transitional zones (National Research Council, 1995).

Wetlands are not "wet" at all times, so that the designation of a particular site as a wetland requires some skill. To determine if a site is a jurisdictional wetland in the United States (U.S.A.C.E., 1987), a wetland delineation team needs to be armed with a knowledge of wetland indicators, including hydric soils (USDA, 1991; Faulkner and Patrick, 1992), plants (Reed, 1988), hydrology (USACE, 1987; Lyon, 1993), and regional legal qualifiers. Hydric soils are the primary indicator because, even if the soil is seasonally dry or the plants have been removed or are senescent, soils retain clues of past flooding such as colored streaks and blotches, signs of low oxygen conditions that develop in flooded soils. Some soils are wet only briefly, so vegetative and flooding indicators assume more importance (Environmental Defense Fund and World Wildlife Fund, 1992).

*continued*

The actual boundary of a wetland is difficult to pinpoint (Gosselinket al., 1981; Adams et al., 1987; Mulamoottil et al., 1986). This is largely because our need to construct precise property boundaries does not fit the spatial and temporal hodgepodge of environments that make up the real-world ecotones between wetlands and uplands. The logical placement of the wetland boundary is at the outermost limit of the transition zone (National Research Council, 1995).

1985; Rood and Mahoney, 1990; Sparks et al., 1990; Sparks, 1992; Johnson, 1994; Ligon et al., 1995; Middleton, 1995a). Water flow has also been identified as the key hydrologic factor in the management and restoration of nonriverine wetlands such as fens in the United Kingdom (Gilman, 1982) and in the Everglades "river of grass" (Cohn, 1994). This idea has been largely ignored in restoration, the implications of which will become clear as this chapter unfolds.

The flood pulse concept describes the predictable seasonal changes in the ebb and flow of water from the stream channel into floodplains to which the biota are adapted (Junk, 1982; Junk and Howard-Williams, 1984; Junk et al., 1989; Bayley, 1995) and upon which the wetlands are dependent (National Research Council, 1982). The major biological activities of production, decomposition, and consumption in the riverine ecosystem are driven by the floodplain, so this link is critical (Grubaugh and Anderson, 1988; Sparks et al., 1990). Figure 1-1 illustrates the flood pulse concept from a vegetative perspective between the channel and floodplain of riverine cypress swamps in the southeastern United States. In the summer, as water levels draw down, seed germination and seedling recruitment follow the edge of the receding water. The vegetation is most productive at this time. In cypress swamps, most leaves drop within a month or two in the autumn (Brinson et al., 1980; Middleton, 1994a).

In the winter and spring in cypress swamps, flood pulsing is in a high phase due to rain and/or snowmelt and low evapotranspiration (Mitsch, 1977). This flood pulse is associated with the spatial movement of plants, animals, and detrital materials. Fish adapted to the flood pulse follow seasonal water pulses from the channel to the floodplain (Welcomme, 1974; Sparks, 1992), in synch with the spatial movement of organic matter and insects (Bayley, 1995). At this time, 70–100% of the annual growth of fish occurs (Welcomme, 1974). Other types of biota have adapted to the flood pulse. In the Amazon, certain insect species migrate with the flood pulse; juvenile millipedes (*Cutervodesmus adisi* Golovatch) migrate to tree trunks

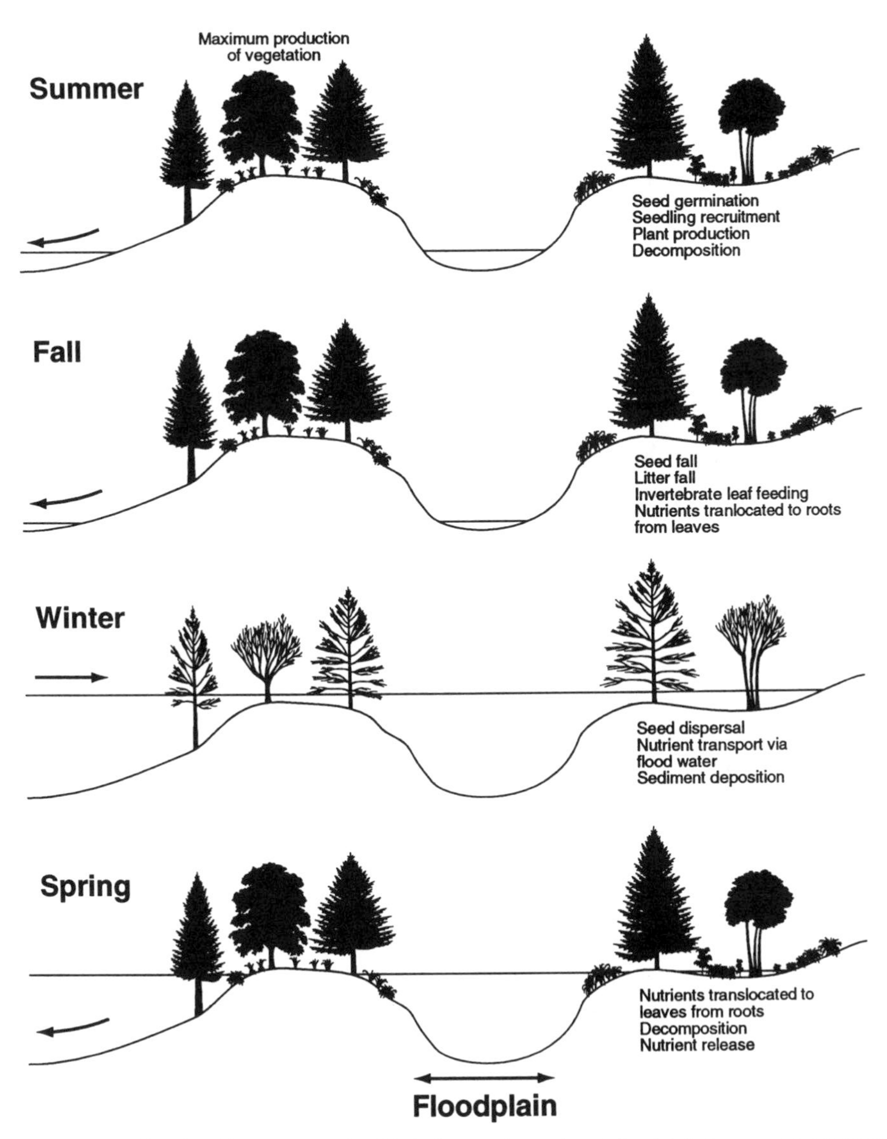


***Figure 1-1.*** *Flood pulsing across a braided stream channel and floodplain in cypress swamps (adapted from Bayley, 1991, as derived from Junk et al., 1989).*

in the flood season and then, during drawdown season, to the forest floor for reproduction (Adis et al., 1996).

Water links detritus and consumers on the floodplain during the flood season. High rates of primary production support consumers, which, in turn, are supported by high rates of decomposition, so that the timing and spatial distribution of detrital decay across the floodplain influence the nature of these relationships (Junk, 1983). In the tropics, decomposition rates are typically high in the early weeks of the aquatic period (Middleton et al., 1992); in temperate areas, decomposition is faster in flooded than in unflooded areas (Shure et al., 1986; Molles et al., 1995). In riverine forests and swamps, as in Amazonian floodplains, most fallen leaf material decomposes in less than one year (Middleton, 1994a; Akanil and Middleton, 1997), so that detritus does not accumulate (Junk and Furch, 1991). Detritivory by invertebrates rapidly breaks down leaves in tropical (Howard-Williams, 1977) and temperate streams (Kirby et al., 1983) and reservoirs (Webster and Simmons, 1978). Because as much as 46% of leaves may become buried, some of the invertebrate detritivory occurs in saturated sediments (Herbst, 1980; Metzler and Smock, 1990; Smith and Lake, 1993). Nutrients are released during the decomposition process and also move during flood times. A study along the Garonne River, in France, demonstrates that, depending on the structure of the channel and floodplain, forests can act as either sources or sinks for nutrients such as C, N, and P (Pinay et al., 1992).

The flood waters also link other components of the system across the floodplain via the movement of hydrochorous **dispersal** of swamp seeds and spores (diaspores) partially decomposed organic leaf material, and sediments. High flood flows are critical to the dispersal and recruitment of cypress swamp species (Middleton, 1995a), western riparian species such as velvet mesquite (*Prosopis velutina*) (Stromberg et al., 1991), and cottonwood (Scott et al., 1997).

In the spring, as water levels recede in cypress swamps, seeds are stranded at the highest water levels and then germinate as water levels draw down successively across elevations. With protracted drawdown, regeneration of woody species such as cypress (Middleton, in review), herbaceous species (Middleton, 1995a,c), and liverworts (Conrad, 1997) can occur. The flood pulse is essential in the functioning of floodplain wetlands.

The flood pulse concept, with some adjustment, can potentially be applied to dryland situations in which the pulses are not seasonally regular but instead erratic. Species along rivers in such climates also develop particular life history strategies to adapt to the complicated nature of an unpredictable flood pulse, that is, variable duration, magnitude, and timing. Unfortunately, while the flood pulse concept remains largely descriptive, there is a need both for models to describe the potential impacts of regulation (Walker et al., 1995) and for templates of restoration.

Some authors have not regarded flood pulses as disturbances (Bayley, 1995); however, these fit the definition from a vegetative perspective. Flood pulses in riverine wetlands, as well as water level fluctuations in other wetland types, produce annual and interannual forces that destroy plants and subsequently allow regeneration from buried seed (Harris and Marshall, 1963; van der Valk, 1981; Keddy and Reznicek, 1986a; Finlayson et al., 1989; Leck, 1989; Keddy, 1990; Middleton et al., 1991). Flood pulsing along rivers is important in the regeneration and maintenance of dominant riverine species in monsoonal wetlands in Australia and India (Figs. 1-2a, b; Finlayson, 1991a, b; Finlayson et al., 1989; Middleton et al., 1991, Midleton, accepted). In North America, alterations to the natural pulsing of water have led to regeneration problems in cottonwood (Rood and Mahoney, 1990; Scott et al., 1993b; Johnson, 1994; Stromberg and Patten, 1996; Scott et al., 1997), willow (Stromberg and Patten, 1988), and cypress forests (Middleton, unpublished).

More attention should be paid to the importance of flood pulsing and disturbance in restoration. These ideas are completely ignored in the regulatory process. The focus in mitigation is on obtaining surface saturation for a limited number of days during the growing season and then planting the right types of plants to obtain ''in kind'' replacement (Nancy Rorick, personal communication).

## Disruptions of the Flood Pulse

The intentional regulation of **water regimes** (annual and long-term fluctuations in water levels in wetlands; Cowardin et al., 1979) or, from a riverine perspective, the pattern of water discharge or flow in the river (Sparks, 1992) is accomplished by the construction of dams, reservoirs, or levees; diversion for irrigation; and manipulation of banks along rivers and streams (Gregory, 1977). In large rivers, engineering projects may have unanticipated impacts because the dynamic link between floodplains and rivers by the flood pulse, as well as their far-reaching and long-term consequences, are not fully appreciated prior to project construction (Sparks, 1992). The restoration of riverine habitats requires a reversal of these impacts to reintegrate the channel with its floodplain (Ward and Stanford, 1995a).

Human habitat alterations that involve the reengineering of the river and floodplain are referred to as **press events. Pulse events** are cyclic or periodic features of natural systems more limited in scope and defined in duration than press events (Vogl, 1980) and sometimes include human disturbances from which ecosystems quickly recover, such as chemical spills (Niemi et al., 1990; Allan and Flecker, 1993). While recovery from press events may require decades without human intervention, ecosystems recover more quickly from pulse events (Niemi et al., 1990; Allan and Flecker, 1993).

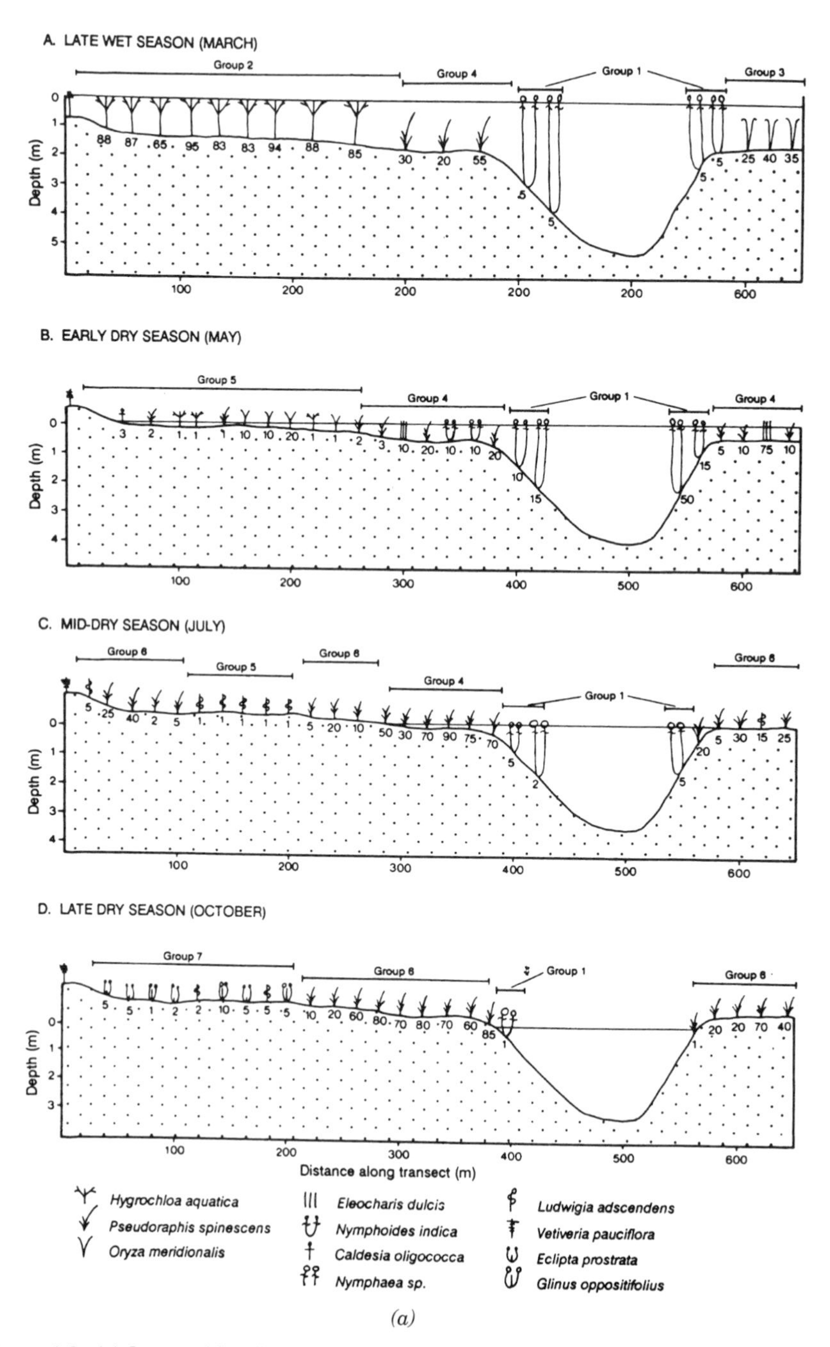


***Figure 1-2.*** *(a) Seasonal flooding and vegetation patterns. (b) Twinspan dendrogram of species and eigenvalues for macrophytic vegetation along the Magela Creek, Northern Territory, Australia (Finlayson et al., 1989; copyright © by Australian government by permission).*

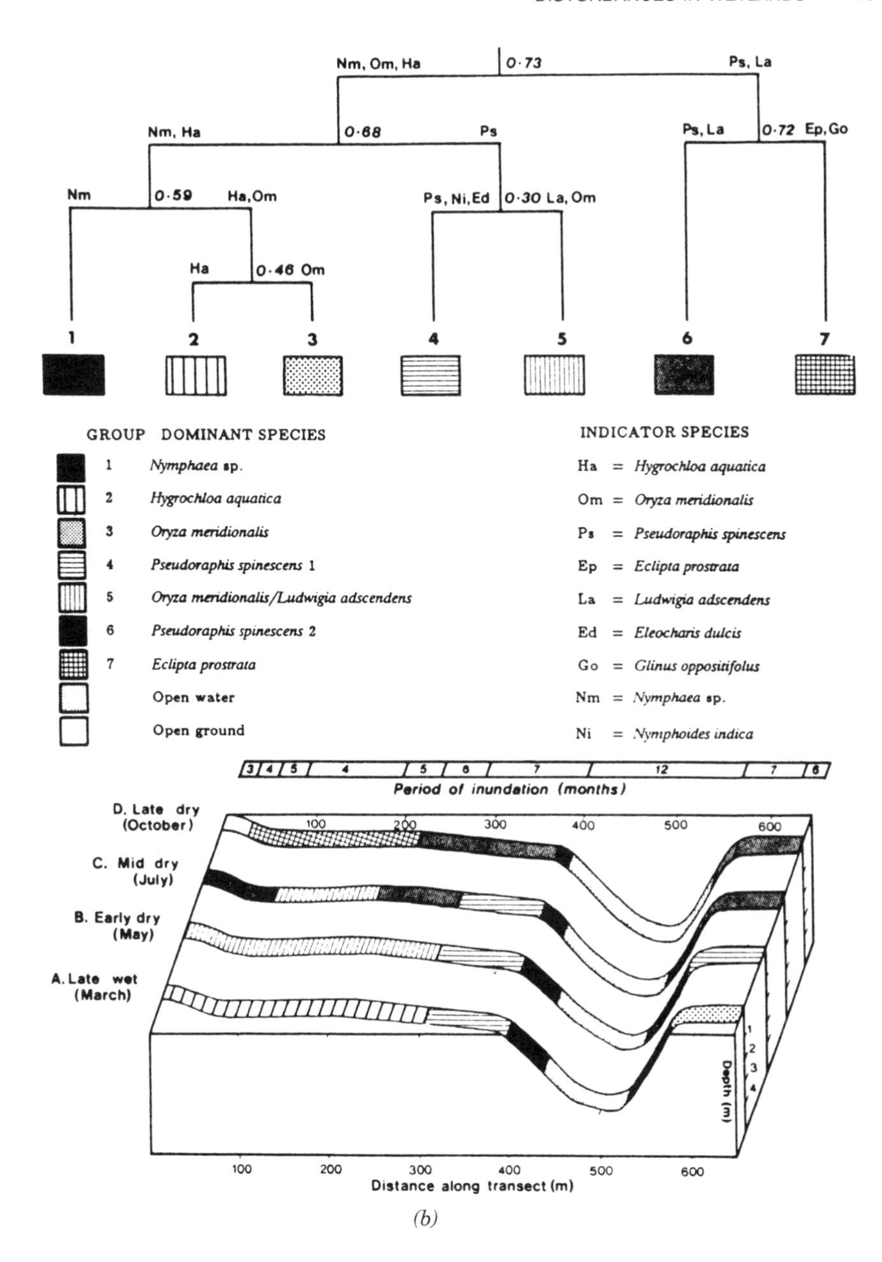


(b)

**Dams** A dam in a river changes the river's biological processes in conjunction with water flow, sedimentation, nutrient cycling, and energy exchange (Sparks et al., 1990; Ligon et al., 1995). The impoundment of water in reservoirs to create hydroelectric power, urban water supplies, and other uses has left few of the world's waterways untouched.

*BOX 1-3*

**ANCIENT ENGINEERS**

While a few dams date back to early civilizations, most of the world's dams were built after the beginning of the Industrial Revolution, 85% of them after 1950 (Smith, 1971). Ancient dams were constructed for reasons that parallel those of today, including rechanneling water for irrigation, navigation, or drinking; storing water for seasonal drought; and protecting settlements against floods. More recently, dams have been used to produce hydroelectric power (Smith, 1971; Schnitter, 1994). Dams are eventually abandoned due to sedimentation or catastrophic bursting.

Warfare over dams has occurred since ancient times. More recently, Serbia allegedly blew up Croatia's Peruca Dam in a cease-fire zone on January 28, 1993. Millions of liters of water rushed into the Cetina Valley below the dam after one of its regulation towers and a 12-m-wide hole opened in the dam itself (Roche, 1993).

A few ancient dams were constructed thousands of years ago; a large dam was built during the Pyramid Age by the Egyptians in about 2600 BC (Box 1-3; Smith, 1971; Costa, 1988; Schnitter, 1994). Most dam construction occurred after the beginning of the Industrial Revolution, with 85% of the world's dams constructed after 1950 (Schnitter, 1994). Currently, impoundments regulate the flow of a large percentage of rivers in Africa, North America, Europe, Asia, South America, and Australasia (20, 20, 15, 14, 6, and 4%, respectively) (Petts, 1984). Estimated dam numbers in the United States vary between 68,000 and 80,000, with 17% of 3.5 million miles of natural rivers dammed (Echeverria et al., 1989).

The impoundment of waters with dams has changed the natural dynamics of flood pulsing along rivers and their floodplains by altering channel characteristics, habitat availability, flow regime, and movement of migratory species such as fish (Ward, 1978). Wetland function is determined less by the presence of water as by its dynamic properties. River alterations including water stabilization, shifted flood timing, and increased or decreased flooding have taken a tremendous toll on riverine communities (Klimas, 1988). The loss of biodiversity along many streams and rivers is attributable to the inability of species adapted to flowing waters to live in the altered

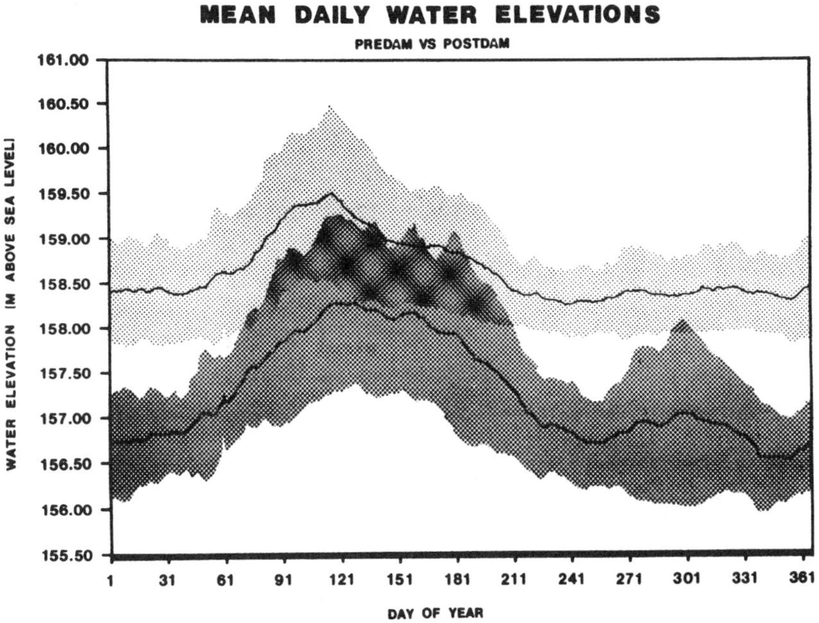

*Figure 1-3.* Pre- and postdam hydrograph generalized for the years before and after the construction of the Lock and Dam 19, upper Mississippi River, Burlington, Iowa upper curve represents before dam construction, the lower after, (from Grubaugh and Anderson, 1988, in Sparks et al., 1990; copyright © by Environmental Management by permission).

environments produced by dams and reservoirs (Hackney and Adams, 1992; Powers et al., 1996; Williams, 1997).

Dams change the peak flow and discharge rates of water (Petts, 1977). After dam construction, mean water elevations below the dam are typically but not always lower. In the case of the flood pattern at Burlington, Iowa (Fig. 1-3), the dam is part of a sequence of dams designed to maintain constant water levels for navigation (Fig. 1-3; Grubaugh and Anderson, 1988). The result of this water control activity has been to reduce the amplitude of the flood pulse and also to cause permanent flooding on the floodplain and a reduction of primary and secondary production (Sparks et al., 1990). The larger the dam, the less the river reflects the predam regime of discharge, water temperature, and sediment transport, that is, the more unnatural the river becomes (Petts, 1977).

The effects of impounding water can be divided into upstream in-channel, downstream in-channel and floodplain effects (Fig. 1-4). After construction, the reservoir becomes a long, multibranched water body that is deepest adjacent to the dam (Darnell et al., 1976). The major effects of impoundment in the reservoir are increased volume of water and constant

Dam Effects

| Class | Effects: In-Channel Upstream | Effects: In-Channel Downstream | Effects: Floodplain Downstream |
|---|---|---|---|
| **Flow Regime** | | | |
| ► water volume | ► increased volume | ► altered volume; flow variability increased<br>► flow increased if channel straightened | ► regular inundation eliminated |
| ► temporal distribution | ► constant inundation | ► altered flows; typically increased volume during normal low flow, decreased volume in high flow | ► decreased duration of inundation (when present) |
| **Water Quality** | | | |
| ► temperature | ► thermal stratification develops; varies from pre-dam temperature, warmer in fall | ► altered temperature regimes; lowered temperatures in summer, warmer in fall | ► altered if flooded |
| ► dissolved oxygen (DO) | ► altered diurnal cycle or lower DO | ► lower DO | ► altered if flooded |
| ► sedimentation | ► increased | ► decreased | ► increased if farmed |
| ► turbidity | ► increased | ► decreased | ► increased if farmed |
| ► nutrients (nitrogen and phosphorus from ag) | ► increased | ► increased | ► increased |

***Figure 1-4.*** *Impoundment effects on the characteristics of floodplain, and upstream and downstream in-channel habitats (adapted from Petts, 1984, Karr et al., 1986, and other sources).*

Dam Effects

| Class | In-Channel Upstream | In-Channel Downstream | Floodplain Downstream |
|---|---|---|---|
| | Effects | | |
| **Nutrient Dynamics** | | | |
| input source | altered; allochthonous instream input while formerly autochthonous streambank vegetation in small streams/rivers | autochthonous, cut off from floodplain sources | autochthonous, cutoff from past role as source of organic matter for channel |
| temporal pattern of input | little end of growing season input | altered somewhat | totally cut off from predam role |
| volume of input | reduced | reduced, depending on robustness of vegetation after alteration | negligible |
| decomposition | altered species, rate and timing | altered species, rate and timing | altered species, rate, timing and interconnection with channel cutoff |
| **Habitat Profile** | | | |
| substrate | increased sedimentation | channel erosion and streambank instability | floodplain dynamics controlled by flooding; sedimentation altered |
| water depth | deeper | altered; typically shallower in wet season, deeper in dry season | no water |
| vegetation | | | |
| type | changes from riparian to submersed and algal | highly variable; sometimes changes to flood-intolerant species | eventually changes to flood-intolerant speices |
| production | decreased | decreased | decreased |
| woody debris | decreased | decreased | decreased |
| dispersal | decreased | decreased | decreased |
| channel dynamics | channel eliminated | pools and riffles altered<br>channel often straightened | n/a |

***Figure 1-4.*** *(Continued)*

Dam Effects

| Class | Effects: In-Channel Upstream | Effects: In-Channel Downstream | Effects: Floodplain Downstream |
|---|---|---|---|
| **Biotic Considerations** | | | |
| trophic structure | altered species composition and abundance | altered species composition and abundance | altered species composition and abundance |
| vegetation composition | riparian species eliminated | altered | often, flood-intolerant species replace flood-tolerant ones; flood-tolerant do not recolonize well if altered by humans |
| fish ecology | shift to lotic species | population numbers decline | major role as fish nursery and spawning ground eliminated |
| insect functional group (scrapers, shredders, etc.) | shift to lotic | altered species composition and abundance | altered species composition and abundance; doesn't recolonize well if altered by humans |
| human encroachment | fishing/boating in reservoir | logging, grazing | logging, grazing, farming, urbanization |

***Figure 1-4.*** *(Continued)*

inundation, which result in the total replacement of the riparian vegetation by submerged or algal communities. Reservoirs trap up to 95% of the sediment load transported by the river (Leopold and Maddock, 1954), and, not surprisingly, sedimentation is the primary reason that dams are eventually abandoned (National Research Council, 1992).

Downstream of dams and reservoirs, peak flows are diminished (bankfull discharge); this is the one factor that disrupts channel processes the most. The periodicity of peak flow varies, but in the United Kingdom it has a return interval of 1.5 years (Petts and Lewin, 1979). The in-channel effects of dams vary somewhat, depending on the type of structure (e.g., large or small hydroelectric dam, or dams for irrigation or water supply) and the climatic situation, as these both affect the volume and variability of flow from the dam (Petts. 1984; Williams and Wolman, 1984). Channel metamorphosis occurs because of the altered relationship of water discharge and sediment load (Schumm, 1969; Kellerhals and Church, 1989).

Because sediments are trapped behind the dam, sediments transported downstream from the dam decrease from pre-dam conditions (Joglekar and Wadekar, 1951; Williams and Wolman, 1984; Muñoz and Prat, 1989), leading to channel scouring, that is, downcutting and erosion in the channel (Petts and Lewin, 1979; Hickin, 1983; Petts, 1984). Downstream from a hydroelectric dam built along the Oconee River on the coastal plain of Georgia in the 1950s, the river has downcut the stream bed. As a result, the inundation of the floodplain has been reduced because the lateral migration of the flood pulse has been hampered by the incision of the channel. Channel incision results in the simplification and stabilization of braided channels. The best approach to remedy this problem is to reengineer predam geomorphologic processes rather than adding structures to the channel (Ligon et al., 1995).

Downstream of hydroelectric dams, flow is typically higher during the normal low flow of the dry season and lower during the normal high flow of the wet season (Kirchner and Karlinger, 1983; Hadley et al., 1987; Muñoz and Prat, 1989) because of controlled releases from dams or return flow from irrigation (Williams and Wolman, 1984). In temperate climates in the summer, the water in the reservoir is cooler (depth related) than under predam conditions, so that the water released downstream is also cooler (Petts, 1984); in the fall and winter, the water in the channel downstream is warmer (Ridley and Steel, 1975). Deeper reservoirs alter water temperatures more than shallower ones, with marked decreases in annual and diel temperature variation (Ward and Stanford, 1985a). Dissolved oxygen is usually lower downstream from a reservoir than in streams without reservoirs (Camargo and de Jalon, 1990), but the peaking of hydroelectric operations causes fluctuations (Dwyer and Turner, 1983). In altered streams and rivers where flood pulsing is managed to mimic natural flow variation, the biota of the stream is benefited (Powers et al., 1996).

The vegetation response to the changes in stream flow below dams is highly variable. Riparian vegetation eventually can increase by up to 95% 30 years after the construction of a dam (Williams and Wolman, 1984). Because flood frequency was reduced along the Colorado River downstream from the Grand Canyon Dam, marshes formed there (Stevens et al., 1995). In contrast, after damming and/or diversion, white willow forests along the upper Rhine (Dister et al., 1990) and *Populus* (poplar) forests in the western United States collapsed (Stromberg and Patten, 1988; Rood and Mahoney, 1990; Milhous, 1994), except in situations with adequate moisture (Johnson W. C. 1994).

Downstream from dams, vegetation often changes to flood-intolerant species. In Australia, exotic weed invasions are likely to occur along rivers with altered and low flow frequencies (Chesterfield, 1986, in Bren et al., 1987). Also, any change in the annual regime of water flow results in the restructuring of the invertebrate community (Cummins, 1979).

The effect of dam construction is most profound on the floodplain. Dams eliminate the regular inundation of the floodplain and thus cut the link between the stream channel and the floodplain via flood pulsing for the movement of biota (fish, seeds, insects), resulting in poor recolonization (Middleton, 1995a); (Middleton, in review). If flooding does occur on these impacted floodplains, it normally has a shorter duration than in pre-dam conditions. From a biotic perspective, the floodplain is nearly cut off from its predam role as a source of organic matter for riverine fishes. After damming, human land usage (e.g., logging, grazing, farming, urbanization) usually expands on the floodplain (Box 1-4).

While dams provide many useful human services, the arguments against building any new dams intensify. In developing countries, the advantages brought by new dam construction, such as hydroelectric supply, water for drinking and irrigation, and tourism, are possible only after people are displaced from the intended reservoir site (Bardach, 1972; National Research Council, 1982). In the projects at the Volta and Aswan dams in Africa, the uprooted people were so unhappy with the arrangements provided (smaller houses built of different materials) that they left the new site. Dams in the tropics create slow-moving, stagnant water, suitable environments for disease vectors such as snails, mosquitoes, and tsetse flies, as well as exotic nuisance plants (Msangi and Ellenbroek, 1990).

If dams break, sudden, violent releases occur (Costa, 1988); the Buffalo Creek disaster of 1972 and the Teton Dam failure of 1976 were catastrophic (Federal Emergency Management Agency, 1993, and National Research Council, 1992, respectively). Dam failure is a product of poor construction,

*BOX 1-4*

**THE FLOODPLAIN DRIES OUT**

Much of the human suffering due to floods in the world is caused by encroachment of the floodplain. Disastrous floods displacing thousands of people in recent years have overflowed onto occupied floodplains on nearly every continent, from rivers from the San Joaquin in the United States to the Brahmaputra/Ganges/Meghna in Bangladesh (Brammer, 1990).

The World Bank and the Bangladeshi government are planning to engineer a drying out of the floodplain in Bangladesh (Brammer, 1990), a country that lies almost entirely on a delta (Winkley et al., 1994). Some say that floods are worsening in Bangladesh because deforestation has increased in the Himalayas since 1950 and/or because sea levels are rising off the coast due to the global warming associated with the greenhouse effect. However, erosion due to deforestation has not increased since the 1950s (Ives, 1989, 1991), and there is little evidence of sea elevations rising (Rogers et al., 1989; Brammer, 1990).

The annual floods, which typically cover 20% of Bangladesh, have economic benefit in that they renew the soil; record crops grew there in the moisture, fertile soil, and algae bestowed by the large floods of 1987 and 1988 (Ives, 1991). However, while the floods renew the soil, they make capital investment difficult (Pearce, 1991).

The recommendations of the Eastern Waters Study for flood alleviation in Bangladesh include the construction of upstream water storage and hydroelectric dams, limited embankment, underground storage for dry season crops, and drainage improvement. This group also recommends a study of how the local people currently "flood proof" or live with the floods—for example, flood warning systems, raised roads, and refuges.

High embankments, the higher the more dangerous, pose a hazard to poor people who live directly behind them. When the British arrived in Bangladesh in 1757, the Bangladeshis already had a low flood embankment system in place. Today the people are carrying off the clay in the embankments to build the burgeoning cities and towns (Pearce, 1991).

There is growing grass-roots opposition to the flood action plan. At least 5 million people will be displaced. The project may endanger the

*continued*

country's fisheries because most of the fish in the rivers there spawn and rear young on the floodplain, and embankments would block the access of the fish. Furthermore, 80% of the protein consumed in Bangladesh is derived from fish (Sklar, 1993).

At least some Bangladeshi women appreciate the flood season as a convenient time to visit family members by boat. As the flood waters rise over the floors of their homes, the women still cook atop their beds but their otherwise prodigious work schedule is curtailed (Shaw, 1992, in Roggeri, 1995).

old concrete (lasts for 50–100 years; National Research Council, 1992), or undersized materials (Eckenfelder, 1995).

While dams provide flood control under normal conditions, in large flood events huge quantities of water must be released (Cairns and Palmer, 1993), often onto altered floodplains with poor water storage capacity. Land drainage for agriculture also increases the peaks of flood flow (Bailey and Bree, 1981).

Dams present other difficulties. Power plants create flow fluctuations that can endanger downstream river recreationists and result in bank erosion. While the public views hydroelectric dams as permanent electricity-producing stations, after the reservoir fills with sediment (and sometimes hazardous chemicals), the dam is no longer useful for hydropower generation. Yet, in the United States, there are no bonding funds to provide for ecological restoration in the predictable event of dam decommissioning.

***Dam Removal*** In the United States, dams are licensed for 50 years only (National Research Council, 1992), so that many that were built earlier in the twentieth century are coming up for relicensing. In Sweden, many 30-year-old dams will be up for relicensing soon. Studies of the effects of dams there have shown permanent changes in the biodiversity of riverbanks and concerns for the impacts of dams have already led to a decision to increase flow through the hydroelectric dams by 5% (Williams, 1997).

Many of the 75,000 dams in the United States built between 1900 and 1949 (Federal Emergency Management Agency, 1993) are unsafe and/or no longer useful. Some of these will be removed (Table 1-1). However, the literature on reservoir restoration is scant and discouraging. Cairns (1988) suggests that restoration to prealteration conditions is usually impossible but that a self-sustaining system can sometimes be created. It is disheartening that, because there are so many environmental problems associated with dam removal, defunct or problematic dams are often retained (Shuman, 1995).

**TABLE 1-1. Past and Proposed Dam Removals in the United States**

| Dam | River | State | Proposed or Date of Removal |
|---|---|---|---|
| Washington Water Power Dam | South Fork of the Clearwater River | Idaho | Removed in 1962 |
| Grangeville and Lewiston Dams | Clearwater River | Idaho | Removed in 1963 |
| Sweesy Dam | Mad River | California | Removed in 1969 |
| Newaygo Dam | Muskegon River | Michigan | Removed in 1969 |
| Fort Edward Dam | Hudson River | New York | Removed in 1973 |
| Woolen Mills Dam | Milwaukee River | Wisconsin | Removed in 1988 |
| Columbia Falls Dam | Pleasant River | Maine | Removed in 1989 |
| East Machias Dam | Not given | Maine | Removed ~ 1987–1992 |
| Temaquit Dam | Not given | Maine | Removed ~ 1987–1992 |
| Salling Dam | AuSable River | Michigan | Removed in 1992 |
| Edwards Dam | Kennebec River | Maine | Proposed |
| Elwha and Glines Canyon Dams | Elwha River | Washington | Proposed |
| Rodman Dam | Ocklawaha River | Florida | Proposed |
| Savage Rapids Dam | Rogue River | Oregon | Proposed |
| Elk Creek Dam | Rogue River | Oregon | Proposed |
| Condit Dam | White Salmon River | Washington | Proposed |
| Kettle River Dam | Not given | Minnesota | Proposed |

*Source:* Shuman (1995) and Anonymous (1992).

An alternative approach to dam removal for restoring a flood pulse in rivers is to provide controlled releases of water to simulate flood flow conditions. To simulate spring flooding, a surge of water was released in March 1996 from a federal hydroelectric dam built in 1963, 15 miles above the Grand Canyon (Stevens 1997; Vaselaar, 1997). While it did not wash away exotic carp, catfish, and tamarisk, as hoped, the simulated flood did build new sandy beaches, bars, and backwaters. While this procedure represents both a political and a policy breakthrough, the government calculated a $1.8 million cost in lost electrical generating capacity at the dam (Stevens, 1997).

Dam removal almost always results in the release of sediment plumes in the downstream reach (Stocker and Williams, 1991). When the Newaygo Dam along the Muskegon River, Michigan, was removed, 40% of the impounded sediment was immediately released downstream. The sediment wave moved at the rate of 1 mile per year for the first 9 years; it is predicted that it will flush out of the system in 80 years (Simons, 1991).

Sediments held in dams sometimes are chemically contaminated; the sediments held by the Fort Edwards Dam along the Hudson River, New York, were contaminated with PCBs and have complicated the 1973 removal of the dam (Tofflemire, 1986). Today, the PCBs have moved downstream in the river and have volatilized from exposed sediments. The PCBs have broken down much more slowly than anticipated, and it is feared that the poor reproductive success of eagles along the river is attributable to this contamination (Revkin, 1997).

The negative consequences of dam removal can be so substantial that other alternatives are pursued. Along the Elwha River in Washington, hydroelectric dams have interfered with the movement of salmon, but the dams are estimated to cost $64 million to remove. As part of the dam removal proposal, tunnels could be built to divert the water from the river while sediment is removed from the reservoirs. Fish passages are being considered as a cheap alternative to dam removal (Stocker and Williams, 1991).

At least one successful dam removal along a formerly impounded portion of the Milwaukee River near West Bend, Wisconsin, sets a positive precedent for future dam removals (National Research Council, 1992). The Wisconsin Department of Natural Resources (WDNR) ordered the removal of the Woolen Mill Dam and the subsequent restoration of the impoundment for reasons of public safety (Nelson and Pajak, 1990). The WDNR involved the community in a ten-year project to remove the dam and to restore a flowing river with a riparian zone in the gaping space that was once the 1.5-mile former reservoir. Within six months of the start of the project, most of the silt and sand that had filled the reservoir had moved downstream, leaving a coarser substrate and an improved fish habitat. Fisherman now catch smallmouth bass, walleye, and northern pike there. While long-term monitoring is necessary to determine the biological impact, the project is a qualified success (Nelson and Pajak, 1990).

Dam removal is nearly impossible after people have moved onto the floodplain. The disaster left by Hurricane Fran in 1996 fully demonstrated the extent to which people have altered floodplains (Lathbury, 1996). Even after the devastation following a flood, people are not likely to move off of the floodplain (Mitchell, 1974). While the National Flood Insurance Act of 1968 encouraged human occupation, it also restricted new development and made provisions for the conservation and restoration of floodplains (National Research Council, 1992; Lathbury, 1996).

Destructive floods can provide the catalyst for rethinking urban and agricultural usage of the floodplain. The sensible people of both Valmeyer, Illinois, and Rhineland, Missouri, voted to move to higher ground after their towns were destroyed by the Mississippi River floods of 1993–1994 (Cortner, 1993, and McGuire, 1997, respectively). Planning for the restoration of wetlands, the elimination of levees, and changes in flood insurance policy is occurring following these floods (Interagency Floodplain Management Review Committee, 1994; Hey and Philippi, 1995; Upper Mississippi River Summit, 1996). After the 1997 flood in Yosemite National Park, discussions began to rebuild the park's infrastructure to accommodate future floods (Golden, 1997).

***Channelization*** From an engineering standpoint, **channelization** is a broad category of channel alteration, the consequences of which are little studied from an ecological perspective. Among the categories of channelization are

1. **resectioning** by widening or deepening the channel,
2. **realignment** or shortening the channel via an artificial cutoff,
3. **diversion** or diverting flow around an area to be protected,
4. embankment with a linear levee, bund or dike to prevent channel water from overflowing onto the floodplain,
5. **bank protection** or channel stabilization,
6. channel lining (concrete in urban areas),
7. culverts and dredging, and
8. cutting and removal of channel obstructions to reduced the roughness and thus increase water flow rate through the channel (Brookes, 1985.

Channel change by human engineering is referred to as *direct change*, whereas *indirect, human-induced changes* include afforestation, deforestation, precipitation modification, road construction, interbasin transfer, urbanization, and reservoir construction (Park, 1977).

Channelization results in long-term changes to the morphology of the stream and subsequent stream instability (Shields and Hoover, 1991) because decreased sinuosity results in sharper pulses of flow (Brinson et al., 1981b), along with increased sediment transport (Hickin, 1983). In the United Kingdom, 35% of the upland rivers showed instability from 1870 to 1950, much of it attributable to changes in land use (e.g., dams, channelization). The maximum erosion rate was 2.8 m per year for two to three years along the River Exe, Devon (Hooke and Redmond, 1989). Twenty-mile Creek, Mississippi, was channelized in 1910, 1938, and 1966. After 1966, the channel bed degraded by 2–4 m, and its cross section enlarged 1.4–2.7 times. Weirs were constructed along the channel in the 1980s to restore stability (Shields and Hoover, 1991). Similarly, after channelization, channel widening occurred along the Cimarron River in southwestern Kansas because of high peak floods and below-average precipitation (Schumm and Lichty, 1963). Along some portions of the tributaries of the Cache River in southern Illinois, downcutting has followed channelization and has contributed sediments to the river (Sengupta, 1995). While rehabilitation of floodplain is very difficult where downcutting has occurred, at least in the Big Creek portion of the Cache River, remeandering of the channel is being considered (U.S.A.C.E., 1996). Channelized rivers often become somewhat more sinuous over time (Hooke and Redmond, 1989), but channelization greatly impairs the ability of the river to migrate across the floodplain and few channels can regain their original sinuosity after straightening (Downs, 1994).

As urban areas expand, the natural hydrology of the region is completely reorganized to accommodate it. Discharge into rivers and streams is higher near urban areas after the drainage system is revised to rid streets of storm water (Newson, 1986). Modern suburbs have large drainpipes and culverts that lead directly to streams. Generally, there is more massive erosion around trees and structures near cities. Flooding near cities has been linked to channel modifications, but little effort has been made to study this impact (Roberts, 1989).

Indirect changes to the watershed affect stream flow. Logging increases runoff and flow and decreases water quality. The effects of logging on stream morphology can be quite direct when the stream channel itself is altered to facilitate the movement of logs (Meehan et al., 1985). Water retention and flood abatement vary with land use; they are highest in conifer forests and lowest in agricultural land (Ward, 1978). Along a stream in South Africa, in-stream flow increased after clear cutting for two years, but there was no increase in sedimentation (Scott and Lesch, 1996). Wetland drainage in the watershed typically increases streamflow, but the magnitude of this change varies by locality (Newson, 1994). For example, wetland drainage increased streamflow in northern, central, and southern Illinois by 33, 50, and 70%, respectively (Demissie and Khan, 1991).

Because channelization disrupts the horizontal sinuosity of the stream channel, the types and functions of the species change as the habitat becomes more homogeneous. Along channelized first-order streams, bank cover is eliminated, so the amount of allochthonous organic input in the stream is diminished (Neuhold, 1981).

The effects of channelizing the Kissimmee River north of Lake Okeechobee have disrupted the Everglades by reducing the amount of water flowing through the "river of grass" (a term coined by Douglas 1947). In the 1960s, the 106-mile Kissimmee River became a 52-mile canal (Cohn, 1994).

Because channelization can affect the ability of the channel to migrate, the regeneration dynamics of many floodplain species is hampered. Lateral channel movement creates point bars and oxbows (Shankman, 1993), which are suitable surfaces for plant regeneration. Cypress, black willow, and tupelo establish along abandoned meanders and oxbow lake margins due to drawdown (Fig. 1-5; Shankman and Drake, 1990).

***Embankment*** Levees, dikes, and bunds effectively cut off the transfer of flood water between the river channel and floodplain, except in dramatic

***Figure 1-5.*** *Cypress swamp in southern Illinois (photograph by Beth Middleton).*

flood events when these structures are breached. These **embankments** are built to dry out the floodplain and reduce flooding on the floodplain (see Box 1–4), but regionally, these force the water from natural storage sites on the floodplain and increase flooding downstream (Leopold and Maddock, 1954). In addition to devastating towns and ruining crops for the year, the force of the levee breaks associated with the Mississippi River floods of 1993 and 1994 damaged farm fields alternately by erosion (Tipton and Schlinkmann, 1993) and coarse debris deposition (Cortner, 1993).

After floods (Fig. 1-6), while many people call for more dams and levees, some are rethinking policies initiated 50 years ago. In response to the 1997 floods in California, some reconstruction may set back levees to allow a wider floodplain and access for the river to the floodplain.

Farmers are now looking at the expense of rebuilding levees and, in

***Figure 1-6.*** *Flooding on the floodplain along Squaw Creek, Ames, Iowa, May 1990 (photograph by Beth Middleton).*

some cases, opting for the river (Christenson 1997). In the Danube Biosphere Reserve of Romania, levees are being broken that were built in the mid-1980s to create grain fields in the delta (Simons, 1997).

Along the lower Yellow River in China, levee breaks have a long history. Between 602 B.C. and 1949 A.D. there were 1593 levee breaks. To control flooding, the Sanmenxia Reservoir was completed in 1960. Now the lower river is eroding, and the reservoir is filling rapidly with sediment (Zhou and Pan, 1994).

The disruption of the flood pulse created by levees along the Illinois River near St. Louis, Missouri, threatens the extinction of a rare composite, *Boltonia decurrens*. After the 1993–1994 floods, *Boltonia* populations increased greatly, particularly at sites with the most flooding. The sedimentation, moisture, and reduced competition created by flooding increased the populations of this species (Smith et al., 1998). Similarly, crocodile populations plummeted along the Chambal River in India after a series of dams were built in the 1960s (Sharma and Singh, 1986). Levees and other alterations can impact communities, endanger species, and hamper restoration projects.

Embankments are created on the seaward side of salt marshes or mangroves to prevent tidal inundation. This way, ponds are created for aquaculture or salt collection (Hutchings and Saenger, 1987). Walls are also created to desalinate the landward side. These areas then can be used for cattle grazing or agriculture. This type of ''poldering'' has been done along the coast of Nova Scotia in Canada (Pollett, 1979), in the Netherlands (Bijlmakers and de Swart, 1995), and in central coastal Queensland in Australia (Hutchings and Saenger, 1987). In large freshwater lake systems, impoundments are created by embankment and pumping to control water levels in wetlands for waterfowl; this alters water level fluctuation (Fig. 1-7; Mitsch, 1992) and natural pulsing dynamics that once occurred along the shoreline.

***Stream Diversion and Interbasin Water Transfer*** **Interbasin water transfer**, the diversion of water away from a river for irrigation, drinking water supply, or sewage disposal often has unexpected and far-reaching effects. This practice is particularly common in arid regions around the world. Interbasin water transfer is especially common near cities; in arid South Africa, for example, cities are generally located near mineral deposits or harbors, so freshwater supplies are often transferred from one of the few rivers in the country (Ashton and Van Vliet, 1997). In estuaries, the combined loss of freshwater input from diversion near cities and buffering freshwater wetlands can lead to drastic salinity flux following rainstorms (Welsh et al., 1978). In Hong Kong, on the lower Tsuen River, so much water is extracted from the river that almost no flow continues downstream (Dudgeon, 1996).

The ecological implications of interbasin water transfers have been little considered prior undertaking water diversion schemes (Walker et al., 1995). These transfers change the hydrologic regime and quality of the water, which is likely to cause the loss of biological integrity as endemic floras are invaded by alien and/or invasive species, genetic intermixing of once separated populations, and the spread of disease vectors (Davies et al., 1992).

Typical of rivers used for irrigation, water from the Platte River in the United States is overappropriated in that it is diverted from the river, used for irrigation, returned to the river, and rediverted to make as much use of the water as possible. Some of the water in the South Platte comes from the diversion of water from the Colorado and North Platte rivers (Knopf and Scott, 1990). Water from the Platte River changed the ''Great American Desert'' into a vast agricultural land (Simons and Simons, 1994).

In arid regions, the diversion of water can increase salinity in complex ways (Hammer, 1986). Along the San Joaquin River in California, water taken for the city of Los Angeles and various other interbasin water projects diverts more than one-half of the fresh water to its delta near the Pacific Ocean and has altered the dynamics of fresh water and salt water there (Fig.

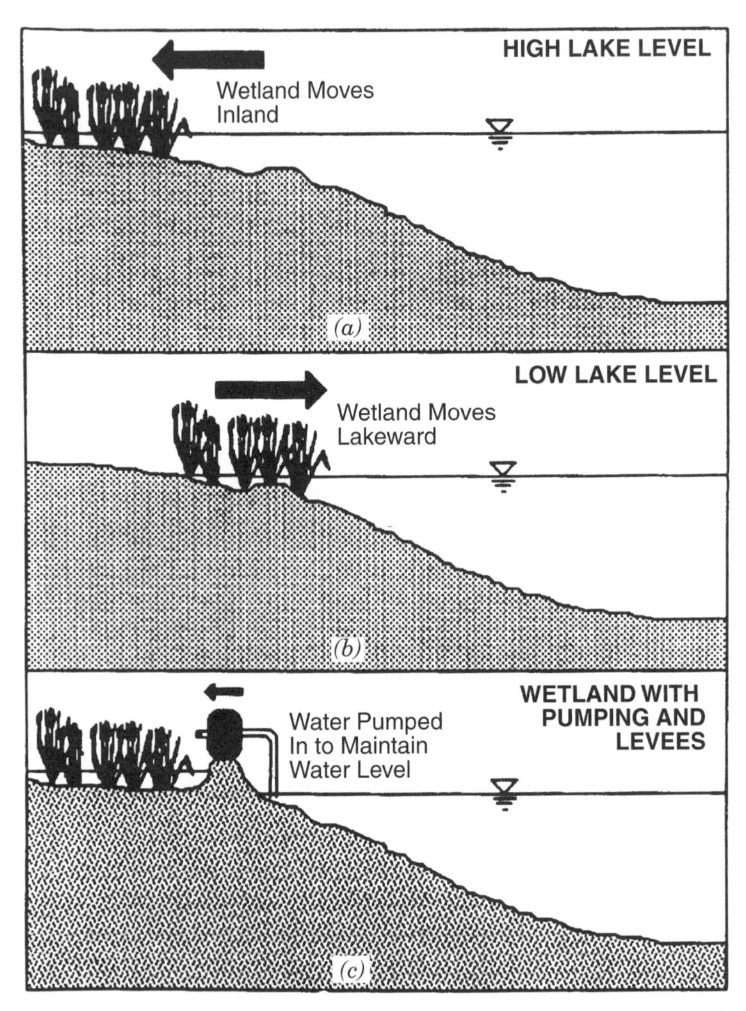


***Figure 1-7.*** *Natural water level fluctuation versus impoundment in a Great Lake wetland (from Mitsch, 1992, with permission from the International Association of Great Lakes Research).*

1-8; Davies et al., 1992). Along the lower Ebro River, Spain, the dynamics of the salt water wedge in the estuary are affected by upstream water diversion for the irrigation of rice and the erratic discharge of fresh water from hydroelectric dams (Muñoz and Prat, 1989; Prat and Ibañez, 1995; Ibañez and Prat, 1996).

Wetland loss has occurred in lower Iraq because of water diversion from the Tigris and Euphrates rivers in middle and upper Iraq (Scott, 1993). Water diversion from the Aral Sea for cotton irrigation has led to the nearly complete collapse of the fisheries and human communities surrounding the sea (Box 1-5; Cunningham and Saigo, 1995).

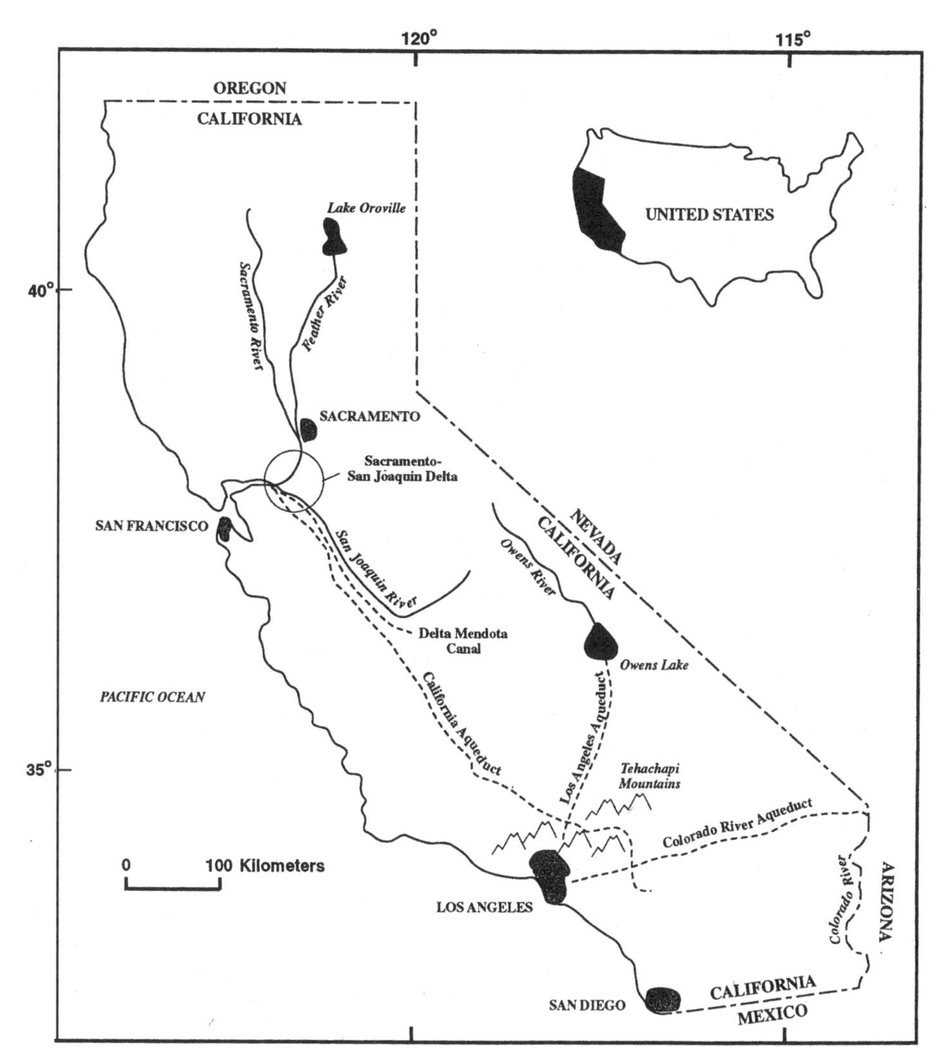


*Figure 1-8. Interbasin water transfers in California (from Davies et al., 1992; copyright © by John Wiley & Sons by permission).*

In Illinois, water diversion from Lake Michigan into the Illinois River via a 45-km canal has a long, volatile history. Before 1845, sewage from the growing town of Chicago was dumped directly into Lake Michigan, but as a result of several typhoid epidemics, waste water was redirected into the Illinois River. Since then, the diversion of this water to the Mississippi River via the Illinois River has been the subject of several federal court cases pitting Missouri against Illinois (Changnon and Changnon, 1996).

*BOX 1-5*

**BAD DAY FOR FISHING: THE ARAL SEA TRAGEDY**

I was standing in the sun on the hot steel deck of a fishing ship capable of processing a fifty-ton catch on a good day. But it wasn't a good day. We were anchored in what used to be the most productive fishing site in all of central Asia, but as I looked out over the bow, the prospects of a good catch looked bleak. Where there should have been gentle blue-green waves lapping against the side of the ship, there was nothing but hot dry sand—as far as I could see in all directions. The other ships of the fleet were also at rest in the sand, scattered in the dunes that stretched all the way to the horizon.

. . . As a camel walked by on the dead bottom of the Aral Sea, my thoughts returned to the unlikely ship of the desert on which I stood, which also seemed to be illustrating the point that its world had changed out from underneath it with sudden cruelty. Ten years ago the Aral was the fourth-largest inland sea in the world, comparable to the largest of North America's Great lakes. Now it is disappearing because the water that used to feed it has been diverted in an ill-considered irrigation scheme to grow cotton in the desert. (Excerpt from *Earth in the Balance.* Copyright © 1992 by Senator Al Gore. Reprinted by permission of Houghton Mifflin Co. All rights reserved.)

Water diversion from a river can eliminate or alter wetland vegetation. Namibia, suffering from a long-term drought, is contemplating a water diversion project from the Okavango Delta, a project with far-reaching potential for damage to freshwater wetland communities (Hanna, 1996). While water diversion may be inevitable, it may be possible to lessen the damage by abstracting the water at the apex of the swamp without constructing a dam or weir, which inevitably results in downstream erosion (Ellery and McCarthy, 1994). Despite the ecological problems associated with water diversion, many people in arid regions view water flowing into the sea as ''wasted'' (O'Keeffe et al., 1990).

Along the Murray River in Australia, meadows of Moira grass (*Pseudoraphis spinescens*) have been invaded by less flood-tolerant red gum (*Eucalyptus camaldulensis*) (Chesterfield, 1986; Bren, 1992). Those portions of the forests that still flood after the diversion flood with less frequency and duration (Thoms and Walker, 1993). Cottonwood forests died after the Los Angeles Department of Water and Power and other agencies diverted almost all of the water from Rush Creek in the eastern Sierra Nevada of the western United States. In the 1980s, heavy snowmelt along with minimal flow releases allowed their partial recovery (Stromberg and Patten, 1988).

## Wood and Channel Morphology

Large woody debris, or **snags**, cause a natural disturbance by altering channel morphology and long-term geomorphology in streams and rivers (Fig. 1-9; Young, 1991; Maser and Sedell, 1994), but it has been removed from many rivers (Maser and Sedell, 1994). While large woody debris produces local impoundments, pools, and backwater areas, other objects such as boulders do not produce equivalent results (Fig. 1-10; Kaufmann, 1987). Because of the nature of the obstruction it creates, wood in streams causes

***Figure 1-9.*** *Woody debris in a stream in Pantnagar, northern India (photograph by Beth Middleton).*

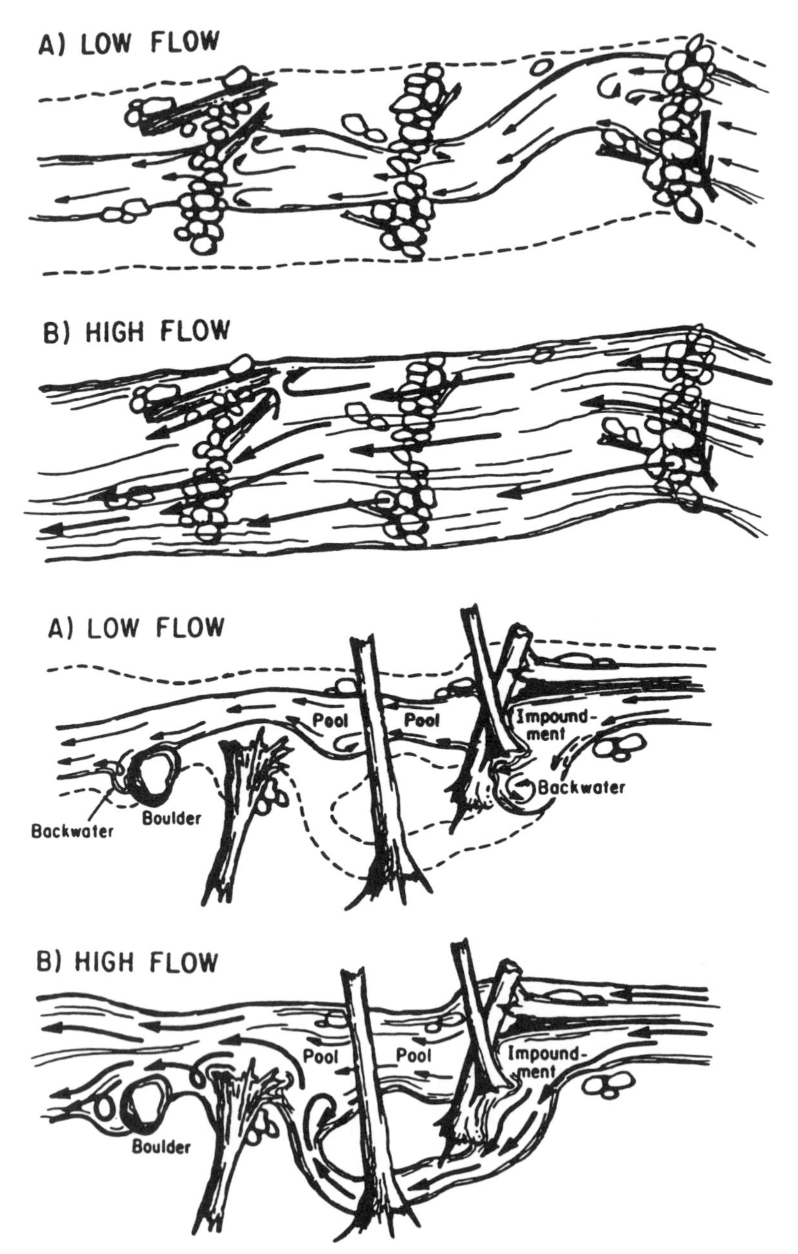


***Figure 1-10.*** *Water flow behavior in stream reaches with large boulders (upper a, b) versus driftwood (lower a, b) during low and high flow. Large driftwood creates protected areas for fauna in high flow (lower b; from Maser and Sedell, 1994, as adapted from Kaufmann, 1987; copyright © by St. Lucie Press by permission).*

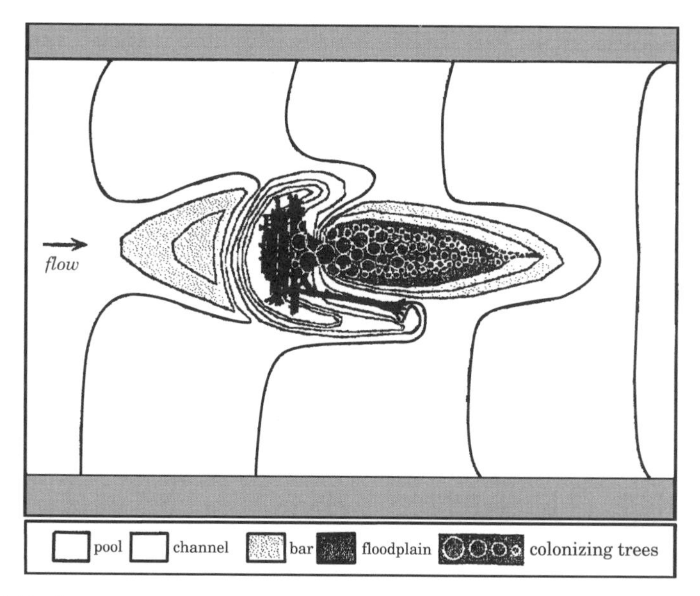


***Figure 1-11.*** *A bar apex jam showing progressive tree colonization in a channel behind an obstruction caused by large woody debris (from Abbe and Montgomery, 1996; copyright © by Regulated Rivers: Research and Management by permission).*

the stream bed to widen as flow deflects to the side of the bed. This results in sediment deposition, pool formation, channel depth variation, backwater creation along the margin of the channel, and the formation of secondary channels (Maser and Sedell, 1994). Root wads in banks, dead or alive, greatly reduce bank erosion, but their function is altered once they fall into the channel (Keller and Swanson, 1979). Water behaves differently in channels with wood because the wood creates flow resistance, variable flow, reduced water speed, increased water depth, and increased flood travel time. Large snags reduce flow more than small snags (Shields and Smith, 1991; Gippel, 1995), so small debris (<1 m in cross section) has less influence on flood conveyance or frequency than large snags (Gippel et al., 1996a).

These changes in water behavior and stream morphometry result in a more heterogeneous habitat for stream invertebrates (O'Connor, 1991), fish, and flora (Maser and Sedell, 1994). Jams created by large, woody debris create opportunities for tree colonization (Fig. 1-11; Abbe and Montgomery, 1996). Salmon decline has been linked to lack of woody debris and/or channel changes, and is essential for fish spawning and rearing habitats. Stream degradation has been blamed for the decline of the salmon industry in Germany along the Rhine (van Dijk et al., 1995) and, in North America, in the Pacific Northwest (Sedell et al., 1991) and Maine (National Research Council, 1992).

While it is logical for woody debris to cause flooding in streams, it is also argued that it decreases downstream flooding by increasing retention time upstream (Swales, 1989; Gippel, 1995). In reality, the effect of large woody debris in the channel varies, so that during floods, large woody debris has less influence on flow than during low flow (Shields and Smith, 1991). In nonflooded conditions, woody debris caused a 21% increase in the local water surface elevation behind the wood dam, while small debris had very little effect (<0.2%; Shields and Smith, 1991; Gippel et al., 1992). Small debris had little influence on flood conveyance or frequency in the Thompson River of Australia (Gippel et al., 1996a).

Debris dams move dynamically through the stream reach on their way to the sea (Maser and Sedell, 1994), but only a few studies have documented the amounts of debris moved by rivers (Table 1-2). In Europe, along rivers with highly managed forests, there is less woody debris than in North America (0–5 t $ha^{-1}$ versus 500–550 t $ha^{-1}$, respectively). A less managed river, the Ain Tributary of the River Rhône in France, has 0.001–200 t $h^{-1}$ of woody debris in the river channel (Piegay, 1993). Some debris dams remain in situ for more than 100 years. About one-third of the debris dams in the New Forest of the United Kingdom shift every year (Gregory and Davis, 1992), with 65% of the debris redistributed every six years (Lienkaemper and Swanson, 1987).

## Debris Removal

Debris is removed from rivers all over the world to improve navigation (Peterson et al., 1987) and fish migration. However, because fish can move around debris, particularly at times of flooding, the contention of improved fish migration is not only unfounded, but debris removal has caused a great deal of fish habitat degradation (Piegay, 1993). Debris is also removed to increase the flow rate and improve drainage (Shields and Smith, 1991; Ministère d l'Environnement et Agences de L'Eau, 1985, in Piegay, 1993).

In the waterway, debris removal alters the dynamics of the stream; for example, it reduces wood, litter, and sediment storage; reduces channel depth variability; increases water speed; and decreases habitat variability (Fig. 1-12). The number of minor channels in the Willamette River of Oregon has been greatly reduced because of woody debris removal and other stream regulation techniques (Fig. 1-13; Sedell and Froggatt, 1984; Benner and Sedell, 1997).

The amount of wood that has been removed from streams around the world is staggering. Table 1-3 shows the number of snags and trees removed from the Willamette River from 1870 to 1930 (Sedell and Froggatt, 1984). In the Murray River of Australia, 22,000 items of debris were extracted from a 233-km stretch between 1976 and 1986 (Gippel et al., 1992).

**TABLE 1-2. Volume Loading and Density of Debris in Relatively Undisturbed Rivers Draining Catchments Greater than 100 $km^2$**

| Site | Drainage Area ($km^2$) | Debris Volume ($m^3m^{-2}$) | Debris Density ($km^{-1}$ | Lower Size Limit (m) | Channel and Riparian Condition | Reference |
|---|---|---|---|---|---|---|
| East Fork Salmon River, Idaho, U.S. | 196 | 0.00025 | — | 0.1 | Undisturbed forested floodplain | Lienkaemper (unpublished) in Harmon et al. (1986) |
| Ogechee River Georgia, U.S. | 7,000 | 0.0148 | — | <0.01 | Undisturbed forested floodplain | Wallace and Benke (1994) |
| Black Creek Georgia, U.S. | 767 | 0.0168 | — | <0.01 | Undisturbed forested floodplain | Wallace and Benke (1994) |
| McKenzie River, Oregon, U.S. | 1,024 | 0.0010 | — | 0.1 | Undisturbed forested floodplain | Keller and Swanson (1979) |
| Pranjip Creek, Victoria, Australia | 787 | 0.035–0.055 | — | 0.01 | Narrow stand of riparian trees | O'Connor (1991) |
| South Fork Obion River, Tennessee, U.S. | 927 | 0.043–0.094[a] | 35–38 | Formations 1 $m^2$ in area | Forested floodplain, previously channelized and cleared | Shields and Smith (1992) |
| 6 streams in Iowa, U.S. | 381–821 | — | 1–14 | Not stated[b] | Previously channelized, grassy | Zimmer and Bachman (1976) |
| 5 streams in Iowa, U.S. | 373–712 | — | 14–34 | Not stated[b] | Undisturbed woodland | Zimmer and Bachman (1976) |
| River Murray, Lake Hume to Yarrawonga, Australia | approx. 11,000–20,000 | — | 94[c] | Not stated[b] | Stand of riparian trees, variable width | Gippel et al. (1992) |
| River Murray at Overland Corner, Australia | Approx. 1,000,000 | — | 15–46[d] | Not stated[b] | Narrow stand of riparian trees; River flow highly regulated | Lloyd et al. (1991) |
| Gouldburn River, Shepparton, Victoria, Australia | 16,125 | — | 2364[e] | Not stated[b] | Relatively undisturbed riparian forest | Anderson and Morrison (1988) |

[a]Debris formations measured without regard for spaces between stems. Value is an overestimate.
[b]Large woody debris (in the form of snags, large branches, or accumulations of logs) was implied by authors.
[c]Deduced from the number of snags removed during a clearing operation. Value is an underestimate because not all snags were removed.
[d]A large, turbid river where only visible emergent debris was counted. Value is an underestimate.
[e]A sonar technique was used that probably resulted in some items being counted more than once. Value is likely to be an overestimate.

*Source:* Gippel et al., 1996a.

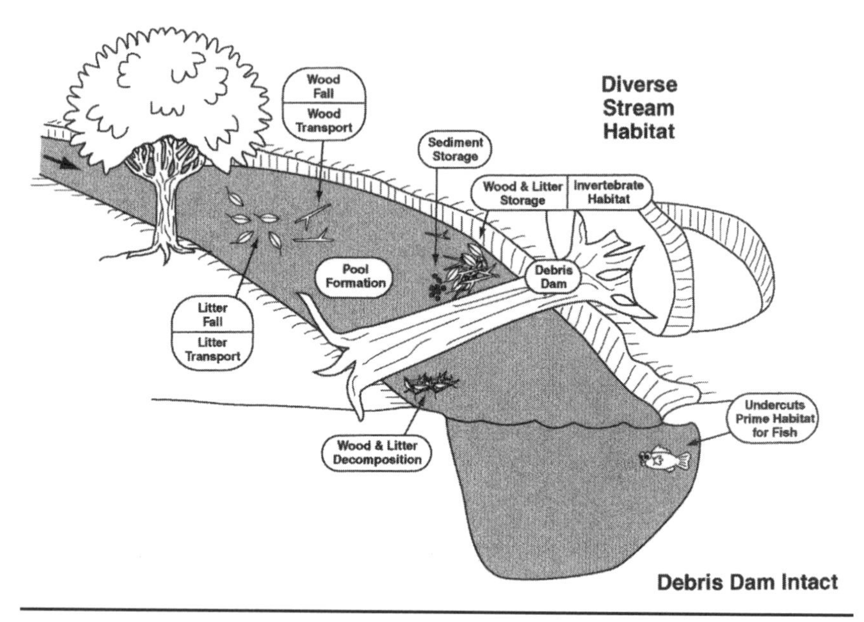


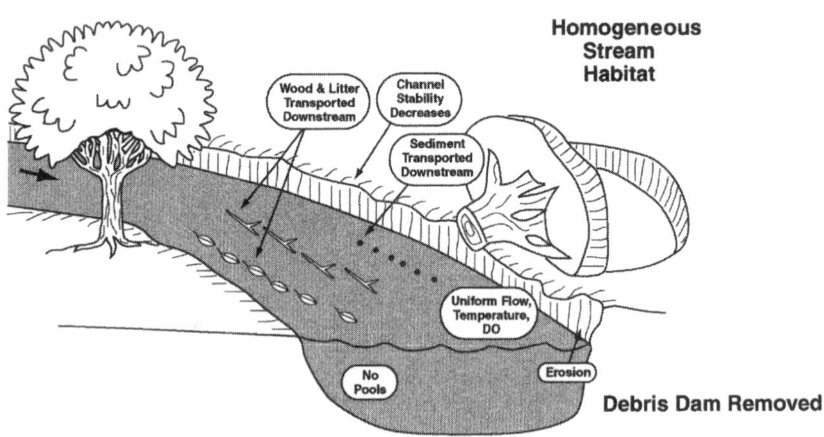


***Figure 1-12.*** *Debris dam and the ecology of streams. Contrast the intact stream with a woody debris dam (above) with the altered stream with debris dam removed (below) (adapted from Gregory, 1992, and Davis and other sources).*

Along the lower 1600 km of the Mississippi River, 800,000 snags were pulled over a 50-year interval (Sedell et al., 1982); this was also authorized by the U.S. Congress for the middle Mississippi River in 1824. Along the Satilla River of Georgia, 15,000 snags were removed from a 260-km stretch of the lower river before the 1940s (Wallace and Benke, 1994).

Beyond the local waterway, the combined effect of debris removal, logging, and damming has affected interconnected ecosystems. Wood is no

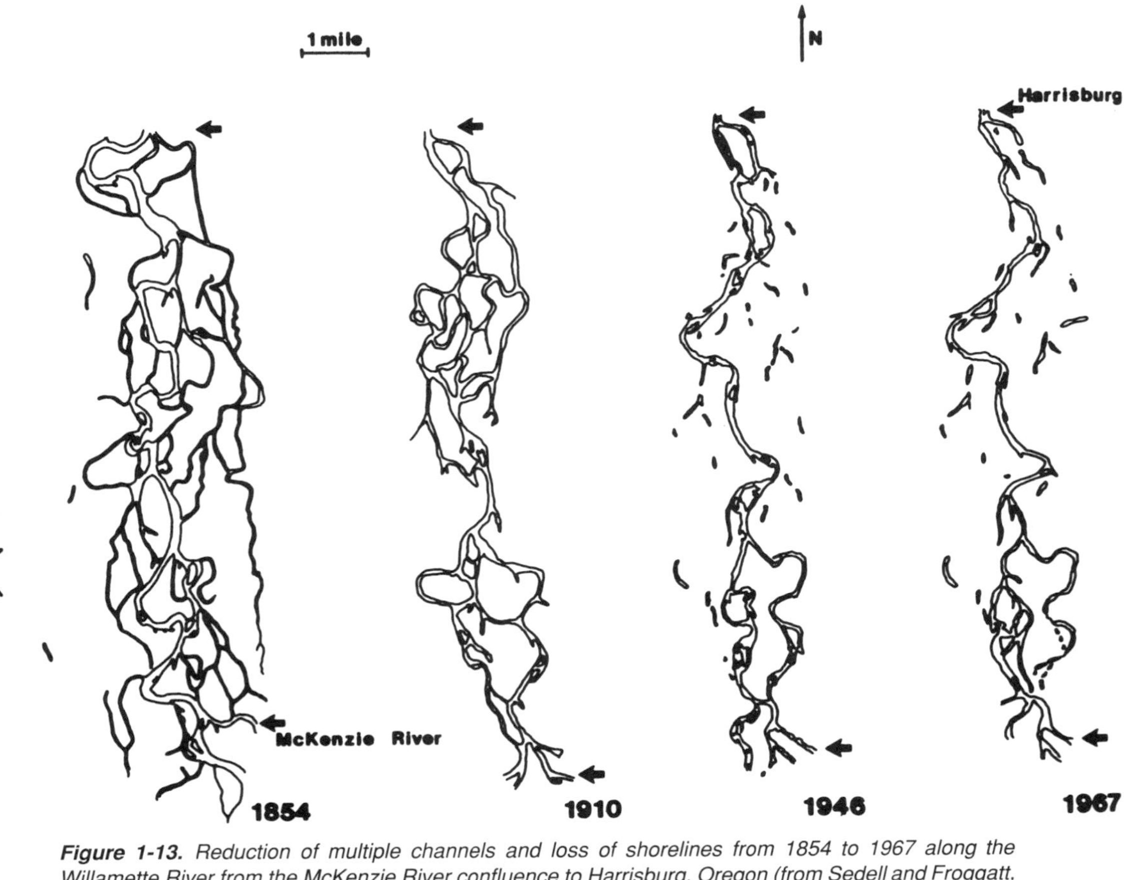


***Figure 1-13.*** *Reduction of multiple channels and loss of shorelines from 1854 to 1967 along the Willamette River from the McKenzie River confluence to Harrisburg, Oregon (from Sedell and Froggatt, 1984; copyright © by Verhandlurgen Internationale Vereinigung für Theoretische and Angewandte Limnologie by permission).*

**TABLE 1-3. Summary of Snags Pulled and Stream-Side Trees Cut Along the Willamette River, Oregon, for the Purpose of Navigation**

| Year | Snags | Streamside Trees | Snags/Streamside Trees |
|---|---|---|---|
| 1870–1875 | 1566 | 17 | |
| 1876–1880 | 4620 | 542 | |
| 1881–1885 | 3900 | 3910 | |
| 1886–1890 | 2556 | 545+ | |
| 1891–1895 | 3735 | 1211+ | |
| 1896–1900 | 7070 | 4520 | |
| 1901–1905 | 2246 | 918 | |
| 1906–1910 | 2701 | 1744 | |
| 1911–1915 | 3055 | 6810 | |
| 1916–1920 | | | 7498 |
| 1921–1925 | | | 250 |
| 1926–1930 | | | 891 |
| 1931–1935 | | | 2130 |
| 1936–1940 | | | 4221 |
| 1941–1945 | | | 1960+ |
| 1946–1950 | | | 836+ |

*Note:* Snags and stream-side trees were not listed separately after 1919.

*Source:* Sedell and Froggatt (1984).

longer carried in quantity to the sea, where it once was a focus of consumer activity. Depletion of driftwood on seashores deposited from rivers upstream has negatively impacted beach processes (Fig. 1-4; Maser and Sedell, 1994).

The restoration of woody materials in streams and rivers is in its infancy (Gippel et al., 1996b), but some attempts to reintroduce wood to degraded streams have been made in the United States (Sedell et al., 1991). The planned reduction of agricultural land in France may result in opportunities for wood restoration in rivers, particularly along less developed rivers (Piegay, 1993). In Australia, the Murray-Darling Basin Commission has stopped desnagging until this policy can be reexamined (Gippel et al., 1996a).

A hindrance to wood restoration in rivers is the lack of knowledge concerning wood dynamics. The majority of the research on the effects of snag removal or desnagging has been done in the Pacific Northwest, Australia, and the United Kingdom (Gippel, 1995). Woody debris surveys have been conducted in a variety of locations (Campbell and Franklin, 1979; Piegay, 1993; Gippel et al., 1996a). However, the minimum amount of wood needed to be replaced in a stream is not known, and there is very little information on how to reintroduce wood into streams. Undisturbed streams can be used as a guide to the target positions and amounts of wood in a stream. The natural recruitment and retention rates are not entirely under-

***Figure 1-14.*** *Woody debris is carried out to sea from streams. Some is deposited on shore during storms and becomes an important part of sand dune dynamics. Pictured is a debris pile on Outer Island, Lake Superior (photograph by Beth Middleton).*

stood; for this reason, the costs and time scales for maintaining wood in stream ecosystems are unclear (Gippel, 1995).

**Beaver** Beaver exert a huge influence over the landscapes they inhabit, creating wetlands (Rudemann and Schoonmaker, 1938; Ives, 1942) and profoundly altering patterns of geomorphology and vegetation (Naiman et al., 1986, 1988; Johnston and Naiman, 1990; Hackney and Adams, 1992). Their temporary extirpation in large portions of Europe and the United States produced a large change in the ecological function of riverine ecosystems.

In North America, before their near-extirpation by trapping around 1850, the number of beaver (*Castor canadensis*) exceeded 60 million (Seton, 1929; Jenkins and Busher, 1979). Their habitats in North America ranged from the Arctic tundra to the deserts of northern Mexico (Jenkins and Busher, 1979). Recently, in the absence of predators and with the introduction of trapping laws, beaver populations in North America have rebounded (Naiman et al., 1991). Thousands of acres are again impounded by beaver, especially in the southeastern (Hair et al., 1978; Hackney and Adams, 1992) and northern United States (Johnston and Naiman, 1987).

Similarly, beaver in Europe (*Castor fiber*) once ranged east to west, from

Britain across the Eurasian continent, and north to south, from the Arctic to the Mediterranean, until hunting and changing land use nearly exterminated them (Hartman, 1996). Beaver were extirpated in Sweden in about 1871 (Jakobsson, 1981,) and then reintroduced successfully in 1922 (Jakobsson, 1981; Hartman, 1996), as well as along several rivers in Russia (Tyurnin, 1984).

Beaver are **keystone** species (Naiman et al., 1984; Naiman, 1988a) in that they fundamentally alter the flow, transport, and nutrient dynamics of the river to influence both primary and secondary productivity (e.g., aquatic insects; Naiman et al., 1984). In uplands, beaver create impoundments with sharp vegetation boundaries along steeply sloping basins, reflecting the original V shape of the stream valley (Fig. 1-15). In wetlands, beaver construct broad impoundments with many vegetation zones within a U-shaped basin. As a result, boundaries of energy and material in upland beaver dams are abrupt but wide and diffuse in wetlands (Johnston and Naiman,1987).

Beaver dams, like other dams, trap sediment (Figs. 1-16 and 1-17; Lowry and Beschta, 1994) and impede the movement of woody debris downstream. Because beaver dams impound water, nutrient transformations by microorganisms in the soil result in methane emissions, which contribute slightly to the accumulation of greenhouse gases (Naiman et al., 1991). It is worth mentioning that other types of natural dams exist as a result of geological activity, including **jokulhlaup** (glacial ice dam: Icelandic), morainal dams, volcanic dams, and landslide dams (Costa, 1988).

The browsing activities of beaver can either maintain the vegetation in sapling stages of aspen and poplar (Vogl, 1980) or change the vegetation from woody to open water or herbaceous species (Johnston, 1994). Because beaver are central place foragers (Orians and Pearson, 1979), their activities are limited to within about 100 m of their ponds (McGinley and Whitham, 1985). Beaver dams are not the equivalent of human dams because they have a very different landscape dynamic. While some beaver dams are occupied continuously for many years, others, abandoned and reoccupied, cycle between flooded and drained stages (Johnston, 1994).

***Fire*** Up to this point, most of the disturbances described that are relevant to restoration are riverine situations. Fire, as well as the other disturbances to be discussed in this chapter, can be broadly applied to include non-riverine situations.

Fire has long been incorporated into restoration plans for prairies (McClain, 1986). However, while fire is the most common and most widely prescribed disturbance in terrestrial ecosystems (Bond and van Wilgen, 1996), its importance has been somewhat overlooked in wetlands (Vogl, 1977) and is badly in need of study (Komarek, 1976).

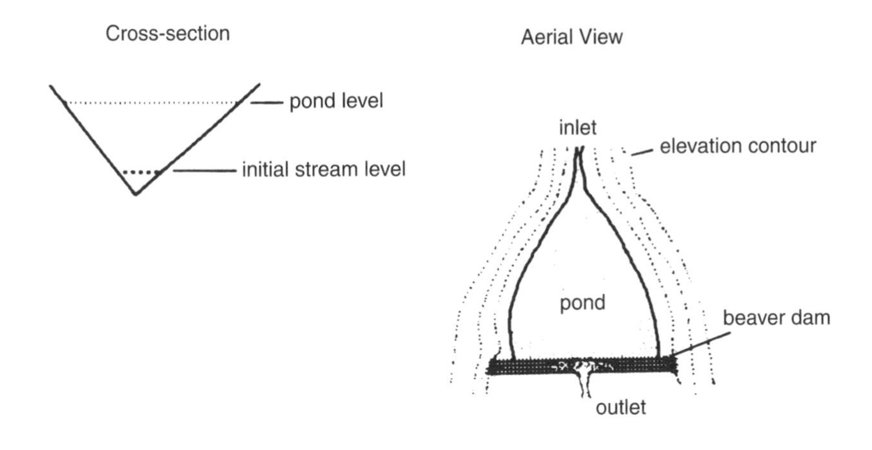


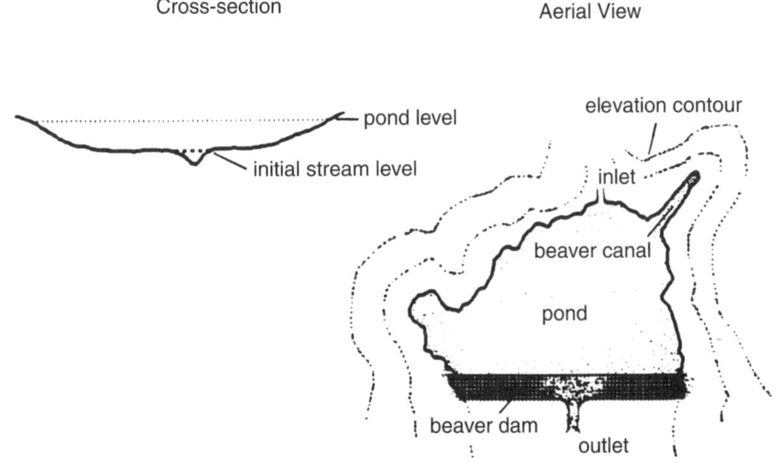


***Figure 1-15.*** *Upland versus wetland beaver dams (from Johnston and Naiman, 1987; copyright © by Landscape Ecology by permission).*

Fire is common in wetlands with seasonal inundation and thus is more likely to be important in their restoration (Buchanan, 1989). Among the wetland types that are fire dependent (Fig. 1-18) or at least burned with some regularity before fire suppression are the wet heathlands in Australia and the United Kingdom (Buchanan, 1989, and Whelan, 1995, respectively), monsoonal wetlands in Australia (Thompson, 1996), subtropical riverine grasslands in Nepal (Lehmkul, 1994), riparian evergreen forests in

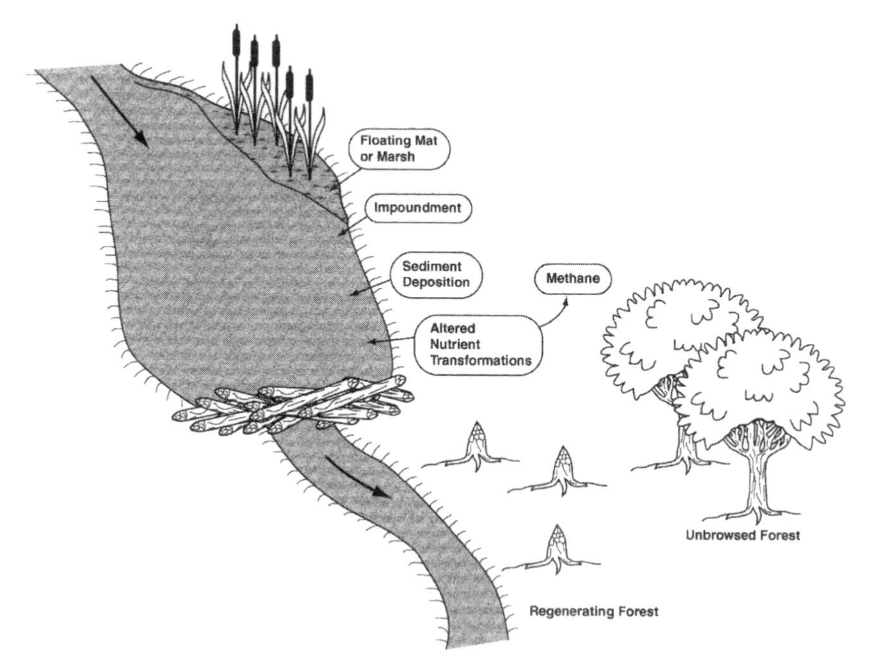


***Figure 1-16.*** *The beaver pond (adapted from Johnston and Naiman, 1987; copyright © by Landscape Ecology by permission).*

Africa (mushitu; Thompson, 1996), herbaceous swamps in Africa (Thompson and Hamilton, 1983) and South America (the Pantanal; Por, 1995), lowland savannas in Malaysia-Indonesia (Stott et al., 1990), tropical lowlands (Budowski, 1996), seasonally flooded palm forests of the llanos of Colombia and Venezuela (Myers, 1990), and peatlands in New Zealand (Clarkson, 1997).

In North America, wetland types that burn regularly include prairie potholes (Kantrud, 1986), tidal fresh marshes (Nyman and Chabreck, 1995), Atlantic white cedar (McKinley and Day, 1979), taiga wetlands (Viereck, 1973), cypress swamps of the southeastern United States (Cypert, 1961; Schlesinger, 1978; Duever, 1984; Duever et al., 1986), cypress domes (Ewel and Mitsch, 1978), scrub cypress (Craighead, 1971), pocosins (Christensen et al., 1981), canebrakes (*Arundinaria tecta*) (Wright and Bailey, 1982), marl prairies (Gunderson and Loftus, 1993), and California fan palm oases (Vogl and McHargue, 1966). Without fire in the North, bogs overtake boreal forests through **paludification** (process in which dry land becomes a bog) (Heinselman, 1975; Post, 1996), and in the South, sawgrass dies out in coastal marshes (Hofstetter, 1974).

South Florida wetlands have well-documented cases of ignition by both

***Figure 1-17.*** *A beaver pond at Leroy Percy State Park in Mississippi (photograph by Beth Middleton).*

lightning and American Indians. These wetlands include seasonally flooded pinelands, sawgrass, wet prairie, tidal marsh, cypress swamp, mangrove, canebrakes (Wade et al., 1980), pond pine (Ewel, 1990), wet slash pine (Phillips, 1996), and marshes, sloughs, and ponds (Duever et al., 1986).

The driest cypress swamps with rapid decomposition and hence less organic matter buildup burn less frequently than wetter ones (Ewel, 1990). Some ponds in cypress swamps may have been formed by peat burning during severe droughts (Duever, 1984). *Spartina*-dominated salt marshes regrow well after experimental burning (de la Cruz and Hackney, 1980).

The literature is divided on whether or not mangrove communities are fire dependent. According to some authors, fire is an organizing force in mangrove communities (Vogl, 1977), second only to hurricanes (Craighead, 1971). In the absence of fire, mangroves move inland to invade subtropical coastal marsh (Wright and Bailey, 1982). However, in lower-lying elevations in Florida, little or no ancient charcoal is found in the peat layers of *Rhizophora/Avicennia* peat (Cohen, 1974), implying that fire may not always be important in mangrove systems.

In riverine systems, the impact of fire on a landscape level has received some attention. While rivers are thought to act as firebreaks, most of the fires burning across white cedar swamps breach the river (Windisch, 1987). Runoff carrying ash increases after fire because streambank vegetation is

*Figure 1-18. Fire in seasonally flooded savanna in the Keoladeo National Park, northern India (photograph by Beth Middleton).*

temporarily destroyed. After fire, streambeds scour, channels recut, and plants regenerate (Vogl, 1977). Sedimentation and the stability of the basin typically increase (Davis et al., 1989; Faber et al., 1989). In chaparral in California, postfire erosion increased on the hillsides (Barro et al., 1989) while dominants resprouted after fire, including white alder, California sycamore and coast live oaks (Davis et al., 1989).

In modern agricultural landscapes, fire dynamics have been disrupted, making it difficult to manage small nature reserves (Hobbs et al., 1993) and restoration sites. Just as in the disruption of riverine dynamics by dam and engineering projects, a fragmented landscape disrupts fire periodicity and the ability of the site to regenerate after fire. The minimum dynamic area (Pickett and Thompson, 1978) constituted by restoration sites may be too small to ensure that an array of successional stages is present in the surrounding landscape.

Lightning fires are common throughout the world in many vegetation types (Gill et al., 1990; Stott et al., 1990; van Wilgen et al., 1990; Clarkson, 1997), with the alleged exception of some tropical and subtropical ecosystems in Central America (Budowski, 1966; Hopkins, 1983; Janzen, 1986; Murphy and Lugo 1986; Koonce and González-Cabán, 1990; Middleton et al., 1997). Fires in New Zealand peatlands are also ignited by spontaneous combustion and volcanic eruption (Clarkson, 1997). Many fires in wetlands are of anthropogenic origin and can be traced to ancient times. American Indians used fire regularly to clear fields (Hughes, 1983), drive animals, and improve blueberry production (Russell, 1982). European explorers commented in their journals on fires witnessed in prairie potholes of the midwestern United States (Kantrud, 1986). The ancient knowledge of burning of Australian aborigines in monsoonal wetlands is incorporated into the management of Kakadu National Park (Yapp, 1989; Thompson, 1996).

Fire in wetlands releases nutrients, opens the canopy and detrital layers, allows earlier warming of the soil (Kozlowski and Ahlgren, 1974), and increases production and diversity (Wright and Bailey, 1982). In prairie potholes, late summer burning kills the root crowns of *Phragmites* (Ward, 1942). The nature of fire varies greatly in different seasons; early dry season fires often go out at night because of dew and so burn less extensive areas (Gill et al., 1996).

Fire is also used to control undesirable vegetation and to remove floating mats of undecomposed vegetation (Mallik, 1990a, 1990b). In North America, burns are widely used in wetlands managed for waterfowl to establish new growth (Chabrek, 1976), expose seed, remove flooded and decomposed organic matter, and create pools and edges for nesting (Payne, 1992). In monsoonal wetlands in India, geese are attracted to young stands that have recently burned (Middleton, 1992), and burning can be used to improve habitats for waterfowl and wildlife (Chabrek, 1976, 1988; Middleton, 1994b). Similarly, waterfowl are attracted to areas heavily grazed by cattle (Dalby, 1957; Chabrek, 1976).

Plants possessing fire adaptations have a number of mechanisms for reestablishing themselves after a fire. The seeds of many species require exposed soils to germinate and remain dormant or have fruiting bodies that do not release seeds until exposed to fire (Vogl, 1977). However, in cold seasons, dormancy may not be broken in seeds if the soil does not warm sufficiently after the fire (Auld and Bradstock, 1996). Other common fire adaptations include extensive root systems, and postfire regrowth from roots or root crowns. Fire-resistant aboveground plant parts include thick bark, fire-protected meristems, absent cambium, strongly vegetative growth, and/or fire-stimulated flowering, seed production, and regrowth (Kozlowski and Ahlgren, 1974). Cypress is fire adapted, and while fires are not common in cypress swamps, without low and regular fire, cypress swamp on drier sites reverts to hardwood forest (Duever et al., 1986).

In created wetlands, controlled burning may help to foster diversity in dense stands of vegetation and increase the hydraulic capacity by reducing organic matter and peat accumulation; however, fire can change the biota and hydrology in ways a manager does not desire (Hammer, 1997). Fire used to control cattail during drawdown in restored marshes stimulates the regrowth of cattail unless the sites are deeply flooded above the stems after burning (Joy Marburger, personal communication). In Australia, where fire ecology is advanced, fire is prescribed in the restoration of all but the wettest wetland types (Buchanan, 1989). At the same time, fire has the potential to burn peat and lower the wetland elevation so that ponds develop (Duever, 1984; Hammer, 1997). Clearly, fire is an important component of wetland restoration (Harker et al., 1993), but it is not always clear at what point in the restoration management plan fire should be incorporated, nor is it likely to be important in all wetland types.

***Windstorms and Hurricanes*** While natural disturbances such as windstorms and hurricanes may be detrimental to a restoration project, these are important events in organizing coastal wetlands. Before the vegetation is established in newly created wetlands, storms can threaten a new project (Broome et al., 1988; Hammer, 1997)—for example, water rising quickly or wave action scouring seeds from the soil surface.

The factors that produce disturbance in a hurricane can be divided into three components: wind, rain, and storm surge. In southern Florida, hurricanes have a return interval of approximately once every three years (Gentry, 1974). Some hypothesize that natural forests in the Caribbean (mangrove and riverine) are in a state of continuous recovery from hurricane (Lugo and Scatena, 1995). Thus, at least in the parts of the world where these are common, hurricanes are part of the disturbance ecology of restored coastal mangroves.

Hurricanes are an important factor in the deposition of woody debris in tropical streams. After Hurricane Hugo, two to three times the nonhurricane levels of woody debris were deposited in streams in the Luquillo Experimental Forest in Puerto Rico (Vogt et al., 1996). Hurricane Hugo reduced the aboveground biomass there by 50%.

After hurricanes such as Hurricane Andrew in Florida or Hurricane Joan in Nicaragua (October 1988), recovery of the vegetation is typically from the regrowth of survivors and from seedling regeneration (Baldwin et al., 1995, and Roth, 1992, respectively). In Florida, after Hurricane Andrew, *Rhizophora mangle* regenerated primarily from propagules. In contrast, *Avicennia marina* and *Laguncularia racemosa* regenerated via epicormic sprouting and had lower mortality rates than *R. mangle* (65.0, 59.5, and 85.1% mortality, respectively; Baldwin et al., 1995). Five years after Hurricane Hugo, 86% of the biomass had recovered through both regeneration and the regrowth of survivors (Scatena et al., 1996). After Hurricane Andrew, while 80–95% of the mangroves died in some areas, small mangroves had a mortality rate of less than 10%. Mangrove propagules were spread widely and in large numbers, but regeneration was poor in areas with newly deposited sediment and increased levels of pore water sulfide (Smith et al., 1994).

While damage to coastal marshes by hurricanes can be quite severe (Louisiana; Chabrek, 1988), after Hurricane Andrew in Florida, inland freshwater marshes appeared little damaged. At least in the short term, only 1–2% of the cypress were damaged (Pimm et al., 1994). However, after Hurricane Hugo in South Carolina, 46% of the cypress showed damage (Loope et al., 1984).

Surging flood waters from storms can remove trees along river banks. A flood in the Medina River in central Texas removed large (>2 m diameter) bald cypress trees, along with 90% of their crown cover. This type of disturbance has a return interval of 100–500 years (Baker, 1988). Macrophytes are more sensitive to sudden summer flooding, so a storm producing sharp water level rises can kill these species (Barrat-Segretain and Amoros, 1995). The importance of water level changes such as these will be discussed in detail in Chapter 4.

***Herbivory*** Herbivores are an integral part of the succession dynamics in many types of natural wetlands, including moose, beaver (already discussed), and elephants (Naiman, 1988a), and logically should be a part of restoration plans. Herbivores remove large quantities of emergent vegetation (Chabwela and Ellenbroek, 1990) and, as a result, provide opportunities for plant regeneration. In flooded situations, cleared areas do not regenerate with emergents until drawdown produced by drought (Middleton et al., 1991; Middleton, *in press*; see Chapter 2 on succession).

Similar to beaver in riverine systems, muskrats play an integral role in the successional dynamics of prairie potholes in North America. Muskrats use the vegetation for both food and shelter building (Fig. 1-19; Ogaard and Leitch, 1981), eventually creating an open water "eat-out." Eat-outs can also be created by geese (Fig. 1-20 and 1-21). This open water condition is maintained until drought occurs. The muskrats die or emigrate during drought, a time when the emergent vegetation (e.g., *Typha, Scirpus*) regenerates from germination from the seed bank. When normal rains resume, the pothole refloods, the emergent vegetation regenerates and muskrats return to begin the cycle again (Fritzell, 1989).

Herbivores are complexly interwoven with their habitats. The lekking behavior of lechwe (*Kobus leche kafuensis*) in the Kafue Flats of Zambia is tied to the rise of water during the rainy season. Nuer and Dinka pastoralists in northern Africa also move with the seasonal patterns of the hydrological regime (Rzoska, 1974). Many herbivores have coevolved with plants, acting as dispersers and influencing plant regeneration patterns (Fig. 1-22; Middleton and Mason, 1992; Middleton, accepted).

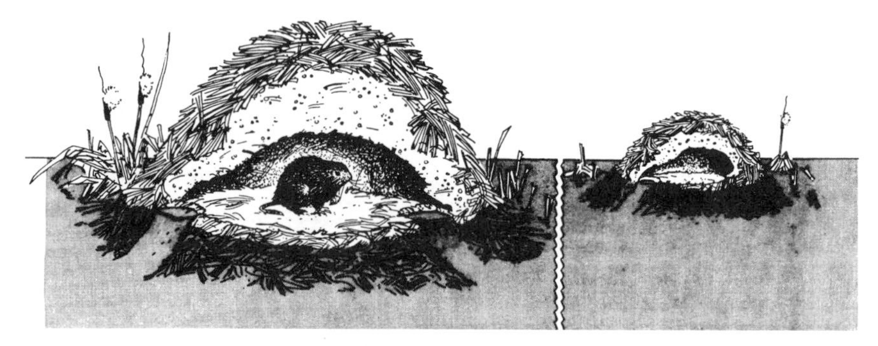

*Figure 1-19. Muskrat lodge in a prairie pothole, North America (from Fritzell, 1989; copyright © by Iowa State University Press by permission).*

Waterfowl influence the species composition and production levels of grassland, sedge meadows (Vogl, 1980), and monsoonal wetlands (Middleton, 1989; Middleton et al., 1991; Middleton, accepted). Grazing tends to increase diversity in wetlands unless the grazing is severe (Kantrud, 1986). Geese can completely "eat out" the vegetation from the monsoonal wetlands during the dry season (Middleton, personal observation) and in Kak-

*Figure 1-20. Eat-out created by geese in the Keoladeo National Park, Bharatpur, India. Pictured are the Barheaded or Indian Mountain Goose (*Anser indicus*) in a monsoonal wetland during the early drawdown season. The geese are standing on a ripe mat of mud, feathers, and goose feces in the early summer heat. Barheaded Geese migrate to the Himalayas for nesting in the summer season (photograph by G&H Denzau by permission).*

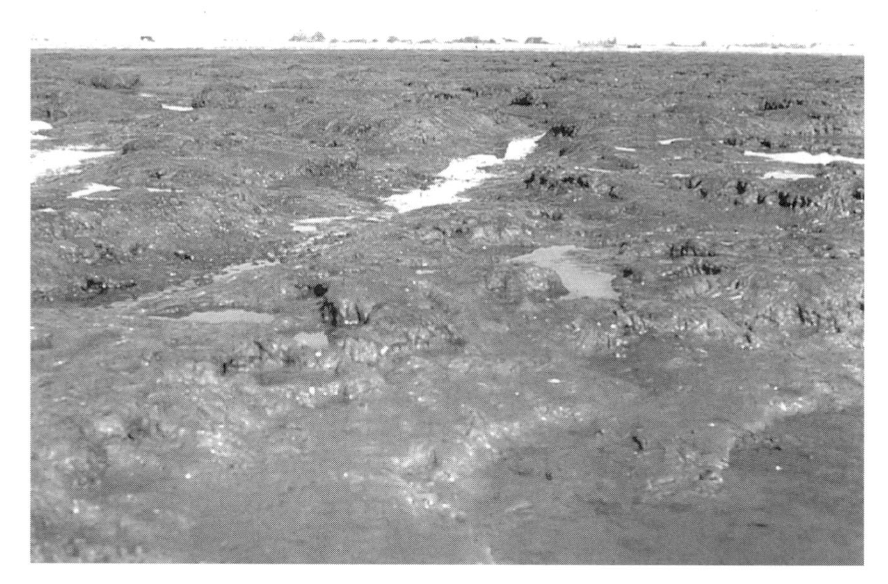

***Figure 1-21.*** *Eat-out created by the Lesser Snow Goose (Chen hyperborea hyperborea) in a salt marsh near Vancouver, Canada (photograph by Beth Middleton).*

adu National Park, Australia, and Florida, where they have been introduced (Yapp, 1989, and Joy Marburger, personal communication, respectively). In North America, muskrats are important in creating openings in prairie pothole vegetation in North America (van der Valk, 1981; Weller, 1981) and deltaic salt marsh along the Mississippi River (Gosselink, 1984). In Europe, both muskrats and coypu have been introduced (Britton and Crivelli, 1993). In Louisiana, introduced nutria in conjunction with prolonged flooding are causing wetlands along the Atchafalaya Delta to become more open (Shaffer et al., 1992).

There is nothing less welcome than a flock of Giant Canadian Geese on a newly restored wetland (Hammer, 1997; Daniel Mason, personal communication). However, herbivory does play an important role in community dynamics once the wetland plant species are established, so the role of herbivory in established, restored wetlands needs consideration.

***Domestic Grazing*** Cattle grazing in wetlands is common the world over. While wetlands can often recover after grazing ceases, even low levels of grazing have a significant impact (Reimold, 1976) due to treading, dung deposition, and grazing (van Wieren, 1991). Cattle grazing is common in emergent wetlands of the Mediterranean (Britton and Crivelli, 1993), salt marshes in the Netherlands (van Wieren, 1991), Great Britain (Beeftink,

1977), and Australia (Hutchings and Saenger, 1987), alpine and montane wetland meadow in central China (Tsuyuzaki et al., 1990), and seasonal wetlands in Iran, Iraq, and Afghanistan (Scott, 1993), South America (Kohlhepp, 1984; Adamoli, 1992; Junk, 1993; Por, 1995), France (Bassett, 1980; Grillas, 1990), the United States (Louisiana; Chabrek, 1972), Australia (Mackay, 1990), and Africa (Fig. 1-26; van Rensburg, 1972; Rzoska, 1974; Burgis and Symoens, 1987; Moss, 1988; Chabwela, 1992). Cattle grazing is also common in North American cypress swamps (Duever, 1984), along streams and rivers (Elmore and Beschta, 1987; Smith, 1989; Sedell et al., 1991; Kauffman et al., 1995), sedge meadows (Bedford et al., 1974), and shrub carrs (White 1965).

In salt marshes, cattle were grazed and hay cut in New England by the earliest Puritans, (Russell, 1982), in England by the Saxons (Dalby, 1957), and in Canada by Acadians near the Bay of Fundy (Mallik, 1990a), and by others in Newfoundland (Roberts and Robertson, 1986). To facilitate haying, the Cowles Bog near the shore of Lake Michigan was shallowly ditched in the early 1900s (Wilcox, 1995). Other wetlands cut for hay include sedge meadows in Wisconsin (Bedford et al., 1974; Middleton, 1978) and monsoonal wetlands in India (Middleton, 1992). These practices either continue or have continued until fairly recently (Nixon, 1982; Russell, 1982; Doody, 1984; Mallik, 1990a).

***Figure 1-22.*** *Sambhar in a wetland at Ranthambhore National Park, India (photograph by G & H Denzau by permission).*

***Figure 1-23.*** *Heavily cut (lopped) tree in Rajasthan, India (photograph by Beth Middleton).*

Wood cutting in wetlands for fodder is common around the world. Branches of trees are lopped off in India for cattle (Fig. 1-23; Berkmuller et al., 1990). Handlers in Nepal cut saplings in riverine forests for domestic elephants (Lehmkul, 1994). Mangroves in southern Iran and the Gulf States (e.g., Qatar) are degraded by excessive cutting for fodder, as well as for browsing by camels (Scott, 1993).

The Camarguias in France have been grazing cattle in the wetlands of the Rhône River for centuries (Mitsch and Gosselink, 1993). To improve palatability for cattle, at least 30% of grasslands in Africa are burned every year, and, along with them, herbaceous swamps (van Rensburg, 1972; Paterson, 1976). In India, cattle grazing is a nearly ubiquitous activity in wetlands. In many parts of the world, cattle or water buffalo grazing either occurs or is proposed for national parks and bird sanctuaries, such as Keoladeo National Park in northern India (Figs. 1-24 to 1-26; Middleton, *in press*) and Palo Verde National Park in Costa Rica. This practice is often defended on the basis of improving the habitat for waterfowl (McCoy and Rodríguez, 1994) or increasing biodiversity (Shukla and Dubey, 1996; Middleton, 1998).

After logging of tropical lowlands, cattle grazing is one of the major causes of lack of reforestation (Buschbacher, 1986; Hecht, 1993). In the Amazon, pastures are abandoned within four to eight years (Buschbacher

***Figure 1-24.*** *Water buffalo in a village wetland in northern India (photograph by Beth Middleton).*

***Figure 1-25.*** *A boy and his dog tending cattle on the Sultanpur Bird Sanctuary, near New Delhi, India (photograph by Beth Middleton).*

*Figure 1-26.* Cattle in a rice field in Madagascar adjacent to Ranamofana National Park. The rice fields are converted wetlands that were formerly depressions in a tropical rainforest (photograph by Beth Middleton).

et al., 1992). After they are abandoned, lowland forest pastures regenerate forest unlike those originally on the sites (Aide et al., 1996). While the reasons for poor rainforest regeneration include seed predation, competition with grasses, degraded soils, and fire, the most important reason for the problem appears to be the inability of seeds to disperse to the deforested site (Aide and Cavelier, 1994).

The effects of cattle grazing vary widely across landscapes. In riverine systems, cattle grazing interferes with seedling regeneration (Mackay, 1990; Scott et al., 1993a) and ultimately decreases the production of woody debris and fish (Maser and Sedell, 1994). Stream types vary in their ability to withstand grazing (Swanson and Myers, 1994). Cattle overgrazing tramples stream banks and causes stream bed widening and shallowing, gradual stream trenching or braiding, degraded fish and insect habitats, and increased water temperature and velocity. Lowered terrestrial leaf input ensues (Behnke and Raleigh, 1978) as tree and/or macrophyte regeneration is diminished. Fencing to exclude cattle does not always result in the immediate restoration of stream width (Kondolf, 1993). Even after 14 years of cattle exclusion in Oregon, not all channel properties had adjusted (Magilligan and McDowell, 1997). In salt marshes, heavy cattle or sheep grazing produces floristically simple vegetation (Doody, 1984). In sedge meadows, tussocks are trampled and flattened from cattle hooves so that *Carex stricta*-dominated tussocks become invaded by blue grass (*Poa pratensis*; Bedford et al., 1974) and dogwood (*Cornus stolonifera*; Middleton, 1978).

Water buffalo grazing, once common in monsoonal wetlands such as those in Australia, produces heavily grazed wetlands with networks of trails and swim tracks, reduction in vegetation, invasion of alien species, increased turbidity, saltwater intrusion, and the reduction of crocodile breeding sites (Finlayson and Von Oertzen, 1993). Water buffalo, which infiltrate farther than cows into wetlands with deep water and soft bottoms, have been introduced into wetlands in Greece, Tunisia, and Italy (Britton and Crivelli, 1993), Iraq (Moss, 1988), Mesopotamia (Scott, 1993), Australia (Yapp, 1989; Finlayson and Von Oertzen, 1993), South America (Junk and Howard-Williams, 1984; Ohly 1987), and South Asia. In the Northern Territory of Australia, herd sizes range from 0.4 to 33.8 km-2 but are highest in coastal floodplains (Graham et al., 1982). In national parks in Australia, it is accepted that water buffalo interfere with conservation objectives, and they are being removed (Finlayson and Von Oertzen, 1993). Wild buffalo herds (*Bubalus bubalis*) are found in isolated pockets in Assam (Gopal and Krishnamurthy, 1993), but before the twentieth century they were common from Nepal to the Indian Archipelago (old name for the Malaysian islands) (Belsare, 1994).

***Farming*** Drainage for agriculture is the single most important reason for wetland drainage worldwide. In forested wetlands, logging is often followed by drainage for grazing and/or farming, both in the tropics (Veldkamp et al., 1992) and in North America (Newling, 1993; Wichman, 1996).

Farming quickly destroys the regenerative ability of certain wetland types. In bottomland cypress swamps of the southeastern United States, the seed bank of native dominants is almost completely destroyed after one year of farming. Many species of cypress swamps, because of their short-lived seeds, are dependent on frequent episodes of dispersal via flooding for natural regeneration (Middleton, 1995a, 1996, unpublished). Wetland species are more resilient in the Prairie Pothole region of North America; up to 60% of the species survive for 20 years (Weinhold and van der Valk, 1989).

Attempts to drain coastal Mediterranean areas for agriculture have failed because salinity increases after surface fresh water is removed (Britton and Crivelli, 1993). Many farmed wetland sites now lie below sea level because of the tendency of peat to decompose in aerated conditions, such as in English fens (Godwin, 1978; Taylor, 1979; Sheail and Wells, 1983).

Many attempts to drain peatlands have failed but have damaged wetlands anyway. Lake Hornborga in the Netherlands has been lowered five times since 1802 in aborted farming attempts. Currently, the elevation of the wetland is sinking so that the restoration of aquatic macrophytes is becoming difficult (Björk, 1974). In the drought of the 1930s, farmers in southern Wisconsin attempted to drain previously unfarmed wetlands, but these sites were abandoned when normal rains resumed. Subsequently, the peat in these abandoned farming sites became oxidized, allowing the invasion of noxious species such as ragweed, nettle, and shrubs (Bedford et al., 1974). In the Red Lake Peatland of Minnesota, ditches were dug in the early 1900s in a failed attempt to drain the area; nevertheless, hydrologic changes near ditches have influenced vegetation. Low-lying water tracks that have dried support more abundant stands of *Carex lasiocarpa*, while species such as *Drosera* sp., *Utricularia* sp., and *Scheuzeria palustris* are eliminated near the drainage ditches (Glaser and Wheeler, 1980).

***Other Human Disturbances*** Changes in groundwater level, often due to groundwater extraction for human consumption, are having disastrous effects on the world's wetlands and on our ability to restore them. Very few studies have explored the implications of this problem. The depletion of groundwater because of drinking and irrigation water extraction in Jordan has caused a series of spring-fed marshes and pools to dry up in the Azraq Oasis (Scott, 1993). In England and the Netherlands, groundwater abstraction (and other changes that alter the patterns of groundwater flow) is the primary cause of species composition shifts in **fen** communities (Fojt, 1994; van Diggelen et al., 1994). Groundwater and surface flooding dynamics can be the determining factor in plant species composition (de Mars et al., 1997). Groundwater abstraction for agricultural production in Zanzibar may not threaten sensitive coastal mangrove communities riddled with natural springs because mangroves are adapted to high salinity (Snowden, 1995), but groundwater abstraction will likely affect the fresh/brackish wetlands inland from mangroves. In western riparian woodland (bosques) of North America, velvet mesquite (*Prosopus velutina*) become stressed as groundwater declines to 15–18 m, and at 18–30 m it is under sublethal stress. *Prosopus* is taller (>12 m) and has larger leaflets where the water table is less than 5 m from the surface (Stromberg et al., 1992).

In North America, commercial cranberry production is another human activity that disturbs wetlands by altering the vegetation in sedge meadows adjacent to the cranberry beds. Because of blowing sand and desiccation near the beds in south-central Wisconsin, otherwise uncommon species such as *Poa pratensis* and *Fragaria virginiana* are increasing in sedge meadows (Jorgensen and Nauman, 1994).

Thermal stress from nuclear power plant effluent damages vegetation.

Most of the floodplain vegetation along the Savanna River was removed downstream of the Savannah River Site because of the combined stress of heated water and associated siltation (Sharitz et al., 1974) between 1954 and 1968. While the site probably was dominated by *Taxodium distichum* before the disturbance (as indicated by stumps), after the cessation of power plant activities in 1968, *Salix nigra* and *S. caroliniana* dominated the site (Muzika et al., 1987). In 1989, undisturbed forested coastal plain was compared to sites disturbed by heated effluents, flooding, and siltation. The disturbed sites had a much higher species richness of shrub and herb taxa, while the undisturbed sites had a higher species richness of trees and vines. The disturbed sites also had stands of very dense trees of species such as *Salix* sp. and *Alnus* sp., while undisturbed sites had larger-diameter trees of *T. distichum* or *Nyssa aquatica* (Firth and Hooker, 1989).

Commercial harvesting of peat has not been common in North America, nor has agriculture in northern peatlands because of the climate. However, for centuries, peat has been cut either directly for fuel, as in the Netherlands (Vermeer and Joosten, 1992) and Ireland, or burned to make steam, as in Finland and Russia (Larsen, 1982).

Highway construction can interfere with the free movement of water and therefore can disturb wetlands (Evink, 1980).

***Figure 1-27.*** *Ice along the shore of Stockton Island, Lake Superior (photograph by Beth Middleton).*

***Other Natural Disturbances*** In general, disturbance in wetlands has received relatively little study considering its important role in these highly dynamic systems (McKee and Baldwin, in press). In addition to those already listed, many other natural disturbances are important in natural and restored wetlands.

Earthquakes created a series of ponds and wetlands for 100 miles on the eastern side of the Mississippi River after the New Madrid earthquake of 1811 (Dennis, 1988). Alternatively, earthquakes can uplift areas as after the 1964 earthquake on the Copper River Delta of Alaska. The entire delta rose by 1.8 to 3.4 m, and the marshland moved seaward by 1.5 km (Thilenius, 1889).

Ice (Fig. 1-27), and particularly wrack deposition after storms, depending on its size and duration, are the most important factors in determining species composition in coastal wetlands such as salt marshes (Hartman, 1983 in Roberts and Robertson 1986). Burial by sediment or wrack is also an important disturbance in other wetland types (McKee and Baldwin, in press). Single trees can be killed by lightning strike, windfall, or senility, as in bottomland forests (Fredrickson, 1979). Other potentially important disturbances include snow, frost, freezing, glaciation, mountain uplift (Vogl, 1980), and insect damage in mangroves (Feller and McKee, in press).

# *Part II*

# *Disturbance Dynamics and Life History Strategies in Wetlands*

# 2

# *Restoration Theory*

*Restorationists typically spend little time dwelling on ecological theory; however, their ideas on the subject direct their approach to restoration. For example, a manager who attempts to maintain a restored prairie wetland without water fluctuation denies the importance of disturbance and/or environmental change in wetlands. This person's views may be slightly more aligned with the Clementsian school of succession, where disturbance is seen as interference in the progressive march of a community toward its climax. However, neither Gleasonian nor Clementsian succession theory suggests that it is possible to maintain a wetland in static conditions. A knowledge of ecological theories can arm managers with powerful approaches in their work.*

*Gleasonian tenets claim that succession, or vegetation change, can be reduced to the responses of individual species to the environment within the constraints of their life histories, that disturbance is an integral part of the process, and that species act more or less independently of one another (Gleason, 1926). Clementsian tenets claim that the wetland metamorphizes as a whole through stages to its climax, that disturbance interrupts this progression, and that species are dynamically linked within each stage (Clements, 1916).*

*Whether to replant a restoration site or to allow the site to revegetate on its own is the heart of the design versus self-design controversy and is illuminated through a discussion of succession theory. These restoration design theories are restatements of Gleasonian and Clementsian viewpoints that have been updated by recent research. Gleasonian reductionist notions underpin the Gleasonian succession model for wetlands (van der Valk, 1981), which is based on the idea that the life history parameters of species are the operational foci of successional change. Odum (1969) fixed trajectories of thermodynamics to Clementsian stages as part of their orderly ecosystem change through time. The purpose of this chapter is to review ecological theories relevant to restoration ecology, including self-design versus design, ecological succession, invasion, and river dynamics theories.*

## THEORY IN WETLAND RESTORATION

### Self-Design versus Design Theory

Many restoration practitioners feel that theory plays little role in their endeavors and that ecologists can learn from restoration experiences (Bradshaw, 1987). Nevertheless, more attention to theory could avoid unnecessary expense and failure. Self-design versus designer theories are the only ones that claim to be conceived especially for restoration, but according to van der Valk (in press), these are just restatements of earlier theories of **Clementsian** versus **Gleasonian succession**.

Theoretical considerations should enter early into the project planning of the restorationist. After restoring the hydrology of a restoration site, a restorationist's next dilemma is whether to replant or to allow the site to revegetate naturally. The decision will likely be dictated less by circumstances than by the manager's thoughts on the relative ability of a restoration site to restore itself over time, a self-design versus design issue, and thus the emphasis in this chapter on theory.

**Self-design**, at least from a restoration perspective, is the idea that over time a restored wetland will organize itself around and eventually alter its engineered components. In this approach, the restorationist may or may not plant wetland species at the site, but ultimately it is the environmental conditions there that determine the vegetative outcome. This strategy deemphasizes interventionist approaches and views wetland development as an ecosystem-level process (Mitsch and Jørgensen, 1989). Self-design is a restatement of the holistic community views of F. E. Clements (1916) and their later reorganization into ecosystem-level approaches by H. T. Odum (1971; van der Valk, in press).

The **designer** wetland approach is not associated with any particular person, but is more likely to be practiced by restorationists trained in landscape architecture, horticulture, forestry, or agronomy. Following reductionist reasoning, the most important factor in the success of the revegetation of a restoration site is an understanding of the life history of the species involved, that is, the nature of their dispersal, germination, and establishment requirements (van der Valk, in press). Both self-design and design theories can be seen as having been reinvented for the purposes of restoration (see Box 2-1). Succession theory will be discussed in detail in the next section of this chapter.

According to the self-design concept, as elaborated by Mitsch and Jørgensen (1989), the nature of the environment dictates the plant species that can be maintained in the restoration site. Even if diverse species are planted at many different water depths, these will sort themselves out and species will establish only in those environments that suit them (Mitsch, 1996; Mitsch, in press). The logical extreme of this approach is not to replant restoration sites at all but, instead, allow them to revegetate naturally. In a test of this idea, Mitsch found that after 3 years little functional difference could be found in a planted versus unplanted created wetland (Mitsch, in press). However, proponents of the self-design concept sometimes encourage species introductions into restoration sites as a way of speeding up the revegetation process (Odum, 1989). Self-designers also stress that the return of wetland function to a restoration site may span 15–20 years, and longer in peatlands, forested wetlands, or coastal wetlands (Mitsch and Wilson, 1996). The key to deciding whether an approach leans toward self-design or design is not whether the restorationist has replanted or not, but rather the relative weight given to ecosystem versus life history processes in the restoration of a site.

Far from opposite to the ideas of self-design in their technical approach, proponents of the designer wetland approach emphasize that, because restoration sites may have depauperate seed banks and/or dispersal limitations, the reintroduction of vegetation may require planting or seeding. Following design theory, the restorationist pays close attention to the life history requirements of the species involved. Designer wetland proponents claim that it is not a matter of time, but rather of intervention, that determines the ultimate outcome of a restoration site (van der Valk, in press).

That the designer model of wetland restoration is a restatement of Gleasonian succession is demonstrated aptly by Figures 2-1a and 2-1b (van der Valk, in press). In the design model, the life history constraints of species are emphasized. From a theoretical perspective, the designer model of restoration (Fig. 2-1b) differs not at all from the Gleasonian succession model (Fig. 2-1a) with the added feature that, because of deliberate manipulation, seeds, seedlings, and adult plants may be introduced into the successional sequence and thus influence the outcome. Human intervention is often stressed in designer approaches in that, without the reintroduction of species into the restoration site by the sowing of seeds, and the planting of seedlings and adults, a site with very low **species richness** might result.

*BOX 2-1*

## SUCCESSION REINCARNATED: RESTORATION SELF-DESIGN/DESIGN THEORY

According to van der Valk (in press), self-design and design restoration approaches are merely restated succession theories dating from the early 1900s.

Design is an approach in wetland restoration that views the life history strategy of species as the important factor in developing vegetation on a restoration site. The view favors engineering and replanting strategies directed producing a wetland type with no fixed endpoint. This strategy emphasizes interventionist approaches (e.g., reengineering of hydrology to encourage dispersal and germination from extent seed bank, replanting) based on predictable outcomes. At its heart, design is a restatement of Gleasonian (1917, 1926, 1927) succession theory as applied to wetlands by van der Valk (1981; in press).

Self-design is an approach to wetland restoration that emphasizes the ability of a wetland, given enough time, to organize itself around engineered components. This strategy deemphasizes interventionist approaches and views wetland development as an ecosystem-level process, a reworking of Odum's (1969) succession approaches (Mitsch and Jørgensen, 1989; Mitsch, in press), which are philosophically allied with Clements (1916) at least in that the ecosystem approach recognizes a higher level of species interlinkage than the reductionist approach of Gleason (van der Valk, in press). Similar to Clementsian succession from this perspective, self-design theory suggests that ecosystem transformation over time is analogous to evolution, with the arrival of species akin to mutation in an organism (Mitsch, in press).

So, should a restoration site be thought of as the sum of its components or as something more than that? The self-design/designer controversy crystallizes around this point and is an idea of importance to restorationists.

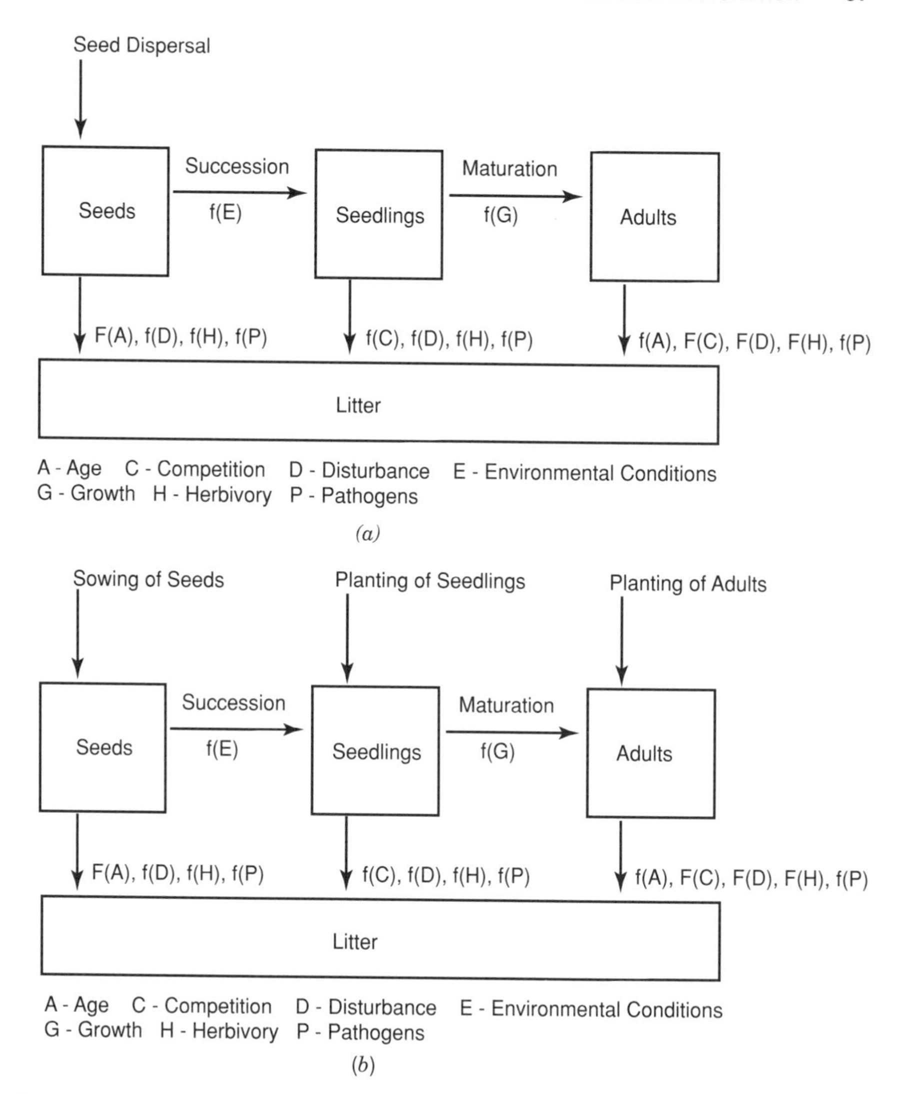


***Figure 2-1.*** *(a) Reductionist succession model. (b) Reductionist restoration model (design theory). Feedback loops and interactions within the system are potentially numerous and not shown (from van der Valk, in press; copyright © by van der Valk by permission).*

Thus, the two restoration design camps differ largely with respect to the emphasis placed on ecosystem-level versus population-level processes. Designer restorationists see vegetation development as the enterprises of individual species, while self-designers view it as a holistic process of the ecosystem. The views are not polar opposites, but they do affect the approach of the restorationist.

A very practical question emerges from this theoretical controversy. Can restored wetlands revegetate without human intervention? The answer varies with the situation, according to the limitations of the extant seed bank and dispersal, and brings out many additional theoretical considerations (see the next section). Also, it should be borne in mind that according to some, this question may overemphasize the importance of vegetation in restored wetlands. Vegetative composition by itself may be a poor indicator of wetland function, and other success criteria should be developed (Reinartz and Warne, 1993; Palmer et al., 1997).

Nevertheless, wetland restoration cannot be successful without the instigation of vegetative succession processes. The role humans can play in the success of vegetative redevelopment is an open question. Relatively little study has been directed toward understanding the role of replanting in reestablishing self-sustaining wetlands from the vegetative perspective.

The success of prairie wetland restoration is somewhat better studied than that of other systems. While a majority of species can survive in extant seed bank for long periods during farming (60% for 20 or more years of farming; Weinhold and van der Valk, 1989), key species are typically missing from the mix. Typically, the best and perhaps only feasible way to reestablish the sedge meadow and wet prairie zones at the periphery of a prairie wetland is by replanting (*Carex* spp.; Galatowitsch, 1996). A prairie wetland composed almost solely of cattail should not be considered a success (Odum, 1988; Galatowitsch and van der Valk, 1994).

One study of small depressional marshes by Reinartz and Warne (1993) in southern Wisconsin looks at the question of whether restored wetlands should be replanted or not, thus illuminating the self-design/designer wetland controversy. Each of the 11 created wetlands on Conservation Reserve Program (CRP) sites had a negligible wetland seed bank at the beginning of the restoration activities because farming had destroyed all but the wind-dispersed wetland species. Note that the seed banks of all wetlands were not checked but that they all had similar agricultural histories (Warne, 1992). After two to three years, an average of 22 species was present in naturally colonized wetlands. More species were present in wetlands where seeds had been introduced (Reinartz and Warne, 1993).

In these wetlands in southern Wisconsin, the species richness of wetland species increased with proximity to the nearest wetland, as well as with wetland age and size, in good agreement with the theory of island biogeography (MacArthur and Wilson, 1967), discussed later in this chapter. After two years, the seeded sites had a higher species diversity and lower cover of cattail than the unseeded sites. Reinartz and Warne predicted that undesirable monocultures of cattail with a fringe of cottonwood (*Populus deltoides*) would develop on unseeded restoration sites. Reinartz and Warne

(1993) probably lean more strongly toward the designer wetland school of thought as they consider the characteristics of species responsible for revegetation and subsequently recommend seeding in small created wetlands to avoid Odum's (1988) ''cattailization of America.''

Ultimately, a restored wetland must be capable of the temporal sequences of vegetation change (succession) characteristic of its type, as described later in this chapter. For vegetation to reestablish after natural disturbance (see Chapter 1), it is necessary for the wetland either to develop a seed and/or propagule bank or, alternatively, to be reintegrated with the landscape so that seeds and/or propagules can disperse to the wetland to fill vegetation gaps after recurring disturbance events. In prairie wetlands, as well as in some other natural wetland types with strong seasonal or climatic water level flux, the reestablishment of emergent vegetation during drawdown is dependent on the presence of a well-developed seed bank (van der Valk and Davis, 1978; Finlayson et al., 1990; Middleton et al., 1991; Haukos and Smith, 1994).

Floodplain wetlands in some cases provide examples of systems that can revegetate naturally. While these wetlands can have very short-lived seed banks, they can be capable of revegetating themselves by dispersing seeds in portions of riverine wetlands where flood pulsing environments are recreated (Middleton, 1995b, in review). Cypress (*Taxodium distichum*) of the southeastern United States reestablishes naturally in swamps at elevations with winter flooding and protracted summer drawdown. In impounded wetlands, a ring of regenerating cypress can often be found at the highest flooding elevations (Klimas, 1987; Middleton, in review). Similarly, cottonwood regenerates along the impounded Upper Missouri River only in the high-medium flood zone (Scott et al., 1993a, 1993b).

The extent to which created wetlands achieve the ability to reestablish their vegetation after disturbance should be part of their functional evaluation. If an allegedly restored wetland cannot revegetate a diverse assemblage of species after natural disturbance, then it is not a functional wetland. For example, while coal slurry ponds in Illinois acquire some wetland function, these have little innate ability to undergo the long-term successional changes required to maintain vegetation. Even after relatively long periods of time (>40 years), seed banks are mostly composed of exotics and annuals, that is, poorly developed in comparison to reference (natural) wetlands in the region (Middleton, 1995c).

A landscape-level criterion that ought to be applied to created wetlands is that wetlands be required to be self-maintaining within their hydrogeologic and climatic setting (Bedford, 1996; Parker, 1997). Once an ecosystem is restored, it should require little further human intervention (Mitsch, 1993). Coal slurry ponds fail this test of self-sustainability. Drawdown, which is required for the long-term maintenance of vegetation, is both undesirable and unfeasible because it produces acid conditions in the coal slurry pond (Middleton, 1995c). The flooding is designed primarily to prevent acid mine drainage, not to create wetlands.

## Clementsian versus Gleasonian Succession

**Succession** and restoration are largely the same phenomenon. Wetland restoration involves the reestablishment of wetland hydrology and environment, vegetation and function. Functional attributes such as maximum biomass are generally a direct result of the reestablishment of the vegetation; thus, an understanding of succession is essential to successful restoration (van der Valk, in press).

The Gleasonian succession model developed by van der Valk (1981) was the first to stress the importance of the life history attributes of species to plant development in wetlands. It is similar to the vital attributes approach developed for Australian forests by Noble and Slatyer (1977), following the earlier inhibition model of Connell and Slatyer (1977). Inhibition is the idea that species establishment is limited by **preemption** throughout succession (Connell and Slatyer, 1977). Very recently, vital attributes have been applied to landscape characteristics (Aronson and Le Floc'h, 1996; Noble and Gitay 1996).

In these models based on vital attributes, the life history characteristics of species are the key to predicting vegetation changes; these characteristics include seed dispersal limitations, longevity, germination, and establishment requirements. Reductionist approaches to succession theory originated with H. D. Gleason (1917), developed as a response to the holistic approach of F. E. Clements (1916). Not surprisingly, Australians, the originators of the vital attributes idea, never followed Clementsian succession because disturbance, particularly fire, was such an overpowering influence in their landscapes (Noble and Slatyer, 1977). For decades, the ideas of Clementsian succession dominated the thinking of ecologists both in the United States and in Europe, at least among ecologists in the Braun-Blanquet school (Naveh and Lieberman, 1984).

The Gleasonian succession model, designed for prairie potholes in North America, predicts cyclic wetland dynamics (van der Valk and Davis, 1978; Fig. 2-2). These vegetation changes (maturation and fluctuation) are a response to recurring disturbances (drought, high water, herbivory) in the wetland. In the dry marsh phase of the cycle, drawdown occurs because of drought at 5- to 30-year intervals. Some portions of the marsh remain flooded and are thus noncyclic. During drawdown, species with an annual life history strategy germinate, set seed, and die. Emergent vegetation is

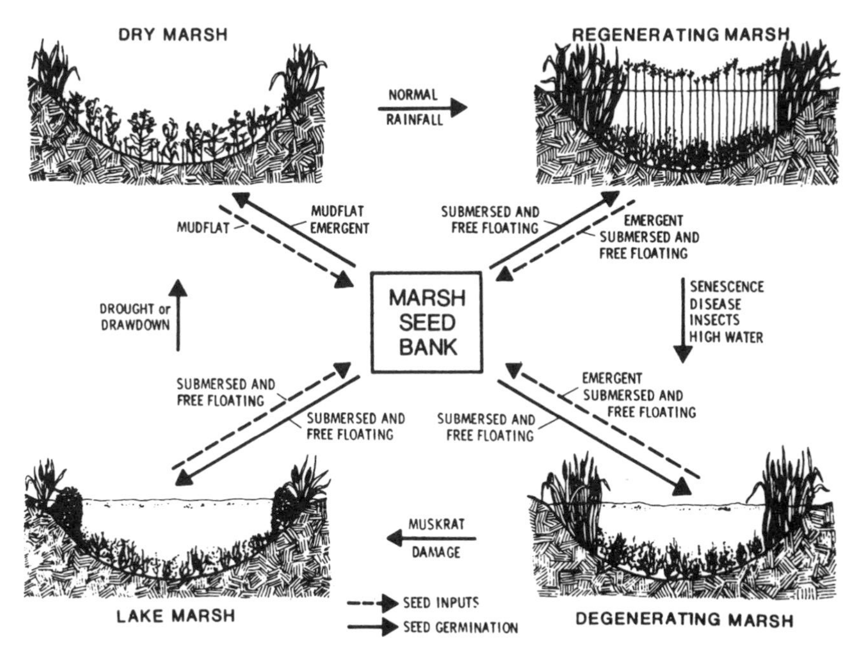


***Figure 2-2.*** *Vegetation succession in prairie pothole wetlands. The cycle is driven by disturbances (drought, flooding, herbivory) and the responses of species based on their life history characteristics (dispersal, germination, establishment; from van der Valk and Davis, 1978; copyright © by the Ecological Society of America by permission).*

able to germinate by seed from the **persistent seed bank** or is spread vegetatively to reestablish on wet mud. Drawdown, as an environmental factor, represents a phase of the "environmental sieve" that allows certain species to enter the wetland vegetation (Fig. 2-3; van der Valk, 1981). Following the definition of Gitay and Noble (1997), groups of species such as annuals or emergents should be referred to as **functional groups**, or organisms with similar responses to disturbances.

When normal rains resume and the pothole refloods, the emergent species dominate the regenerated pothole until high water and herbivores remove the vegetation. At the same time that muskrat activity (Hayden, 1939) and/or high water create successively less vegetated wetland and changes in nesting marsh bird populations (Weller and Spatcher, 1965) invertebrate populations occur (Fig. 2-4; Voigts, 1976). During the degenerating marsh stage, the wetland becomes increasingly open until a lake marsh develops. The wetland remains open until the next drawdown associated with drought (van der Valk and Davis, 1978).

Initial floristics, an applicable related theory for upland situations, predicts that at least for secondary successions, the ultimate vegetation of a

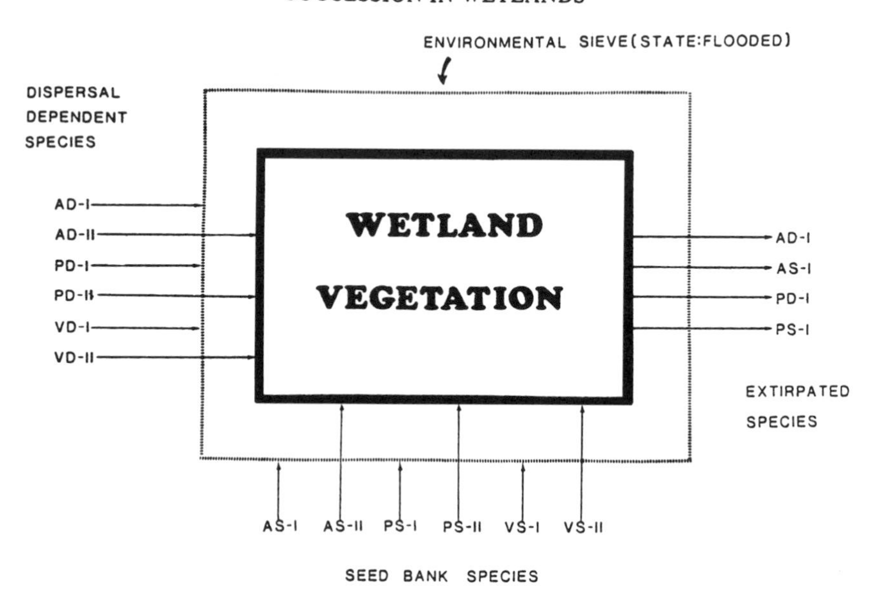


***Figure 2-3.*** *The environmental sieve. Emergent and annual species are unable to germinate under flooded conditions and so are environmentally barred from entering the wetland vegetation (from van der Valk, 1981; copyright © by the Ecological Society of America by permission).*

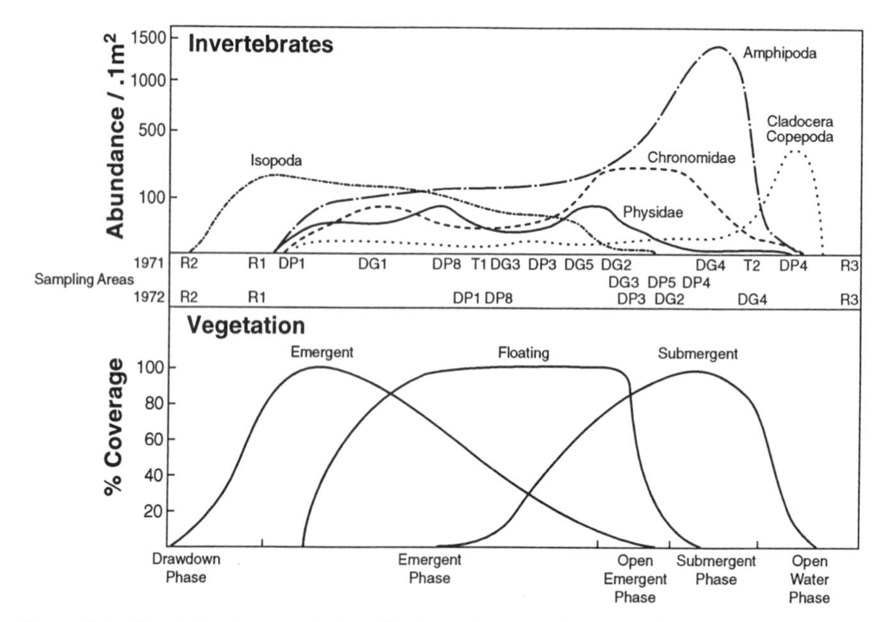


***Figure 2-4.*** *Vegetation-insect relationships in various marsh stages (from Voigts, 1976; copyright © by American Midland Naturalist by permission).*

**a.** Clements/Pearsall - Classical Succession

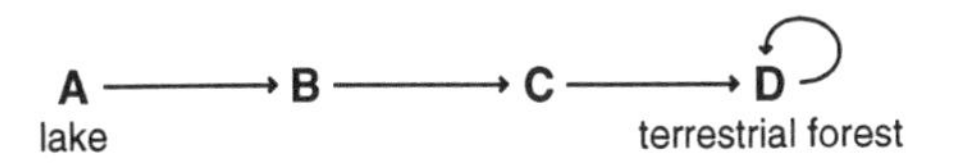


**b.** van der Valk (Gleasonian) Vital Attributes Approach

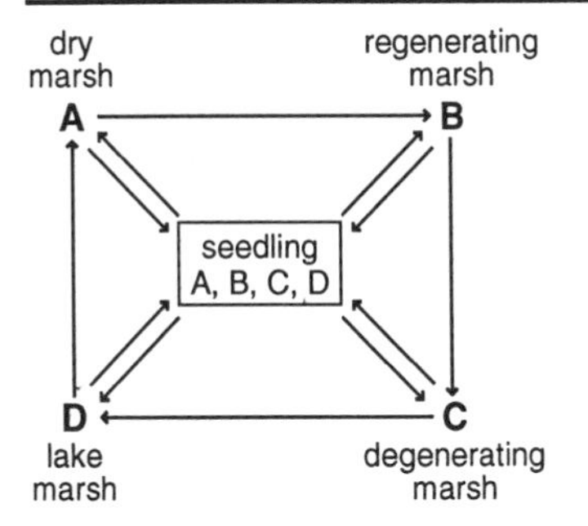


**c.** Egler-Initial Floristics

***Figure 2-5.*** *Replacement sequences with A, B, C, D representing hypothetical dominant species. Lowercase letters stand for subdominants. (a) Classical succession (Clements, 1918). (b) Gleasonian succession (van der Valk and Davis, 1978). (c) Initial floristics (Egler, 1977) (based on Noble and Slatyer, 1977; copyright © by the Ecological Society of Australia by permission).*

community is present within those species that initially colonize the site (Fig. 2-5c; Egler, 1977; also see Robinson and Dickerson, 1987). On Alaskan floodplains, both pioneer (willow, alder, poplar) and late-forest (spruce) species are present within the first five years of silt bar formation (Walker et al., 1986). Similarly, along desert streams with flash flooding in Arizona, many taxa recolonize within the first week of the flooding event (Fisher et al., 1982). Many species that become dominant sometime after a disturbance are present immediately after the disturbance (Noble and Slatyer, 1980).

Clementsian or classical succession (Clements, 1916) has been applied to aquatic systems by Pearsall (1920; Fig. 2-5a), Gates (1926; Fig. 2-6), Wilson (1935), and Wetzel (1983). The idea was that successional stages replaced each other through both allogenic and autogenic processes until the lake filled in to become a terrestrial system. Each stage or sere produced changes in the environment that made way for the next stage until a climax (Box 2-2). This idea is also referred to as *relay floristics* (Egler, 1977) or *facilitation* (Connell and Slatyer, 1977). The emphasis in the classical model was placed on holistic processes, in contrast to the reductionist notions of the Gleasonian model. Clements held the extreme view that the

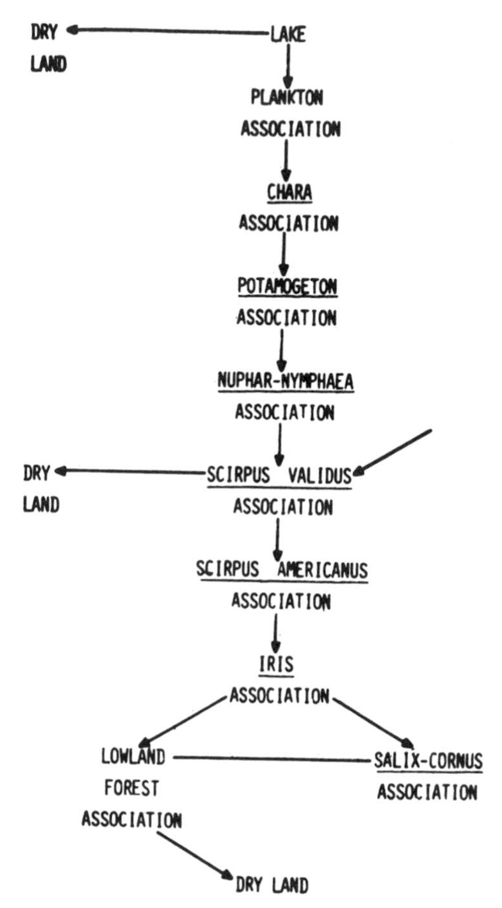


***Figure 2-6.*** *Hydrarch successional sequence proposed for Douglas Lake, Minnesota (from Gates, 1926; copyright © by Botanical Gazette by permission).*

community was the equivalent of an entity or *superorganism* that grew toward a fixed climatic climax determined by macroclimatic conditions (Clements, 1928). Many wetland scientists now suggest that the idea of a regional climax as applied to wetlands is inappropriate (Mitsch and Gosselink, 1993).

Succession can also be defined as a community development process that is predictable and reasonably directional arising from changes in the environment and biomass accumulation (Margalef, 1963; Odum, 1969). In this view, at any given point in the succession, the most successful species combinations are those that divide up available resources to maximize productivity. Species substitutions are made as conditions change through time (Dunn, 1989). The end result is a stable ecosystem with predictable char-

*BOX 2-2*

## DECONSTRUCTING SUCCESSION THEORY

Despite the growing outcry from ecologists, introductory textbooks of biology continue to present to students some of the discarded succession theories of F. E. Clements (1916), that is, that communities progress through a series of stages that culminate in a climax.

The problem begins early in the training of students (Gibson, 1996). The Biological Science Curriculum Study (BSCS) (1992) high school text (green, environmental version), states that "the process by which one type of community replaces another is called succession," followed by "the community that ends a succession, at least for a long period of time, is called the climax community" (p. 619).

Concerning Gibson's (1996) charge that high school students are being taught outdated ideas on succession, McInerney (1996), director of BSCS, countered that BSCS states later in their text: "Recently, many ecologists have challenged this traditional model of succession in which predictable and orderly changes result in a climax community. Most communities are in a continual state of flux, or change, and the types and number of species change throughout all stages of succession, even at the so-called climax stage" (p. 624).

Recently? The first challenges to the Clementsian perspective were voiced about 100 years ago. While it is true that Gleason's (1917) challenges were ignored for years, by the 1950s, researchers such as Whittaker (1953), Raup (1957), and many others had relegated Clementsian notions on climatic climax to the dustbin of ecological theory (van der Valk, in press).

So why are these outdated ideas taught to students? It is a mystery to most ecologists. Unfortunately, educators sometimes get one only chance, via high school instruction, to get the message right. And getting it wrong can have devastating environmental consequences. Clementsian notions likely are at the heart of the confusion of some politicians who see wetland protection as futile, believing that wetlands will eventually fill to become dry land (**hydrarch succession**). Some managers attempt to manage wetlands as static entities, an impossible feat following neither Gleasonian nor Clementsian tenets.

At the college undergraduate level, coverage is no better, at least in

*continued*

most ecology textbooks designed for nonmajors. According to a synopsis of the succession treatments in these texts, Gibson (1996) charges these texts with fostering the ecologically untenable ideas of Clementsian succession. As an example, Cunningham and Saigo (1997) state: ''Succession proceeds from open lake to shallow pond with highly vegetated edges to marshy area with rooted, emergent vegetation and finally to grassland or forest'' (p. 87). Later, the authors interject, ''the process of succession may not be as deterministic as we once thought'' (p. 88). Why teach nonmajors Clementsian succession at all, or, if we do, then why not as a historical feature of ecological thought? Happily, Gibson found coverage in texts designed for majors to be more in keeping with modern views of succession.

acteristics including maximum biomass and symbiosis between organisms (Table 2-1). Odum's model projects Clementsian stages of progressive change to a stable state (climax) in an ecosystem context (van der Valk, in press).

Few studies have attempted to test Odum's model. In one study along Sycamore Creek, a desert stream in Arizona, the trajectory of ecosystem development following flash flooding fit Odum's model poorly (Table 2-2). Only seven attributes fit or partially fit his model. The ecosystems did decrease in production per unit biomass ($P_g/B$) and did increase in total organic matter and reliance on detritus in the food chain. The seven remaining attributes did not fit Odum's projections of ecosystem development over time (Fisher et al., 1982).

Several points also argue against the original concept of Clementsian succession. There are no documented cases of vegetation change from open water to terrestrial forest transition (Box 2-3). Upland is more likely to become bog, as discussed below. Regarding the idea that species are closely tied to one another, palynological evidence indicates that while changes in vegetation over many thousands of years in North America have oscillated with climate fluctuations, these vegetation changes have been complex and dependent on individualistic species responses (Overpeck et al., 1992). In addition, since communities have no DNA—that is, are not organisms in the genetic sense—they have no mechanism to evolve or change as a unit.

With temporal modification, the Gleasonian model can be applied to tropical and subtropical monsoonal wetlands in Australia (Fig. 2-7; Finlayson, 1991a, 1993; Finlayson et al., 1990, and India (Middleton, 1989; in press; Middleton et al., 1991). One of the major differences in the application of the Gleasonian succession model to prairie potholes versus

**TABLE 2-1. Trends Expected in the Development of Ecosystems**

| Ecosystem Attribute | Developmental Stages | Mature Stages |
|---|---|---|
| | *Community Energetics* | |
| 1. Gross production/community respiration (P/R ratio) | Greater or less than 1 | approaches 1 |
| 2. Gross production/standing crop biomass (P/B ratio) | High | Low |
| 3. Biomass supported/unit energy flow (B/E ratio) | Low | High |
| 4. Net community production (yield) | High | Low |
| 5. Food chains | Linear, predominately grazing | Weblike, Predominantly detritus |
| | *Community Structure* | |
| 6. Total organic matter | Small | Large |
| 7. Inorganic nutrients | Extrabiotic | Intrabiotic |
| 8. Species diversity—variety component | Low | High |
| 9. Species diversity—equitability component | Low | High |
| 10. Biochemical diversity | Low | High |
| 11. Stratifiction and spatial heterogeneity (pattern diversity) | Poorly organized | Well organized |
| | *Life History* | |
| 12. Niche specialization | Broad | Narrow |
| 13. Size of organism | Small | Large |
| 14. Life cycles | Short, simple | Long, complex |
| | *Nutrient Cycling* | |
| 15. Mineral cycles | Open | Closed |
| 16. Nutrient exchange rate between organisms and environment | Rapid | Slow |
| 17. Role of detritus in nutrient regeneration | Unimportant | Important |
| | *Selection Pressure* | |
| 18. Growth form | For rapid growth (r-selection) | For feedback control ("K-selection") |
| 19. Production | Quantity | Quality |
| | *Overall Homeostasis* | |
| 20. Internal symbiosis | Undeveloped | Developed |
| 21. Nutrient conservation | Poor | Good |
| 22. Stability (resistance to external perturbations) | Poor | Good |
| 23. Entropy | High | Low |
| 24. Information | Low | High |

*Source:* Reprinted with permission from Odum (1969). Copyright © 1969 American Association for the Advancement of Science.

**TABLE 2-2. Comparison of Trends in Postflood Recovery of Sycamore Creek with Those Predicted by Odum (1969)**

| Ecosystem Attribute | Odum 1969 | Sycamore Creek | Agreement |
|---|---|---|---|
| 1. P/R | Approaches 1 | Rises from <1 to >1 | No |
| 2. P/B | Decreases | Increases rapidly, then declines | Yes |
| 3. P/chlorophyll a | Not considered | Decreases | — |
| 4. $P_n$ (yield) | Decreases | Increases | No |
| 5. Food chains | Grazing→detritus | Detritus→grazing→ detritus | Yes |
| 6. Total organic matter | Increases | Increases | Yes |
| 7. Inorganic nutrients | Increasingly intrabiotic | Increasingly intrabiotic | Yes |
| 8. Species diversity | Increases | Depends upon group | |
| Algae, H' | — | Relatively stable | No |
| Invertebrates, H' | — | Fluctuating | No |
| Invertebrates, richness | — | Relatively stable | No |
| 9. Biochemical diversity ($O.D._{480}/O.D._{666}$) | Increases | Stable | No |
| 10. Size of organisms | Increases | Depends on group | |
| Algae | — | Increases | Yes |
| Invertebrates | — | Remains small | No |
| 11. Life cycles | Increasing length and complexity | Stable (short, simple) | No |
| 12. Role of detritus in nutrient regeneration | Increasingly important | Increasingly important | Yes |
| 13. Nutrient conservation | Improves | Improves | Yes |
| 14. Resistance to perturbation | Improves | Remains low | No |

*Source:* Fisher et al. (1982).

monsoonal wetlands is that monsoonal wetlands typically dry each year (Fig. 2-8) in response to annual drought. Prairie potholes dry much less frequently (Middleton, in press).

In monsoonal wetlands and elsewhere, herbivores remove emergent vegetation because these species typically die or do not regrow when cut underwater (Singh et al., 1976; Sale and Wetzel, 1983; Middleton, 1990; Blanch and Brock, 1994). Cutting emergent species underwater causes oxygen levels in their roots to drop (Sale and Wetzel, 1983; Weisner and Granéli, 1989; Mathis, 1996). Herbivores thus create openings in vegetation (Middleton, 1990) that are revegetated by emergents during drawdown. Ultimately, the intensity of herbivory determines the extent of the open lake stage that develops in the course of the monsoonal year (Middleton, 1992, 1994b). Revegetation in monsoonal wetlands occurs during seasonal drought, but succeeds mostly by vegetative means rather than through germination from the seed bank (Middleton, 1989; in press).

In northern peatlands, classic **hydrosere succession**, or the idea that shallow lakes first infill by organic material and then proceed through stages

*BOX 2-3*

## FOR OVER 20 YEARS, $10000 REWARD UNCLAIMED

In an unusually bold move for an academic, ecologist Frank Egler (deceased) put up a bounty for evidence supporting Clementsian succession (Egler 1977; relay floristics). The independent panel has never awarded the money.

A group of individuals did once challenge the claim but were unsuccessful because they could provide no evidence that the so-called invading species became established before or after the ''invaded'' vegetation. The researchers showed Dr. Egler the site with both shrubs and trees but were not themselves sure if the shrubs and trees had invaded the site first. These challengers could not demonstrate even a single stage of succession (John Anderson, personal communication).

### *$10,000 Challenge*

*Challenge*, to any believer in ''plant succession to climax.''

1. *Frank E. Egler*, hereby and herewith agree to wager any sum up to

**Ten Thousand Dollars ($10,000.00)**

against an equal amount, the money to be donated to a non-profit organization scientifically investigating the subject of Vegetation Change under natural or seminatural conditions, thru a period of more than 25 years, if any such believer will produce the evidence, either from the published scientific literature, or from unpublished research.

I stipulate that such research must support the *Belief* that natural and seminatural Vegetation change is a cause-and-effect phenomenon of ignoring and outgoing populations of plants, involving at least five stages, as indicated in diagrams published by me, in the sequence referred to as classical ''Relay Floristics.'' Any contender will give advance notice in writing. He will prepare to submit all evidence in writing within six months of that time to a Committee of Six Judges composed of ecologists Roland C. Clement, William H. Drury, William A. Niering, Ian C. T. Nisbet, and any two others they may appoint. The decision by the judges will be reached within six months of the date of submission of the evidence.*

*I have made it known for many years that I would award a prize of one thousand dollars to any ecologist who would supply published or unpublished data to support the succession-to-climax Belief. I still have the money.

The present challenge is modeled after an analogous challenge made by Harry Houdini a half-century ago. Mr. Houdini spent much of his life exposing and duplicating the phenomena of me-

*continued*

diums at seances, including such behavior as clairvoyance and mental telepathy. (The wish is still with us, now dignified by being called "psychic research." See Time magazine for March 12, 1973, pp. 110, 112, on the interest of the Stanford Research Institute, one of the world's finest think-tanks. I do think that all such phenomena should be scientifically investigated!) Houdini wagered a similar sum "if any Spiritual will produce a medium presenting any physical, so-called psychical manifestation that I cannot reproduce or explain as being accompanied by natural means." Many compulsions to "believe" rise perennially in the human breast. Not all of them are a danger to human society.

*Egler's challenge for evidence of Clementsian succession. (Redrawn from Egler, 1977; copyright© by permission of Aton Forest.)*

## Wetland Succession

### Monsoonal (1 year cycle)

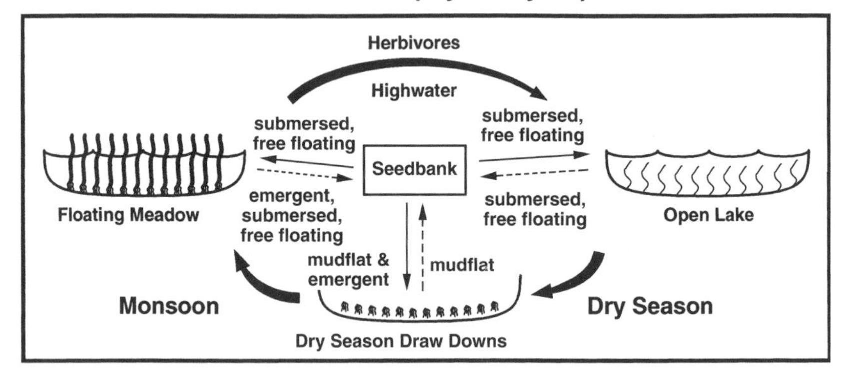


### Prairie Pothole (5 - 25 year cycle)

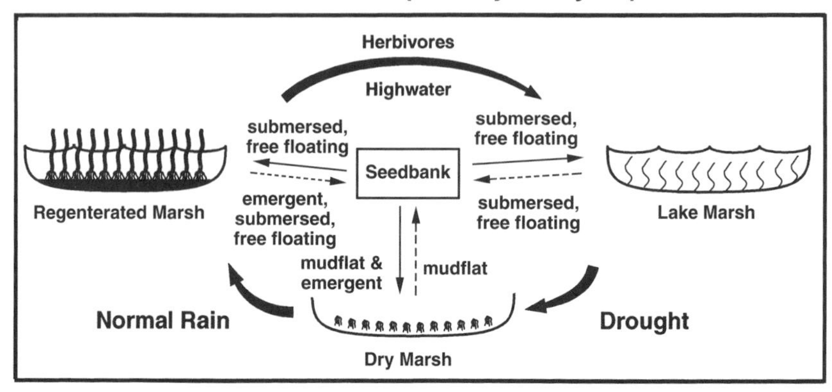


***Figure 2-7.*** *Generalized model of wetland succession in wetlands with fluctuating water regimes (from Middleton, in press, adapted from van der Valk, 1981; copyright © by the Ecological Society of America by permission).*

*Figure 2-8. Drawdown in a monsoonal wetland, Keoladeo National Park, Bharatpur, India (photograph by Beth Middleton).*

to a mature upland forest climax, also has been rejected (Klinger, 1996b). There are no studies that support the idea that **bogs** become upland. Instead, **paludification** can cause upland to become bog (Katz, 1926; Zach, 1950; Sjors, 1963; Heinselman, 1970; Damman and French, 1987; Crum, 1988; Klinger et al., 1990; Klinger, 1996a). Shallow lakes may fill with organic material to become bogs via **terrestrialization** (process of infilling of a shallow lake to become a bog). Here the progression goes no further toward upland forest; these terrestrialized bogs remain bogs, unless disturbed, and do not become upland. Klinger (1996b) argues that the ombrotrophic bog is a climax state resulting from either paludification or terrestrialization (Fig. 2-9), although other authors suggest that a bog or **muskeg** (in Canada, a bog) climax is unsupported because there is either no direction or more than one direction to the changes in these peatlands (Sjörs, 1963; Heinselman, 1970).

Palynological evidence also suggests that the endpoint of vegetational change in British wetlands is a bog, and a transition from fen to oak wood is not supported by any stratigraphic evidence (Fig. 2-10; Walker, 1970). Heinselman (1970) concluded, based on his study of the Lake Agassiz Peatlands in Minnesota, that the classical model of bog filling was not supported but instead that there had been a gradual ''swamping'' of the landscape via paludification and water table rise.

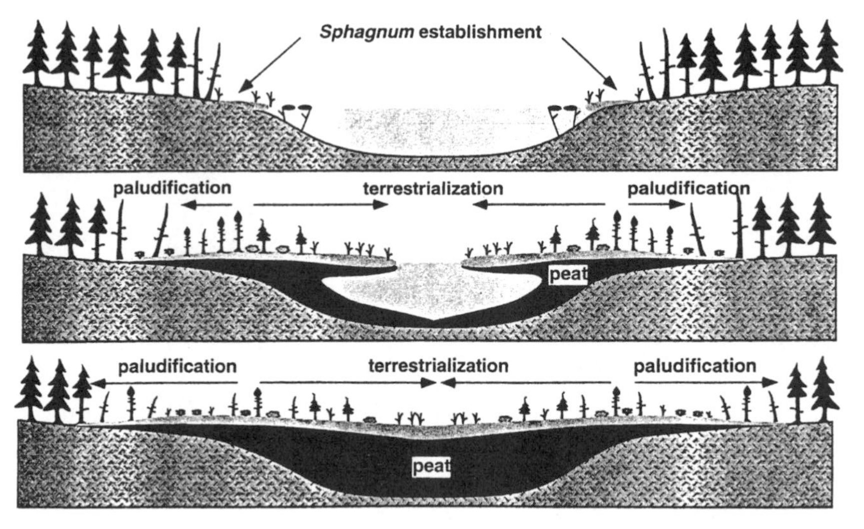


***Figure 2-9.*** *Bog climax model with bogs resulting from both paludification and terrestrialization. Key stages include (a) establishment of peat-forming mosses (Sphagnum)* along shorelines, (b) peat accumulation via both paludification and terrestrialization, and (c) continued peat accumulation resulting in the formation of an ombrotrophic bog (from Klinger, 1996b; copyright © by Regents of the University of Colorado by permission).

Postglacial sea level rise, not autogenic succession, is dominating wetland processes along the coasts of North America. Palynological evidence suggests that some hardwood forests in the Dismal Swamp of southeastern Virginia were replaced by freshwater marshes and cypress-gum (tupelo) swamps as sea levels rose (Whitehead, 1972). Salt water intrusion has accompanied increased tidal flooding in the Cape Fear Estuary of North Carolina and has been related to the presence of chloride in growth rings of cypress growing in some low-lying areas (Yanosky et al., 1995). In the recent past (~50 years), mangroves have been expanding landward into areas formerly dominated by cypress; this is likely due to both sea level rise and a reduced flow of fresh water from inland sources (Alexander and Crook, 1974). Similarly, in coastal wetlands of the Northeast, arrow-grass (*Triglochin maritima*) is replacing black grass belts as sea levels rise 2.5 mm per year (Warren and Niering, 1993).

In Lake Naivasha, Kenya, the primary impetus for vegetation change in the last 9200 years has been climatic and unrelated to Clementsian succession. As determined from microfossils, primarily diatoms, the lake was much larger from 9200 B.P. to 5650 B.P., a time when the climate was much wetter than subsequently. After that, aquatic macrophytes were common as the lake became smaller (even drying briefly). In the last 3000 years, the lake has varied in size but at times has been smaller than the modern lake (Richardson and Richardson, 1972).

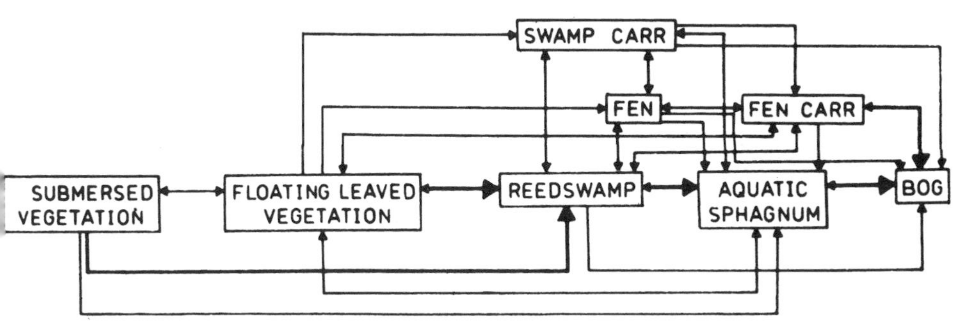


*Figure 2-10.* Successional sequences reconstructed from stratigraphic and palynological studies of postglacial hydroseres in Britain (from van der Valk, 1982, as modified from Walker, 1970; copyright © by International Scientific Publications, Jaipur, India, by permission).

Some authors have erroneously tried to interpret zonational patterns around lakes, ponds, or bogs as belts of varying ages and therefore as evidence of Clementsian succession (van der Valk, 1982; Niering, 1987, 1988; Brown, 1992). Many studies have shown that zonal change reflects responses of vegetation to changes in water level but has nothing to do with long-term lake filling, as predicted by Clementsian succession (Niering, 1987). Clearly, no progression of a change toward a dry land habitat can be seen in Cedar Creek Bog, Minnesota, where the position of the edge of the bog mat did not change for 33 years (Buell et al., 1968). Similarly, in Beckley Bog, Connecticut, the trees throughout the bog seemed to be of the same age. All of the trees from the outermost zone (red maple) to those closest to the open water (dwarf spruce) had about 90 growth rings (Egler, personal communication, in Niering, 1987).

River floodplain areas change gradually due to autogenic and allogenic processes (sedimentation/erosion) that might at first glance support classical succession concepts, but it is important to remember that rivers are governed by processes that cause them to change suddenly (Ellenberg, 1988). In the long term, one can visualize the sweeping of the channel of a river across the floodplain as being like the movements of a hose on a lawn. This movement of the river channel can remove mature forests as the channel's position switches over long time frames (Niering, 1994). A floodplain forest, such as those found in the Amazon, will revert to a terrestrial non-flooded forest only via global climatic change or human alteration (Junk, 1986).

In riverine systems such as those in the western United States, disturbance is too frequent for a theoretical climax to develop (Youngblood et al., 1985). In response to flooding, the river course may change dramatically and, along with it, the vegetation (Ellenberg, 1988). The Loisach River in the "Pupplinger Au" south of Munich, Germany, dramatically changed its

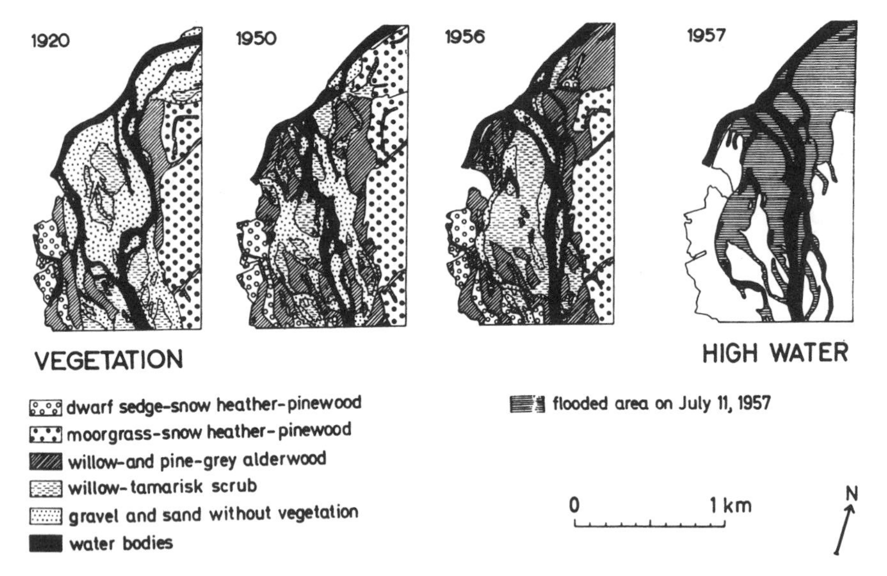


*Figure 2-11.* Changes in stream course (and vegetation) between 1920 and 1957 in the Pupplinger Au south of Munich, Germany (from Ellenberg, 1988, after Seibert, 1958; copyright © by Cambridge University Press by permission).

course as a result of flooding events between 1920 and 1957 (Fig. 2-11; Seibert, 1958).

While public policy concerning protection and management treats wetlands as static entities (Larson et al., 1980), there is ample evidence that wetlands are changing constantly and not in the direction of upland forest. Species in wetlands cannot be managed in a static environment because they evolved in stochastically changing ones. This idea must come to the forefront of current thinking (Hart, 1990; Haufler, 1990; Niemi et al., 1990; Szaro, 1990) in wetland management and restoration. An understanding of succession processes and their relationships to disturbances is essential in the restoration of wetlands (Niering, 1994).

## Invasion Theory

Natural plant invasion is a critical issue in restoration, and the ecological literature on this topic is growing. Restoration sites can be limited in the quantity and quality of plant materials available to regenerate vegetation after hydrological restoration (Reinartz and Warne, 1993; Galatowitsch and van der Valk, 1995, 1996). The restoration of vegetation can occur through either replanting or reliance on the natural ability of species to disperse to the site. Relying on natural processes to restore vegetation has some advantages; for example, it is less costly, and natural genotypes and spatial patterns are more likely to develop. An understanding of **colonizers**, or species first to arrive at an unoccupied site (Bazzaz, 1986), and the nature of their dispersal across landscapes can contribute to our understanding of restoration ecology.

Restored wetlands are linked regionally through the dispersal of seeds or propagules via air, water, or animals. Whether or not aquatic propagules arrive naturally at a particular site depends entirely on the strength of this linkage. For example, one reason restored prairie wetlands may have few seeds of either sedge meadow or wet prairie species is that regional flooding in prairie wetlands is now a rare event (van der Valk, in press).

Similarly, abandoned farm fields in former cypress swamps in the southeastern United States have a very poor ability to regenerate swamp species after farming activities are abandoned. The reason is that their linkage to flood water is typically cut off by the flood control engineering that created the farmable land in the first place. Few seeds of the dominant species of cypress swamp survive beyond the first year or so of farming, due partially to the rapid loss of seed viability on the surface of the soil (Middleton, 1995a). However, portions of farm fields naturally revegetate with cypress swamp species at elevations linked by winter flooding. These sites also must not be impounded so that they experience summer drawdown; under these conditions, they are readily invaded by cypress swamp species. Natural regeneration in these cypress swamps is limited spatially to sites with winter/summer water flux, that is, flood pulsing (Middleton, 1995a; unpublished).

A key theory of importance in the invasion of restoration sites by plants focuses on the idea of safe sites. A **safe site** is a zone with conditions that allow a seed to germinate, that is, a site where the seed can escape pregermination predation and overcome dormancy (Harper et al., 1965). The seed can break dormancy given the proper stimuli (e.g., light/dark), environment, and adequate resources (e.g., water and light). The safe site must also be free from ''specific hazards'' including predators, competitors, toxic soil constituents, and preemergence pathogens (Harper, 1977).

Many theories give partial explanations for plant invasion, but Johnstone's (1986) invasion window theory describes all categories of safe sites based on barriers and selectivity for invasion (Fig. 2-12). Barriers can be botanical (the presence of another plant or preemption), nonbotanical, selective (not the same for all plants), or nonselective. Nonselective removal of a barrier opens an invasion window to create a safe site. For example, the removal of a plant in an area opens an invasion window that allows a new species to occupy a safe site (temporary safe site; Johnstone, 1986). In a restored wetland, if the seed has already arrived at the restoration site,

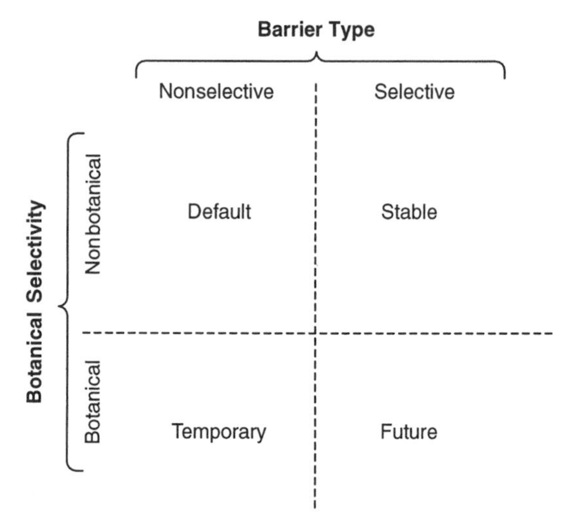


*Figure 2-12.* Safe sites (default, stable, temporary, and future) and their invasion potential based on barrier type and selectivity (from Johnstone, 1986; copyright © by Biological Reviews by permission).

it can enter the unvegetated site through a "temporary" window as long as the nonselective temporary barrier to invasion has been removed, that is, the water is drawn down in the wetland and the plants have not yet begun to grow.

The beauty of the Johnstone theory in characterizing safe sites is that it can be used to describe all types of invasion. Another scenario is that a seed can move across a geographic barrier, such as a continent, to enter a restored wetland. This window is permanently open (stable) and is crossed with a certain level of probability—for example, exotic species that must cross a continent to invade (Figs. 2-13, 2-14, 2-15). Examples of such invasions are numerous, including the invasion of *Lythrum salicaria* in North America and *Mimosa pigra* in Australia. *L. salicaria* arrived in North America in the early 1800s and since then has spread from east to west in glaciated wetlands (Fig. 2-16; Stuckey, 1980). *M. pigra* is an invasive species of tropical wet/dry wetlands in northern Australia (Cook et al., 1996). These are two of a great many geographic jumps by aquatic plant species, mostly assisted by humans (Cook, 1985). One study in Australia showed that tourists could potentially contribute to exotic weed invasions, as seeds were found on vehicles entering Kakadu National Park in northern Australia (Lonsdale and Lane, 1994).

Ultimately, the interconnectedness of the landscape is a prime determinant of the natural ability of species to disperse and invade new areas. While humans can assist this process in restoration, dispersal remains the most important long-term ally in the colonization of species in wetlands.

***Figure 2-13.*** *Exotic* Gladiola *from South Africa (foreground) in the nature reserve at the botanical garden, Perth, Australia (photograph by Beth Middleton).*

***Figure 2-14.*** *Exotic* Eichhornia crassipes *from South America in the Keoladeo National Park, India. The floating plants are gathered by boat in an attempt to eradicate the species from the park (photograph by Beth Middleton).*

***Figure 2-15.*** Eichhornia crassipes *(photograph by Beth Middleton).*

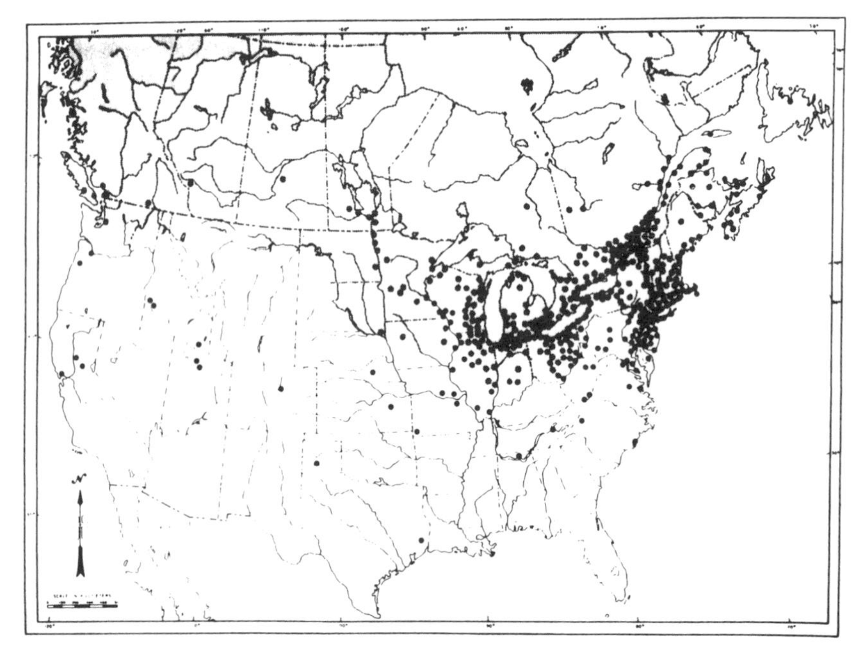

***Figure 2-16.*** *Known distribution of the exotic* L. salicaria *in North America (from Stuckey, 1980; copyright © by Bartonia; reprinted by permission of the Philadelphia Botanical Club).*

In fragmented landscapes, dispersal via corridors can be a particularly important component in the maintenance of both plant and animal populations. Dispersal corridors between such habitats may decrease the probability of local extinctions and the detrimental genetic effects associated with isolation (inbreeding, depression, and random genetic drift) (Noss, 1993). The level of connectivity of habitats may be crucial in the long-term persistence of species in regional landscapes (Turner, 1989; Merriam and Saunders, 1993).

Small, narrow, or fragmented habitats may have certain limitations for the invasion of biota. Degraded habitats, depending on the degree of their anthropogenic influence, may be very difficult for species to cross (Sisk and Margules, 1993). Though the importance of uncrossable barriers in wetland species dispersal is poorly known, the lack of sedge meadow and wet meadow species in restored prairie potholes and the poor natural revegetation of cypress swamps along altered rivers (see above) clearly indicate that such human-induced barriers may thwart our ability to restore habitats.

The edge or **ecotone** of a **landscape element** may in itself inhibit, facilitate, or neutrally influence the passage of species. Dispersal is dependent on the ability of species to cross ecotones, and the flow of seeds or propagules between landscape elements is facilitated by wind, water, animals, herbivory, and other factors (Pickett and Cadenasso, 1995). In the case of wetlands, aquatically dispersed species can navigate long distances by **hydrochory** from intact communities with seed sources to gaps created either by local plant death (e.g., fire, windthrow) or by human disturbance (e.g., farming, levee construction) (Fig. 2-17; Middleton, unpublished). Riparian ecotones can act as dispersal corridors for both plants and animals (Naiman and Décamps, 1997).

Ecotones can be divided into primary and secondary types, with the former located between true terrestrial and aquatic ecosystems and the latter linked to anthropogenic disturbance. Along the River Trent in the United Kingdom, the highest species richness was found in ecotonal areas along the stream corridor, particularly if these areas were moderately disturbed (Large et al., 1994). Because disturbance facilitates invasion, ecotonal areas between wetland and upland are often prime sites of exotic invasion (Ewel, 1986). *Melaleuca quinquenervia* readily invades the ecotone between pine and cypress forests in south Florida (Myers, 1984). Figure 2-18 documents the original location of introduction of *M. quinquenervia* in Florida and its approximate spread by 1983 (Myers, 1983).

Because restoration usually attempts to recapture the original community's character, the presence of exotic species interferes with these efforts (Simberloff, 1990). Exotic plant species are a problem particularly along

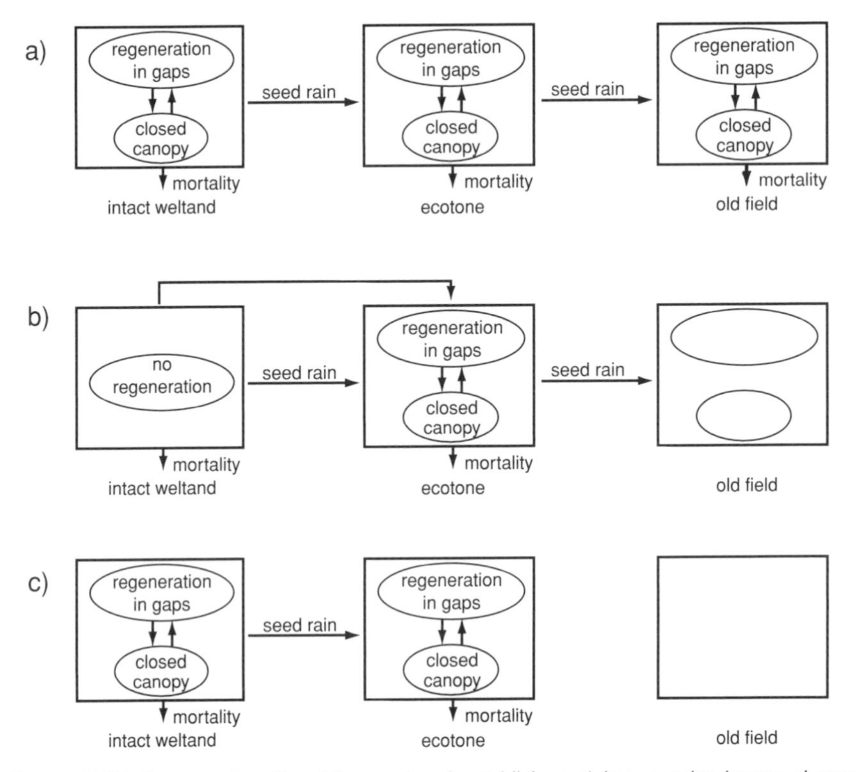


*Figure 2-17.* *Regeneration flow (dispersal and establishment) between landscape elements including intact wetland, old field, and their ecotone. (a) Flood pulse restoration restores regeneration processes in old fields. (b) The site is impounded. (c) The site is not restored. The impounded site (b) has no regeneration except in a narrow band in the ecotone (from Middleton,* in review, *as based on Pickett and Cadenasso, 1995; copyright © by Science by permission).*

river corridors where invasion of species is facilitated by land use and human modification, life history traits of the invasive species, landscape connectivity, and patchiness along the river (Décamps et al., 1995). In Central Europe, *Impatiens glandulifera* seeds and *Reynoutria sachalienensis* fragments have spread along rivers (Pyŝek and Prach, 1994). Similarly, mangroves not native to Hawaii have displaced the former dominant, *Hibiscus tiliaceus*, along the ocean shore (Simberloff, 1990).

In primary successions, as on many restoration sites, immigration and extinction rates, as per Island Biogeography Theory (MacArthur and Wilson, 1967), may control species richness patterns (Rydin and Borgegård, 1991). Any inhospitable terrain can isolate a habitat and render it difficult for organisms to reach the "island" (Simberloff, 1974; Erwin, 1991). In this approach, restoration sites can be thought of as islands (*sensu* Diamond, 1975). Following the theory of island biogeography, naturally revegetated

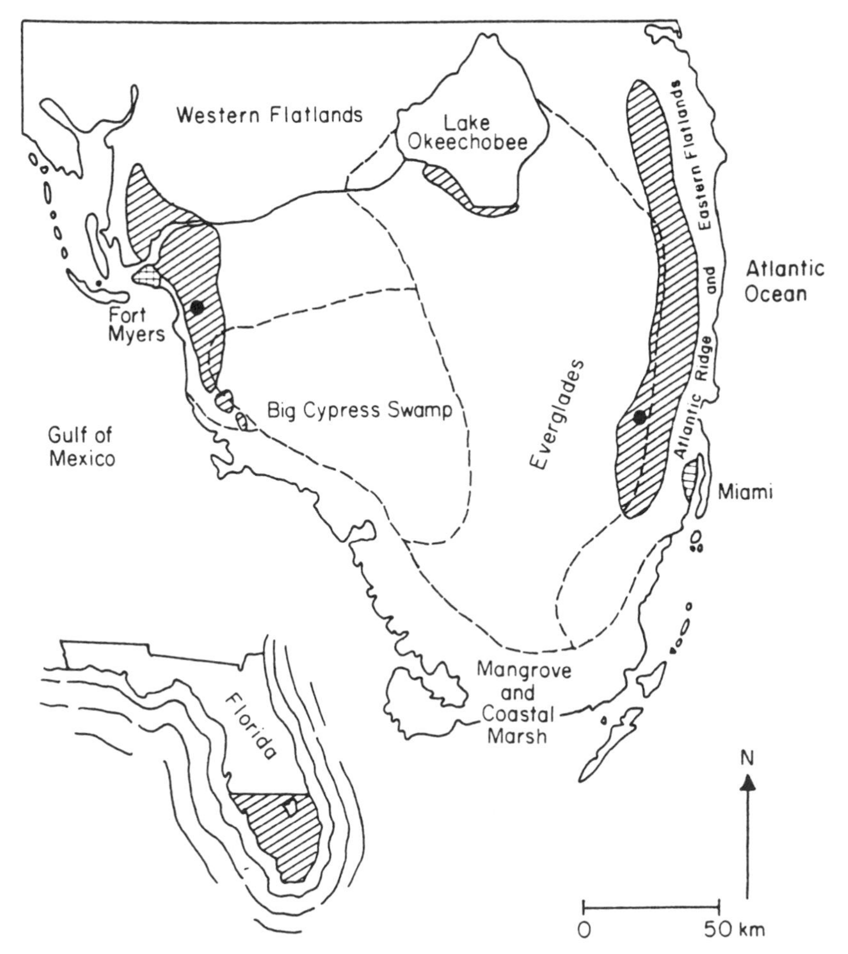


*Figure 2-18. Sites of original introduction and subsequent invasion of* M. quinquinervia *in southern Florida (from Myers, 1983, as adapted from Davis, 1943; copyright © by Journal of Applied Ecology by permission).*

restoration sites that are small or distant from other wetlands would have fewer species than other sites (Fig. 2-19a).

The theory of island biogeography has been very useful in predicting species richness patterns in wetland restoration sites. Godwin (1923) found that older English ponds and those interconnected by rivers had the most species. In created wetlands in southern Wisconsin, Reinartz and Warne (1993) concluded that native wetland species richness increased with wetland age, size, and proximity to the nearest wetland. Similarly, Galatowitsch and van der Valk (1996) found more species in restored prairie potholes that were large and/or near other wetlands.

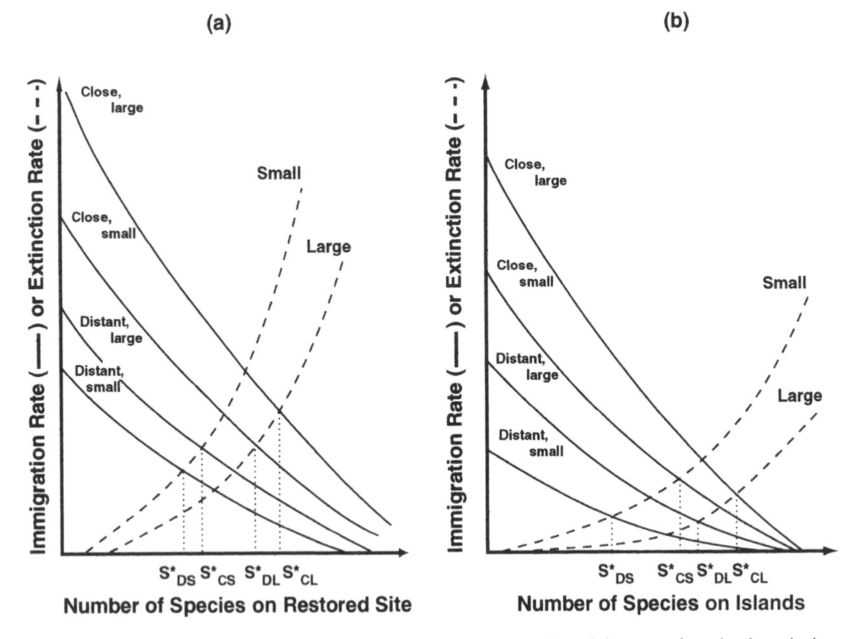


*Figure 2-19. (a) Theory of Island Biogeography. The species richness of a site is a balance between immigration and extinction on small versus large and close versus distant restoration sites to a "mainland" of intact vegetation. S* is the species richness at equilibrium, with S = small, L = large, D = distant, and C = close restoration sites (based on Begon et al., 1990, from MacArthur and Wilson, 1967; copyright © by Princeton University Press and Blackwell by permission).*

*(b) MacArthur and Wilson's (1967) Theory of Island Biogeography. The rate of species immigration to an island plotted against the number of resident species on the island for large and small islands and for close and distant islands (declining curves). The rate of species extinction on an island plotted against the number of resident species on the island for large and small islands (rising curves). The balance between immigration and extinction on small and large and on close and distant islands. In each case S* is the equilibrium species richness (S = small, L = large, D = distant, C = close; vertical dotted lines). (Based on Begon et al., 1990, from MacArthur and Wilson, 1967; copyright © by Princeton University Press and Blackwell by permission).*

Over time, an equilibrium number of species can be anticipated, depending on the relationships of the island's size and distance to the mainland (Fig. 2-19b). Because the number of species at the site's equilibrium is subject to long-term change, its state should be thought of as a quasi-equilibrium (Simberloff, 1974). The arthropod species richness on experimentally denuded mangrove islands rose to a certain level and then declined slightly to a "noninteractive" level (Simberloff and Wilson, 1969). Similarly, after islands were created in Lake Hjälmaren, Sweden, species richness on large islands increased to an equilibrium of 95, with immigration balancing extinction (Rydin and Borgegård, 1988). In Panama, in a reverse situation to that of Lake Hjälmaren, Gatun Lake levels rose in 1913 to isolated islands of less than 1 ha in former core areas of forest. By 1980, the diversity of trees on these islands generally had declined in comparison to that of the mainland. Tree mortality on these small islands was high because of wind exposure and lack of mammals. Animals aid in the spread of seeds in the core forest (Leigh et al., 1993).

## River Theory

Wetlands that lie along river and stream corridors have an additional set of theories relevant to restoration, including the River Continuum Concept (Vannote et al., 1980), the Serial Discontinuity Concept (Ward and Stanford, 1983, 1995b), the Flood Pulse Concept (Junk et al., 1989; Sparks et al., 1990), and other stream ecosystem theories. All of these theories focus on either the linear or lateral flow of water in the landscape from the stream channel. These approaches focus less on the biota than on physical characteristics such as carbon flow, nutrient dynamics, and biogeochemical linkages of terrestrial/lotic systems (Sedell et al., 1989).

The **River Continuum Concept** proposes that functional and structural attributes shift along stream reaches (Fig. 2-20; Vannote et al., 1980). The concept was developed from an insect rather than a plant perspective. Accordingly, it predicts that in headwater, invertebrate shredders rely on allochthonous sources of debris from stream-side trees or grasses. By midreach, autotrophic processes becomes increasingly important to the food chain, while allochthonous input of debris is reduced in comparison to the headwater. Because the vegetation is the most diverse, the midreach is the zone of the highest invertebrate diversity. Downstream where the channel becomes wide and meandering, the habitat is less variable, and thus the invertebrate diversity is predicted to be lower than at midreach.

One study of plant species richness changes along the Adour River in southern France, showed species richness patterns compatible with the River Continuum Concept (Fig. 2-21). Species richness was low in the upper reaches in the mountains because of the narrow valley and extreme habitat conditions, medium-high in the piedmont zone corresponding with a drop in stream gradient (Décamps and Tabacchi, 1994), high in midcourse where the habitat conditions were most heterogeneous and the disturbance was intermediate (Tabacchi, 1992, in Décamps and Tabacchi, 1994), and low in the lower course, with prolonged floods and homogeneous substrates (Décamps and Tabacchi, 1994). In another study of northern rivers in Sweden, while species richness was typically highest in midreach, impounded midreaches of rivers had lower species richness than would be predicted by the River Continuum Concept (Nilsson and Jannson, 1995).

In large rivers, such as the Amazon, the River Continuum Concept fails

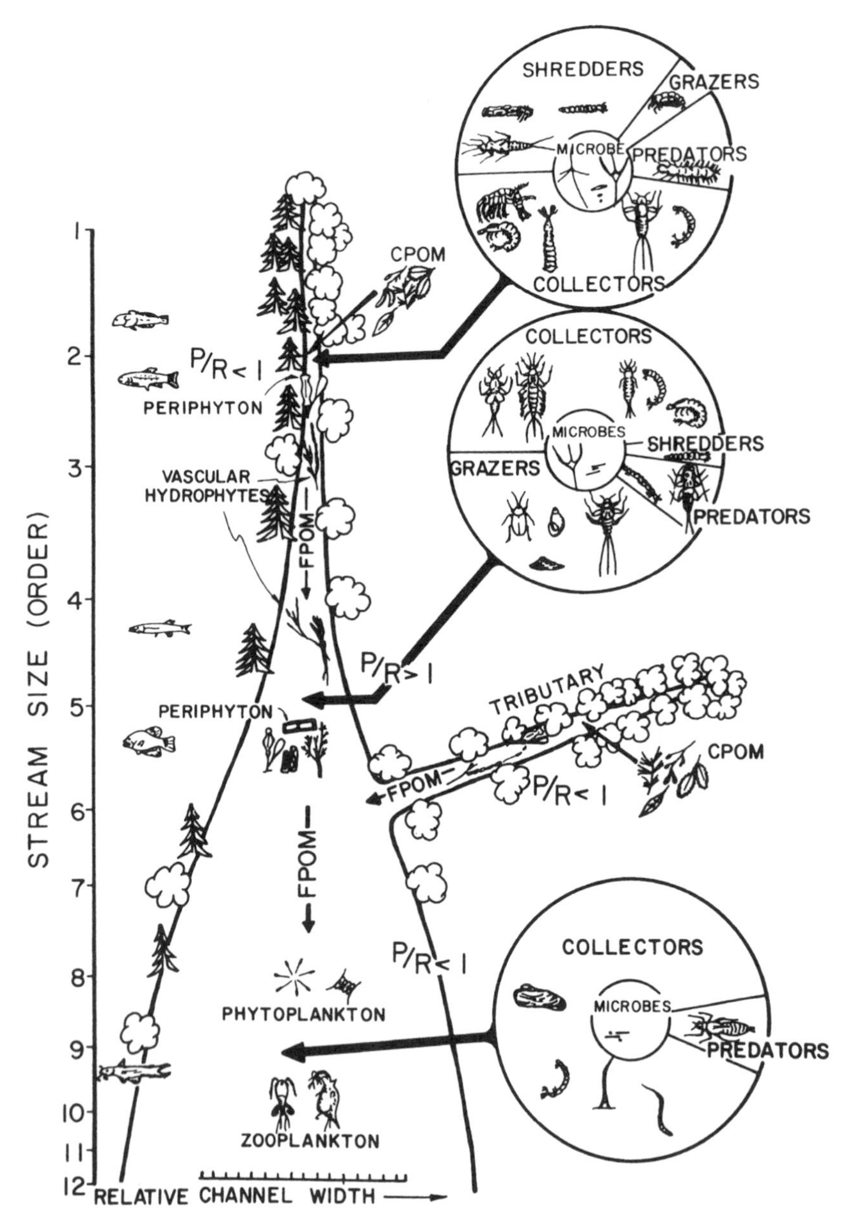


***Figure 2-20.*** *River Continuum Concept showing the relationship between progressive stream size and functional and structural relationships. From a plant perspective, species richness is predicted to be highest in the low-midreach, where habitat diversity is highest (based on Vannote et al., 1980; copyright © by Canadian Journal of Fisheries and Aquatic Sciences by permission).*

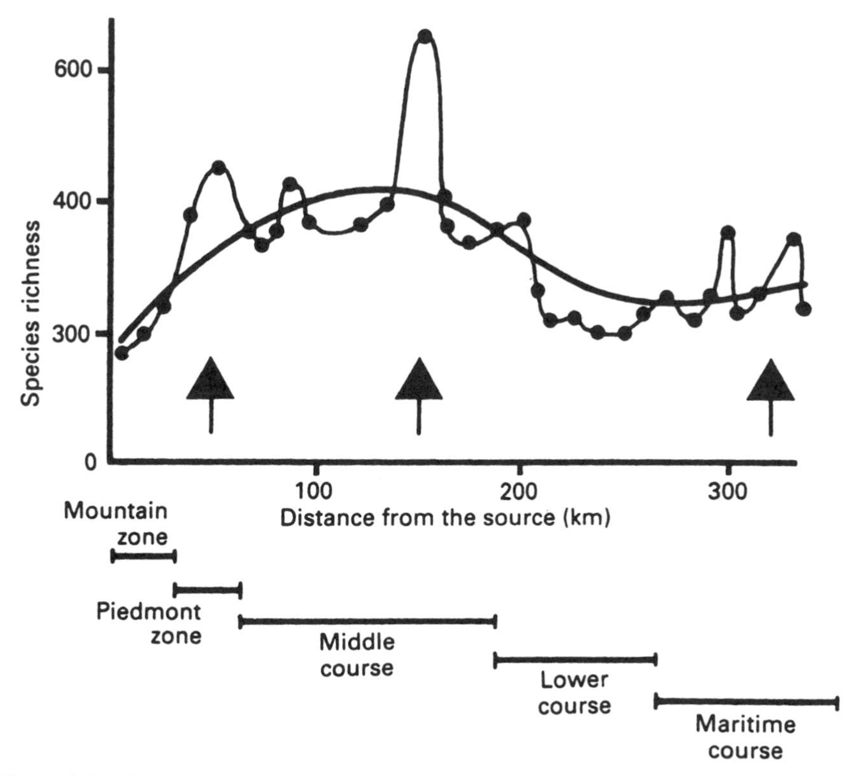


*Figure 2-21.* Species richness of the riparian plant community along the Adour River, southern France (from Décamps and Tabacchi, 1994; copyright © by Blackwell Scientific Publications by permission).

to address the importance of the floodplain environment (trees, groundwater interactions) with the river system. Thus, the flood pulse concept of Junk et al. (1989), may lead to a better conceptualization of the dynamics of large rivers than the River Continuum Concept (Sedell et al., 1989). A number of theories subsequent to the River Continuum Concept have been proposed, mostly to address lateral processes (Cummins, 1979; Cummins et al., 1984, 1995).

Of the river ecosystem theories, the most important from a restoration standpoint is probably the Flood Pulse Concept because, as discussed extensively in Chapter 1, overbank flow is important in the timing and spatial dynamics of seeds, and of propagule movement and regeneration across the floodplain (Scott et al., 1993a, 1993b; Middleton, 1995a).

Channel migration across floodplains isolates ponds, creating new surfaces for plant invasion and succession (Shankman and Drake, 1990; Shankman, 1993). Lateral channel migration occurs across the floodplain in lower stream reaches where rivers and streams have not been channelized

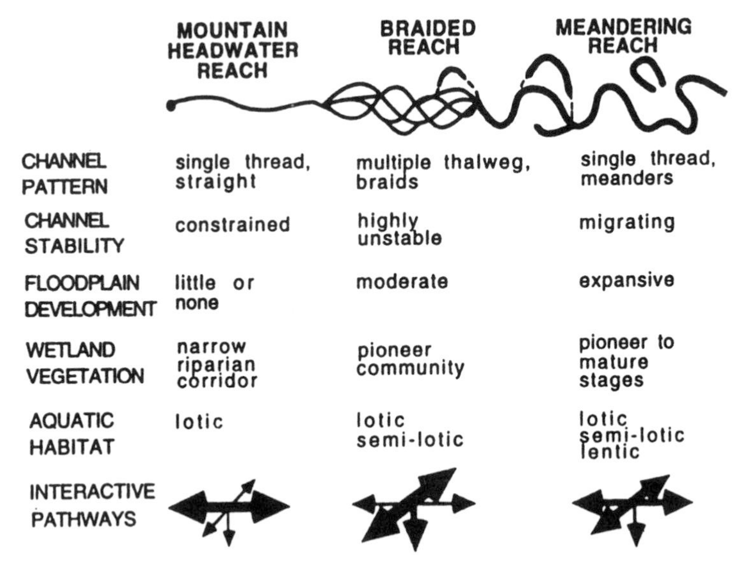


*Figure 2-22.* *Serial Discontinuity Concept in various portions of the stream reach. Arrows indicate the relative strength of the interactions along longitudinal, vertical, and lateral (horizontal, and vertical and oblique arrows, respectively) (from Ward and Stanford, 1995b; copyright © by Canadian Journal of Aquatic and Fisheries Sciences by permission).*

and/or dammed (Fig. 2-22). Because upstream reaches are already relatively stable, according to the **Serial Discontinuity Concept**, dams alter downstream reaches more than upstream ones (Ward and Stanford, 1983, 1995a).

The **Intermediate Disturbance Hypothesis** predicts that sites with an intermediate level of disturbance will have the highest amount of species richness (Connell, 1978). Because rivers and their biota are subjected to disturbances via changes in discharge, these could provide a good test of the hypothesis (Townsend, 1996). Intermediacy must be assigned to the portion of a disturbance defined by its magnitude, frequency, and size, but these quantifiers of disturbance are not equivalent in the level of disruption created (Pickett and White, 1985). Bornette and Amoros (1977) assess levels of disturbance along the Rhône River, France, by examining substrate particle sizes, fine sediments indicating more sheltered situations. However, disturbance levels are seldom objectively quantified in research studies.

Some ecosystem ecologists view the processes associated with the stream or river and its floodplain as a series of discrete patches or communities rather than as a gradient (Frissell et al., 1986; Naiman et al., 1988b) and thus as being more closely related to patch-dynamics concepts (Pickett and White, 1985). Because sharp boundaries can occur either naturally or due to anthropogenic causes along rivers, boundary (ecotone) concepts in ecosystem processes can increase the understanding of these systems. While

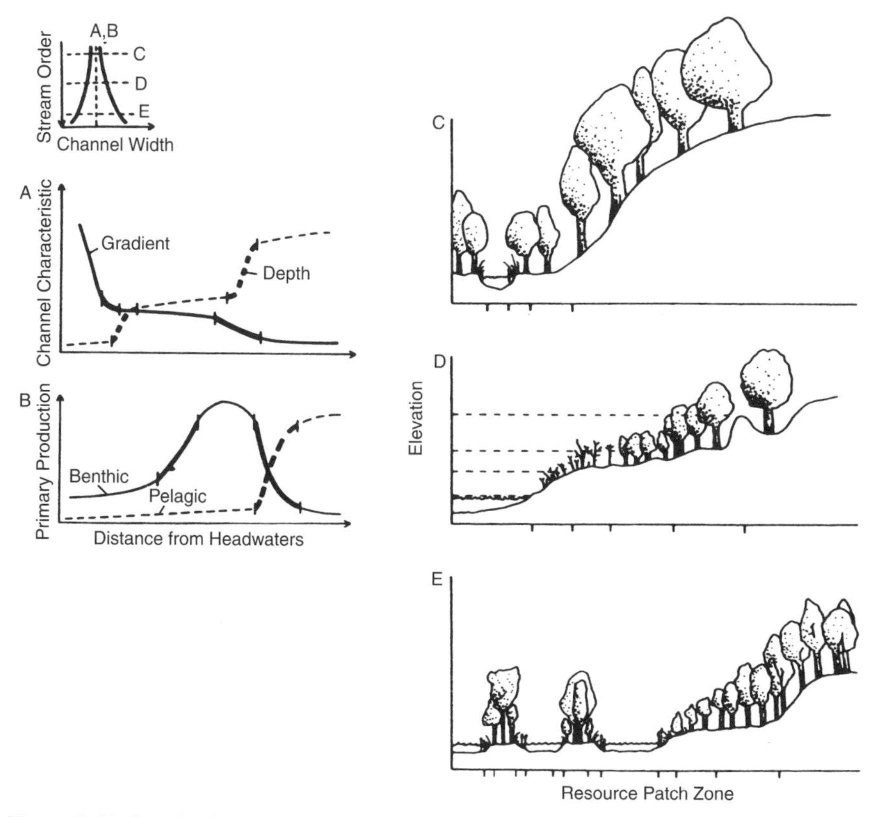


***Figure 2-23.*** *Longitudinal versus latitudinal boundaries in streams. Longitudinal boundaries (ecotones) such as transition zones (bold lines) occur over long distances in (A) channel characteristics and (B) primary production. Lateral boundaries occur over shorter distances but are complicated by longitudinal shifts in channel dimensions (C–E). Dashed lines refer to flood levels (from Naiman et al., 1988b; reprinted with permission of the Journal of the North American Benthological Society by permission).*

latitudinal boundaries along streams may be sharp, longitudinal ones may extend along several stream orders (Fig. 2-23). Channel characteristics such as gradient and depth generally change slowly with stream order, while the width of the floodplain/channel interface is more dynamic laterally, though not independent of longitudinal position (Naiman et al., 1988b).

# 3

# *From Seed to Adult: Missing Links in Restoration*

*The varied life history strategies of plant species coupled with the limits imposed by the created environment set the stage for the plant composition of the restored wetland. Within the same species, seeds, seedlings, and adults often have different requirements, underscoring the necessity of dynamic change in both the regeneration and long-term maintenance of species. Thus, agents of change (e.g., water fluctuation, fire, herbivory) are an integral part of the functional environment for plants. In wetlands, water fluctuation is the most important driving force in both the establishment and extirpation of species. Flood pulsing along unaltered rivers and water fluctuation around lakes and ponds provide the hydrodynamic setting necessary for plants to complete their life cycles.*

*Water fluctuation is a controlling agent in the life processes of wetland plants. Adult stages of many wetland species can be tolerant of periods flooding, particularly during the nongrowing season, though the limits imposed by water regime vary widely by species. Paradoxically, flooding can be essential in spreading seeds onto the floodplain, even while the seeds of most emergent species germinate poorly underwater. For these species, seed germination and seedling recruitment are enhanced by a lengthy drawdown during the growing season. Flood tolerances vary, but even the most water-tolerant species cannot live without oxygen indefinitely. This duality of environmental conditions imposed by flooding versus drawdown favors various life history stages in emergent species. In wetland restoration, this underscores the need for a dynamic water regime to recruit and maintain vegetation successfully.*

*Factors other than water regime can be equally important regulators in the establishment and recruitment of species. Salinity tolerance in seed germination, seedling recruitment, and adult success often vary widely among species and among genotypes within species.*

*This chapter will review models of life history dynamics (sieve model, Grime's plant guilds) and their importance to wetland restoration. Of particular interest to wetland restorationists is a compendium of the tolerances of world plant species for particular water regimes and other environmental characteristics. Specific seed-related information is given for dispersal, seed banks, germination, and dormancy, as well as for seedling recruitment and adult establishment.*

## LIFE HISTORY MODELS IN RESTORATION

The key to the restoration of vegetation in wetlands is an understanding of the life history requirements of plant species. The outcome of the restoration project from a vegetation perspective is entirely dependent on the limitations imposed by the environment on the species involved. Ultimately, vegetation change in plant communities is a product of the environment, including disturbances imposed upon plant species during different stages of their life histories (Grime, 1979; White, 1979). At the same time, because so many restoration sites are both degraded and disjunct from a regional pool of target species, knowledge of dispersal and colonization is vital in the success of projects (Palmer et al., 1997).

Not surprisingly, the life history requirements of species set against environmental fluctuations over time form the basis of many models of community change. A knowledge of some of the basic theories surrounding vegetation change in wetlands can greatly help restorationists to restore wetlands.

Van der Valk's (1981) Wetland Sieve Model is based on the vital attributes approach developed by Noble and Slatyer (1980) to predict the long-term effects of fire on Australian eucalypt forest. Grime's (1977) R-C-S (ruderal, competitive, stress-tolerant) strategies describe the adaptive strategies of species. As applied to aquatic species, the model characterizes the water level and competition (disturbance and stress, respectively) that these species can tolerate (Menges and Waller, 1983; Kautsky, 1988; Murphy et al., 1990).

In the Wetland Sieve Model, the environment functions like a "sieve," allowing only certain species either to become established or to be maintained in a wetland (see Chapter 2, Fig. 2-3). For the model, the key features of the life history of plants include (1) potential life span, (2) propagule longevity, and (3) propagule establishment requirements. The life span of plants may be described as annual (A), perennial (P), or perennial with vegetative growth (V). Propagule longevity may be long (S; seed bank) or short (D; dispersal dependent). Propagule establishment may occur in drawdown (D) or flooded (F) conditions (van der Valk, 1982). These life history constraints also can predict the responses of species to environmental conditions in restored wetlands.

While the model was designed for prairie pothole wetlands of the United States, it has also been applied successfully to monsoonal wetlands in northern Australia (Finlayson, 1988). Typically, all but the deepest channels of monsoonal wetlands draw down seasonally, so that distinctly different communities arise annually (Fig. 3-1). The composition vacillates between species with life history characteristics tolerant of flooded versus drawdown conditions. Figure 3-2 illustrates the predictive capabilities of the model. An ASI is an annual species with short-lived propagules (seeds) that establish during drawdown. Under flooded conditions, these species do not germinate and hence do not become adults. Eventually, a species with an ASI life history strategy is extirpated from a permanently flooded wetland.

Various authors have applied Grime's R-C-S strategies to wetland species. Herbaceous wetland guilds of floodplain forests are categorized by their life history strategies as ruderals (R), or stress tolerators (S), competitors (C), based on their responses to stress and disturbance (Menges and Waller, 1983; Hills et al., 1994). In some studies, species are categorized by juvenile traits that is, seed germination and seedling growth rates (Shipley et al., 1989). Stress slows the growth of plants, and disturbance removes standing vegetation. Ruderals are annual species with fast growth rates, short life spans, and high seed production (Grime, 1977, 1979). These species are neither very stress tolerant nor competitive. To complete their life cycles in frequently flooded areas, they grow and set seed quickly between flooding episodes (e.g., *Polygonum punctatum*, Fig. 3-3; Menges and Waller, 1983). Stress tolerators have physiological or morphological adaptations to survive flooding (e.g., certain *Carex* spp.). Competitors grow tall and flower late in the season in unstressed, undisturbed conditions of high light (e.g., *Helianthus tuberosus*; also see Fig. 3-4). In restored wetlands, another example of the competitor life strategy is *Typha* sp. *Typha* often becomes dominant in sites with high fertilizer input, as in restored wetlands in the Everglades of Florida (Newman et al., 1996).

In the square model, Grime's R-C-S model is expanded to include an additional stress-tolerant strategy, the biomass storer (B; Fig. 3-5a; Kautsky, 1988). Found in conditions of low disturbance but high stress (e.g., limited light or nutrients or high salinity; Fig. 3-5b), biomass storers divert re-

| | DRY SEASON | | WET SEASON | |
|---|---|---|---|---|
| _Pseudoraphis community_ | no standing water<br><br>AS-I _Cyperus_ spp.<br>_Fimbristylis_<br>_Glinus_<br>_Heliotropium_<br>VS-I _Polygonum_<br>_Pseudoraphis_**<br>PS-I _Mimosa_ | flooding<br>→ → → | standing water<br><br>AS-II _Blyxa_*<br>_Hydrochloa_*<br>_Najas_*<br>_Nymphoides_ spp.*<br>_Utricularia_ spp.<br>VS-I _Polygonum_<br>_Pseudoraphis_**<br>VS-II _Eleocharis_ spp.*<br>_Nymphaea_*<br>VD-II _Salvinia_*<br>PS-I _Mimosa_ | drawdown<br>→ → → |
| _Hymenachne community_ | shallow standing water<br>some exposed and moist areas<br><br>VS-I _Ludwigia_<br>_Pseudoraphis_<br>VS-II _Azolla_<br>_Eleocharis_ spp.<br>_Hymenachne_**<br>_Lemna_<br>_Nelumbo_*<br>_Nymphaea_<br>_Urochloa_<br>VD-II _Salvinia_ | flooding<br>→ → → | deeper standing water than during dry<br><br>AS-II _Aeschynomene_ spp.*<br>_Oryza_<br>VS-I _Ludwigia_<br>_Pseudoraphis_<br>VS-II _Azolla_<br>_Eleocharis_ spp.*<br>_Hymenachne_*<br>_Lemna_<br>_Nelumbo_<br>_Nymphaea_<br>_Urochloa_<br>VD-II _Salvinia_ | drawdown<br>→ → → |
| _Oryza community_ | exposed dry areas<br><br>AS-I _Coldenia_*<br>_Commelina_<br>_Digitara_<br>_Heliotropium_<br>_Phyla_*<br>VS-I _Ludwigia_<br>_Pseudoraphis_<br>PS-I _Mimosa_<br>PS-II _Isoetes_ | flooding<br>→ → → | standing water<br><br>AS-II _Aeschynomene_ spp.*<br>_Blyxa spp._<br>_Hygrochloa_<br>_Ipomoea_<br>_Maidenia_<br>_Nymphoides_ spp.<br>_Oryza_**<br>_Utricularia_ spp.<br>VS-I _Ludwigia_<br>_Pseudoraphis_<br>VS-II _Eleocharis_ spp.<br>_Nymphaea_<br>VD-II _Salvinia_<br>PS-I _Mimosa_<br>PS-II _Isoetes_ | drawdown<br>→ → → |

***Figure 3-1.*** *Predicted species changes due to water level fluctuation in floodplains in northern Australia. The dominant species in each community is indicated by ** and the next dominant species by * (A = annual, V = vegetative, P = perennial; S = short-lived propagules, D = long-lived propagules; I = propagules established in areas devoid of standing water; II = propagules established in standing water (from Finlayson, 1991a; copyright © by Association of Wetland Managers, New York, by permission).*

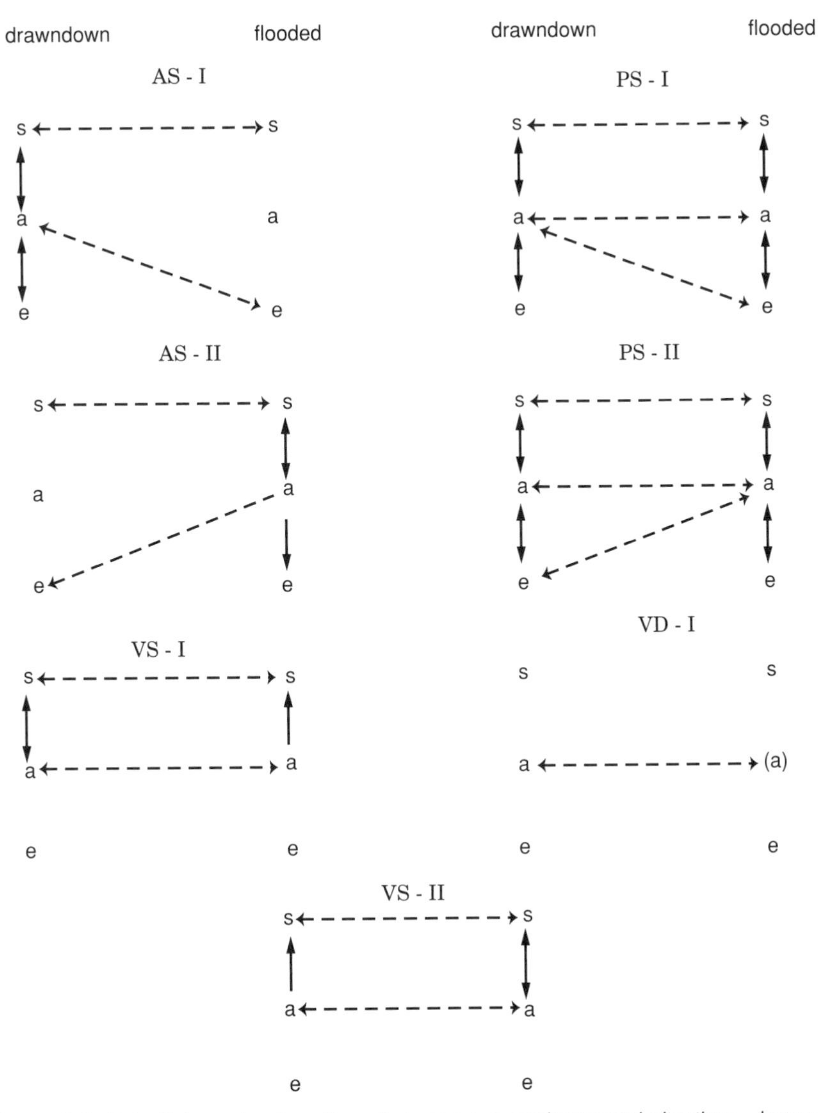


***Figure 3-2.*** *Potential species transitions between two environmental situations—increasing (i.e., flooding) and decreasing (i.e., drawdown) periods in a wetland. Solid lines represent potential transitions within an environmental situation, and dashed lines represent transitions between environmental situations. The species states are: s = present as long-lived propagules in a persistent seed bank, a = mature adults, and e = locally extinct. If establishment is dependent on dispersal from another site, adult populations are indicated in parentheses (from Finlayson, 1991a; copyright © by Association of Wetland Managers, New York, by permission).*

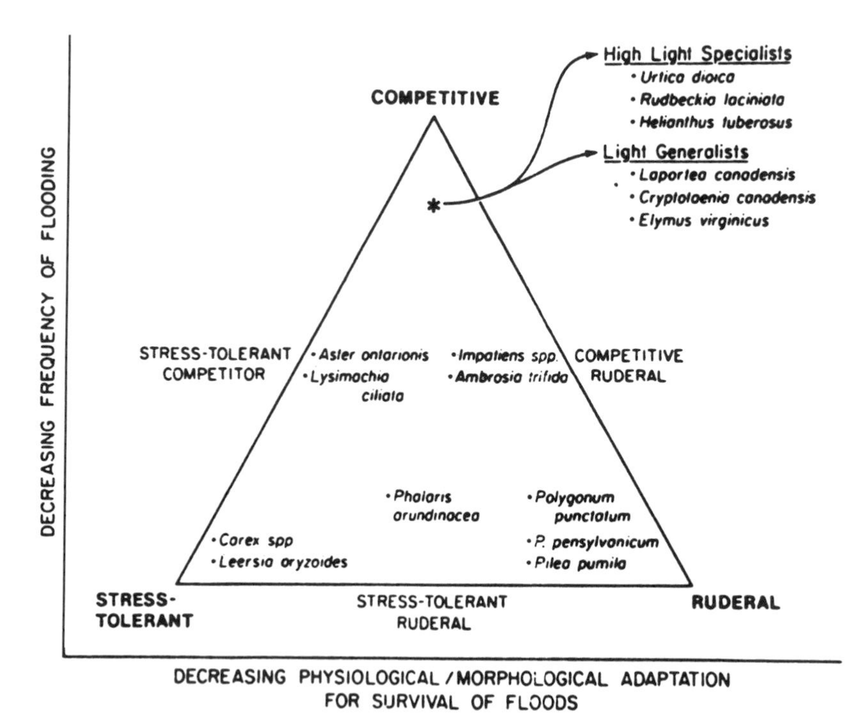


***Figure 3-3.*** *Grime's R-C-S (Ruderal-Competitive-Stress Tolerant) triangle in relation to flooding frequency and adaptation to flooding with representative herbaceous plant guilds (from Menges and Waller, 1983; copyright © by The American Naturalist; reprinted by permission of The American Naturalist and the University of Chicago).*

***Figure 3-4.*** *A "competitor" strategist in Grime's R-C-S triangle, Lilium michiganense* is a tall, late-blooming species of sedge meadows in North America (photograph by Beth Middleton).

sources to storage systems. These have rhizomatous growth and spread vegetatively (e.g., *Zostera marina*; Table 3-1). In related efforts, guild categorizations are based on life history characteristics (Boutin and Keddy, 1993).

## LIFE HISTORY CHARACTERISTICS AND RESTORATION

Life history characteristics of plant species play an important role in the vegetation dynamics of wetlands (Fig. 3-6). Each life history stage, including adult standing vegetation, seedling/sapling, seed, soil seed bank, and vegetative bud bank, as well as vegetative propagation, can have surprisingly different responses to environmental factors. Typically, flood tolerance increases with age as the plant either avoids or adjusts to waterlogging for short to indefinite periods of time (Patrick et al., 1981). For a restorationist to be successful in restoring wetlands, it is essential to understand the changing capability of species throughout their lives to tolerate specific environments, particularly the water regime.

### Seed Dispersal

The long-term ability of a restored wetland to regain/maintain species richness is dependent on the constraints of seed and propagule dispersal (Fig. 3-7). The facility with which seeds disperse is dependent on the dispersal mechanisms available to the individual species (van der Pijl, 1982), as well as the level of interconnectedness of the landscape—for example, a flood link of bottomland forest to abandoned farmland. Beyond gravity, transport agents include wind, water, animals, and birds (**anemochory**, hydrochory, zoochory, and avichory, respectively; Fenner, 1985). Most wetland species are dispersed by more than one of these mechanisms; *Acer negundo* is dispersed by all of them (Appendix 1).

Seeds of wind, dispersed species "catch" the wind with appendages such as wings, hairs, or plumes, which increase the seeds' opportunity for lateral dispersal over simple gravity. Light seeds (<0.05 mg) are often wind dispersed, including *Juncus inflexus*, *Calluna vulgaris,* most orchids (Fenner, 1985), and many spores of mosses and liverworts (Barbara Crandall-Stotler, personal communication). *Dust* seeds, as these are also known, can be carried at high altitudes for very long distances (Burrows, 1975a, 1975b) and are among the most successful in invading restored wetlands that are small or isolated (Reinartz and Warne, 1993).

Hydrochory is very common among wetland species (Appendix 1). Some aquatic seeds can float in the water for extended periods of time (e.g.,

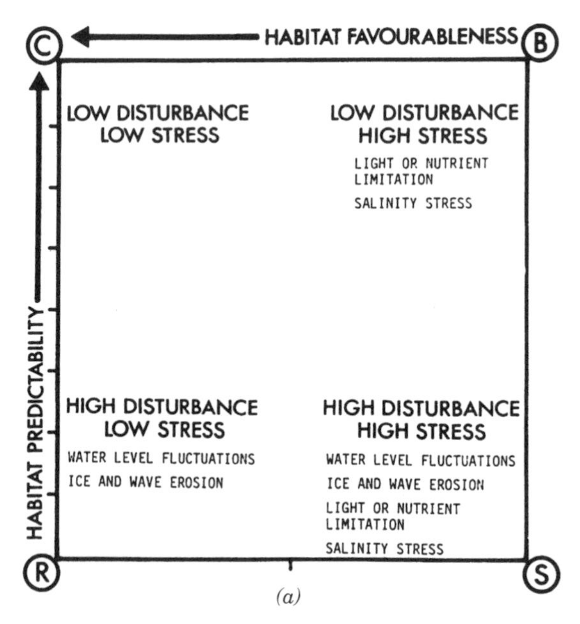


***Figure 3-5.*** *(a) The square model. The corners represent extreme environmental conditions with R = ruderal, C = competitive, B = biomass storer, and S = stunted guilds. The 'B' and 'S' categories are 'S' in the Grime R-C-S model. (from Kautsky, 1988; copyright © by Oikos by permission).*

*Taxodium distichum, Nyssa aquatica*; Table 3-3) and therefore can be dispersed for long distances (Schneider and Sharitz, 1988) with the movement of waves or current in water bodies (McAtee, 1925). Propagules of *Avicennia marina* can be transported a minimum of 10 km and, if submerged, can retain viability for five to seven months (Clarke, 1993). Other species, such as *Salix nigra*, must germinate within 24 hours of dropping from the tree (McKnight, 1965) and, hence, have a much lower potential for long-distance travel via water.

Because hydrologic restoration can reconnect dispersal corridors to restored wetlands from intact sites, species with water-dispersed seeds are among the most successful in reaching restored wetlands, particularly in riverine situations (Middleton, 1995a; in review). Recolonization along rivers depends mostly on the dispersal and deposition characteristics of aquatic species. Along the Rhône River in France, the colonization of *Potamogeton, Groenlandia*, and *Luronium* occurs on the bends of streams where the deposition of vegetative fragments is highest (Barrat-Segretain and Amoros, 1996).

The timing of flooding is crucial to successful dispersal and underscores the importance of proper hydrologic restoration. Seed dispersal can be

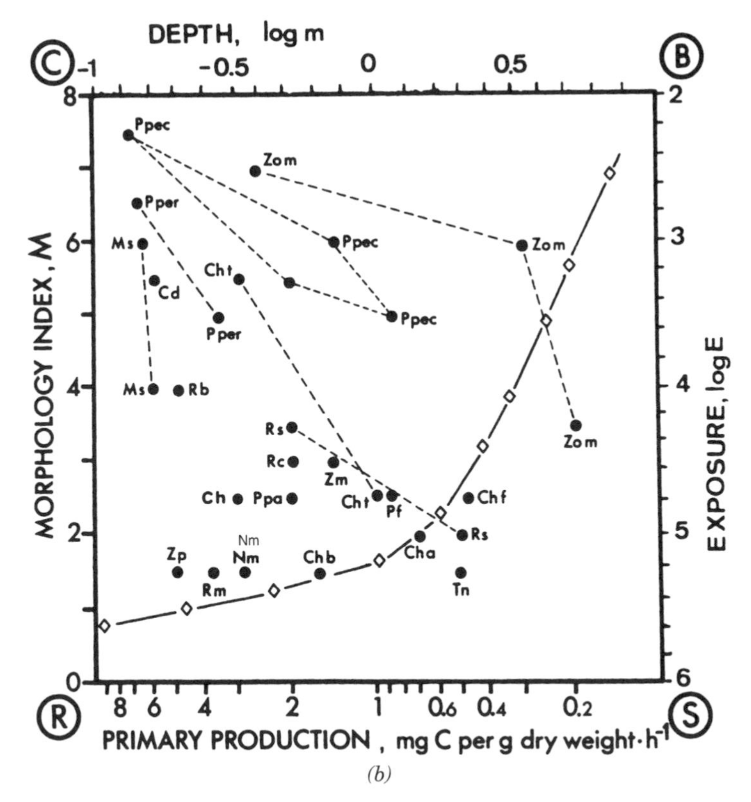


(b)

***Figure 3-5.*** *(b) A square ordination of macrophyte species from the island of Asko, the Baltic Sea. Diamonds show exposure values (log E) plotted against depth (log m) for the most exposed station. Key to the species: Ch,* Callitriche hermaphroditica*; Cd,* Ceratophyllum demersum*; Cha,* Chara aspera*; Chb,* Chara baltica*; Chf,* Chara fragilis*; Cht,* Chara tomentosa*; Ms,* Myriophyllum spicatum*; Nm,* Najas marina*; Pf,* Potamogeton filiformis*; Ppa,* Potamogeton panormitanus*; Ppec,* Potamogeton pectinatus*; Pper,* Potamogeton perfoliatus*; Rm,* Ruppia maritima*; Rs,* Ruppia spiralis*; Rb,* Ranunculus baudotii*; Rc,* Ranunculus circinatus*; Tn,* Tolypella nidifica*; Zp,* Zannichellia palustris*; Zm,* Zannichellia major*, Zom,* Zostera marina *(from Kautsky, 1988; copyright © by Oikos by permission).*

highly variable throughout the year. In cypress swamps, live seeds are dispersed primarily in the winter/spring seasons, including cypress, tupelo (Fig. 3-8a,b; Sharitz et al., 1990), and buttonbush (Fig. 3-8c; Middleton, 1995b).

Animals and birds can carry seeds in their guts, on their fur or feathers, or in the mud on their feet (Fenner, 1985). After ingestion, many species of seeds pass through the guts of animals and birds intact (Middleton and Mason, 1992). For this reason, waterfowl in particular can play a role in the vegetative success of restored wetlands over time (Mitsch and Wilson, 1996). More than 75% of the waterfowl shot in the salt marshes of the

**TABLE 3-1. Key Morphological and Physiological Characteristics of Ruderal, Competitive, Stunted, and Biomass-Storing Species in the Square Model**

| | Ruderal | Competitive | Stunted | Biomass Storer |
|---|---|---|---|---|
| | | *Morphology* | | |
| 1. Life forms | Small algae and macrophytes | Large algae and macrophytes | Crust-forming | Large algae and clonal macrophytes |
| 2. Morphology of shoot | Small stature, limited lateral spread | High, dense canopy; extensive lateral spread above and below ground | Small stature, dwarfish | Often extensive lateral spread above and below ground |
| 3. Leaf or thallus form | Various; often mesomorphic delicate thallus | Robust, often mesomorphic | Often small, leathery, or needle-like | Robust, often mesomorphic |
| | | *Life History* | | |
| 4. Longevity of established phase | Very short | Long or relatively short | Long to very long | Long to very long |
| 5. Leaf or thallus phenology | Short phase of leaf or thallus production in period of high potential productivity | Well-defined peaks of leaf or thallus; production coinciding with period(s) of maximum potential productivity | Evergreens, with various patterns of leaf or thallus production | Evergreens, with various patterns of leaf or thallus production |
| 6. Proportion of annual production devoted to sexual production | Large | Small | Small | Small |
| 7. Vegetative diaspores | No | A few types | Various types | A few types |
| 8. Perennation | Dormant seeds or zygote | Dormant buds and seeds or thallus | Stress-tolerant leaves and roots or thallus | Stress-tolerant leaves and roots or thallus |
| | | *Physiology* | | |
| 9. Maximum potential relative growth rate | Rapid | Rapid | Slow to very slow | Slow |

| | | | | |
|---|---|---|---|---|
| 10. Response to stress | Rapid curtailment of vegetative growth, diversion of resources into reproduction | Rapid morphogenetic responses maximizing vegetative growth | Morphogenetic responses slow and small in magnitude | Curtailment of growth, diversion of resources into storage systems |
| 11. Nutrient absorption | Rapid | Rapid | Slow to very slow | Slow |
| 12. Nutrient losses and turnover rates | Large | Large | Small to very small | Small |
| 13. Storage of photosynthate, mineral nutrients | Confined to seeds or zygote | Most photosynthate and mineral nutrients are rapidly incorporated into vegetative structure, but a proportion are stored and form the capital for the expansion of growth in the following season | Storage systems in leaves, stems, and/or roots, or whole thallus | Storage systems in stems and/or roots, or whole thallus |
| | | *Miscellaneous* | | |
| 14. Litter | Sparse, not usually persistent | Copious, sometimes persistent | Sparse, sometimes persistent | Sparse, often persistent |
| (weight loss % dry wt · $d^{-1}$)[a] | 2 (*Chlorella, Ulva*) | 1.5–1.8 (*Myriophyllum, Potamogeton*) | 1.4 (*Ruppia* sp.) | 0.5 (*Spartina*) |
| 15. Palatability to unspecialized herbivores | Various, often high | Various | Low | Low |
| 16. C:N ratios[a] | 8–11 (*Chlorella, Ulva*) | 11–18 (*Myriophyllum, Potamogeton*) | 25 (*Ruppia* sp.) | 21–40 (*Zostera, Spartina*)[b] |
| 17. DON release (μM)[a] | 20–40 (*Chlorella, Ulva*) | 15 (*Myriophyllum, Potamogeton*) | ~10 (*Ruppia* sp.) | 40 (*Zostera, Spartina*)[b] |

[a]Values from Kemp et al. (1984).
[b]Values from Godshalk and Wetzel (1979).

*Source:* Kautsky (1988).

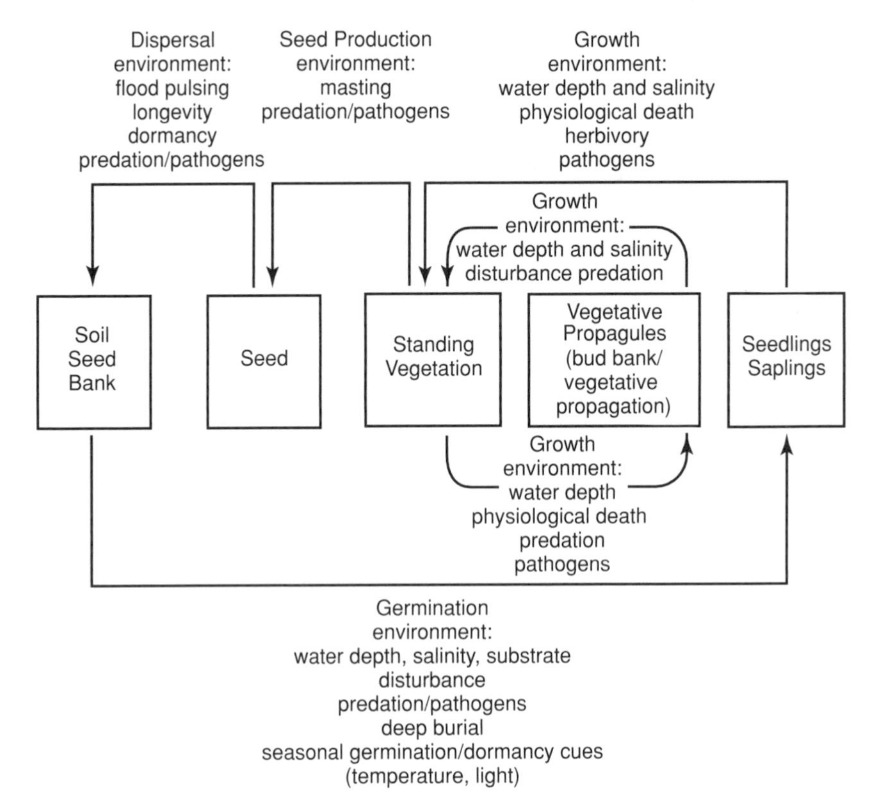


***Figure 3-6.*** *Generalized model of vegetation dynamics as influenced by life history constraints (adapted from Parker et al., 1989, and Leck, 1989; copyright © by Academic Press by permission).*

New Jersey shore collectively carried the seeds of eight wetland species in their feathers or on the mud of their feet (Vivian-Smith and Stiles, 1994). Birds, more than mammals, may carry seeds for long distances during their migrations (Carlquist, 1983).

According to the *foliage-as-fruit* hypothesis, for small seeded species, leaves instead of fruit "reward" the animal for transporting seeds away from the parent plant (Janzen, 1984). In monsoonal wetlands in India, geese transported the seeds of small-seeded species such as *Paspalum distichum*, *Potamogeton indicus*, and *Scirpus tuberosus* (Appendix 1). Wild boar, nilgai (blue bull), and feral cattle also carried the large seeds of species such as *Acacia nilotica* (Middleton and Mason, 1992). Via zoochory, animals can act as local dispersers of seeds to restored wetlands.

Alternatively, plants can be propagated by asexual propagules. For submersed species such as *Potamogeton crispus*, **turions** may be more impor-

***Figure 3-7.*** *Seed traps for anemochory (center) and hydrochory (left and behind), set in a seasonally flooded old field site adjacent to Buttonland Swamp, Illinois (pictured, Shawn Conrad; photograph by Beth Middleton; seed trap design Middleton 1995b).*

tant than seeds, particularly in deep water or where the light is insufficient (Sastroutomo et al., 1981).

## Seed Banks and Seed Dormancy

Restorationists at times are dubious about whether seed banks are of any value in restoration. The value of the seed bank for restoration varies greatly with the type and length of disturbance to which the site was previously subjected. Seed banks may be very short-lived, so that short episodes of farming may almost completely destroy the vegetation (Middleton, 1995a; Middleton, 1996). In contrast, farmed prairie potholes may retain 60% of

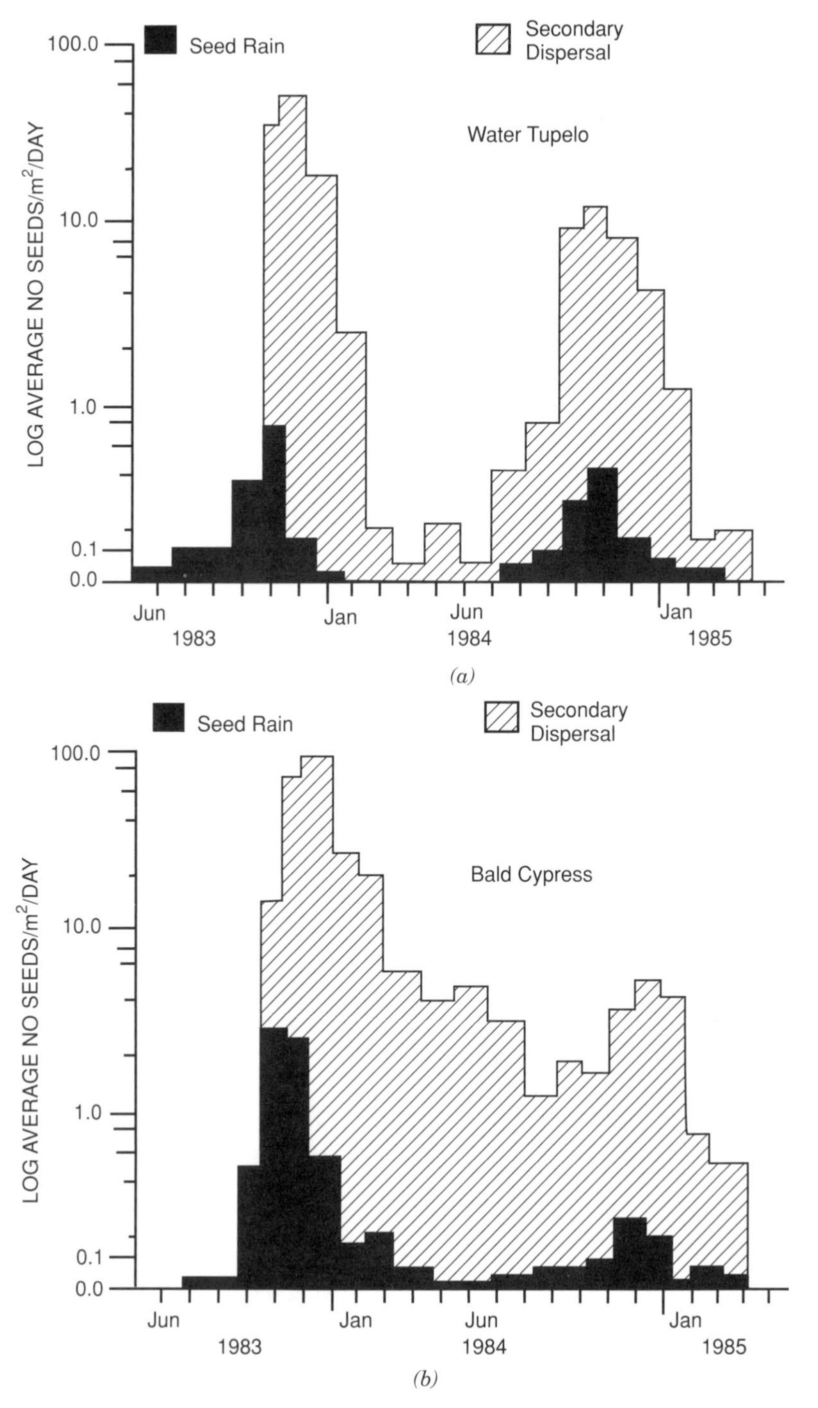


***Figure 3-8.*** *Temporal patterns of seed rain and hydrochory for (a) water tupelo, (b) bald cypress, and hydrochory only for (c) buttonbush. (a and b from Sharitz et al., 1990—copyright © by Lewis Publishers by permission; c from Middleton, 1995a; copyright © by Illinois Water Resources Center by permission).*

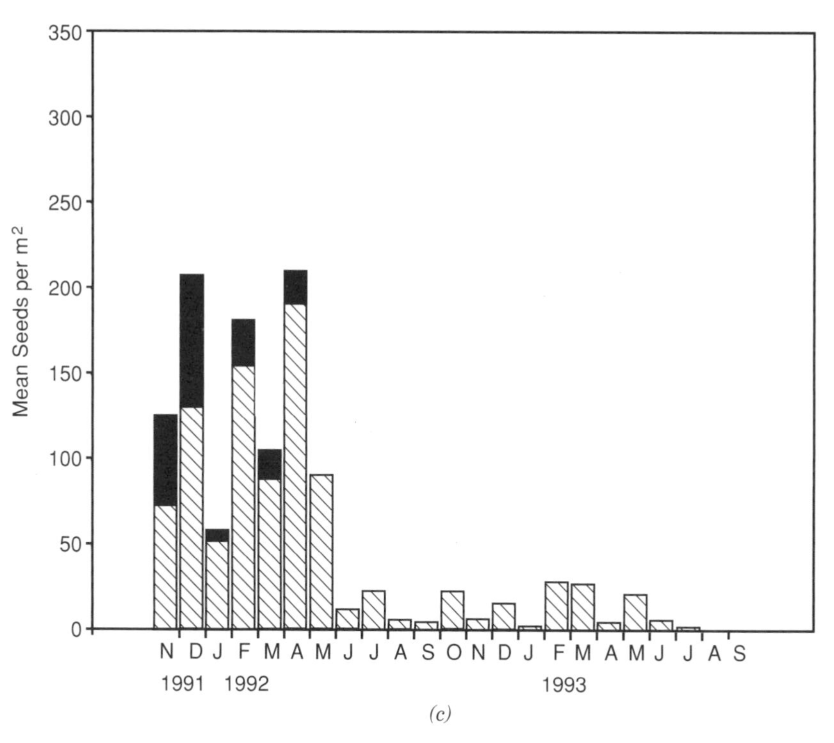


(c)

**Figure 3-8.** *(Continued)*

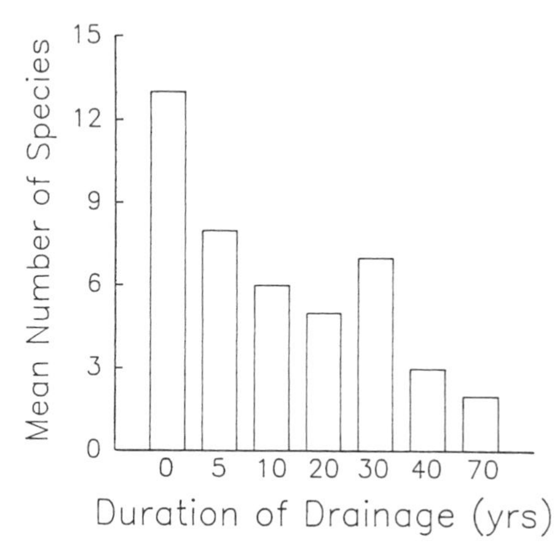


***Figure 3-9.*** *Mean number of wetland species in the seed banks of existing prairie pothole wetlands (0 years) and wetland drained for 5–70 years in Iowa, Minnesota, and North Dakota. Data from Erlandson, 1987 (from van der Valk and Pederson, 1989; copyright © by Academic Press by permission).*

the species for 20 years or more (Fig. 3-9; Weinhold and van der Valk, 1989), and the value of such seed banks should not be ignored by restorationists. Undisturbed seed banks sometimes harbor rare species, as do those of Wilson's Lake, Nova Scotia, where 4 of 18 species were rare (Wisheu and Keddy, 1991). A knowledge of the dynamics of seed banks, and the specific details of dispersal and germination requirements of species, are invaluable in restoration.

If seeds disperse and manage to avoid death by predation and pathogens along the dispersal route, they enter the seed bank (Fig. 3-6), sometimes for lengthy periods of time. After that, germination is ultimately dependent on two factors, dormancy and environment. The nature of seed bank dormancy and longevity are critical factors in restoration. Long-lived seed banks in farmed areas can be a great asset in restoration, but not all seed banks are long-lived.

In terms of overcoming dormancy, seed banks are of four types (Fig. 3-10; Thompson and Grime, 1979). Type I seed banks are dominated by autumn-germinating species with transient seed banks that do not persist through the following summer (e.g., *Dactylis glomerata*). Type II seed banks are dominated by spring-germinating species with **transient seed banks** (Grime, 1981). *Microseris scapigera* seldom retains its viability for more than three months, though it is not a wetland species but rather a species of grassy plains in southeast Australia (Lunt, 1996). Type III seed

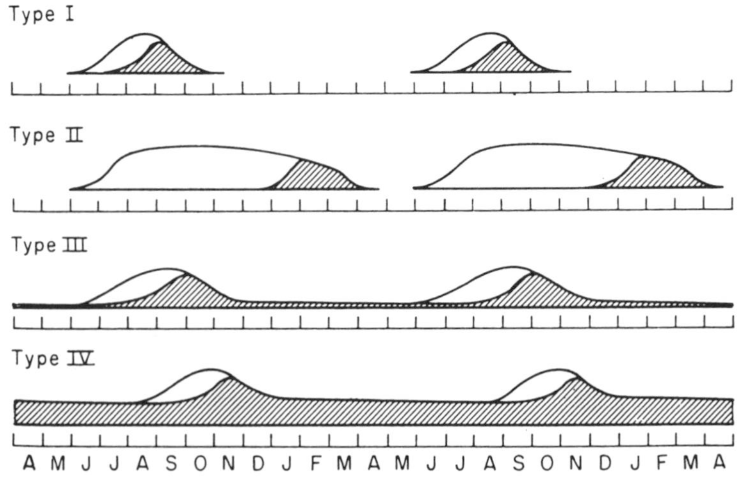


*__Figure 3-10.__ Seed bank types based on germination/dormancy constraints. Unshaded areas represent viable seeds not capable of immediate germination. Shaded areas represent seeds capable of germinating immediately (but usually not able to germinate except in optimal conditions such as suitable laboratory conditions) (from Thompson and Grime, 1979; copyright © by Journal of Ecology by permission).*

banks are dominated by seeds that germinate soon after they are shed, a few of which are incorporated into a permanent seed bank (e.g., *Poa annua*). Type IV seeds banks are dominated by seeds that mostly enter **persistent seed banks**, (e.g., *Calluna vulgaris* and *Juncus effusus*) (Fenner, 1985).

Type III and IV seed bank types are the most common types in wetlands. The seeds of many of the dominant woody species in forested wetlands in the southeastern United States are not persistent due to short longevity, germination, or spring scouring (Schneider and Sharitz, 1986; Titus, 1991; Middleton, 1995a; Middleton, 1996) and thus can be categorized as Type III species. Bottomland forest species can be dependent on frequent dispersal episodes via water to regenerate naturally, thus underscoring the need for hydrologic restoration along waterways (Middleton, 1995a). In contrast, species in northern and prairie pothole wetlands are commonly long-lived and persistent (Leck and Graveline, 1979, and Weinhold and van der Valk, 1989, respectively) and so can be categorized as Type IV species.

After production, some seeds are innately dormant, and must go through a period of after-ripening and then a stage of conditional dormancy until they become nondormant (Fig. 3-12). Strict winter annuals germinate from September through December. Seeds produced in the winter pass through an innate dormancy stage in the spring. Summer annuals germinate from March through June in the southeastern United States (Baskin and Baskin, 1985). Certain other species can germinate anytime, for example, *Cardamine pratensis* and *Ranunculus flammula* in wet Belgian grasslands (Dumotier et al., 1996).

Apart from these rather complicated dormancy scenarios, the germination of seeds from seed banks is constrained by a number of environmental factors (Fig. 3-6). In restoration, drawdown is a key feature enabling many species to germinate from seeds (Appendix 2, Fig. 3-11). The seeds of most tree species will not germinate underwater (Fowell, 1965). In contrast to submerged species, the majority of seeds of emergents also do not germinate underwater (Appendix 2). Careful attention must be paid to the water regime in the initial phases of wetland restoration reliant on either relict seed banks, seed dispersal via wind and water, seed sewing, or seed bank transfer. Early conditions will ultimately affect the vegetation composition of the restoration site.

Within species, seed germination is constrained by salinity levels (Reddy and Singh, 1992; Fig. 3-13). In prairie potholes of the United States, aridity and conductance increase together, moving across North America. Higher conductance is generally found in Saskatchewan and Nebraska than in Iowa (LaBaugh, 1989). For this reason, salinity constraints in seed germination (Galinato and van der Valk, 1986) become an important regional determi-

*Figure 3-11.* Germination from seed banks is dependent on flooding (shown, Greenhouse of the Department of Plant Biology; photograph by Beth Middleton).

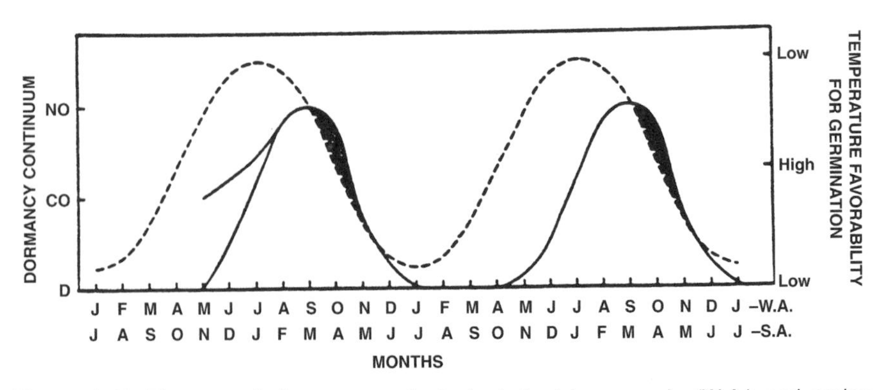


*Figure 3-12.* The annual dormancy cycle in buried winter annuals (W.A.) and spring-germinating summer annuals (S.A.). Shaded areas show when germination is possible. D = dormant, CD = conditionally dormant, ND = nondormant, a = seed conditionally dormant at maturity, and b = seed dormant at maturity. Solid line = dormancy continuum, dotted line = temperature favorability for germination (from Baskin and Baskin, 1985; copyright © 1985 American Institute of Biological Sciences by permission).

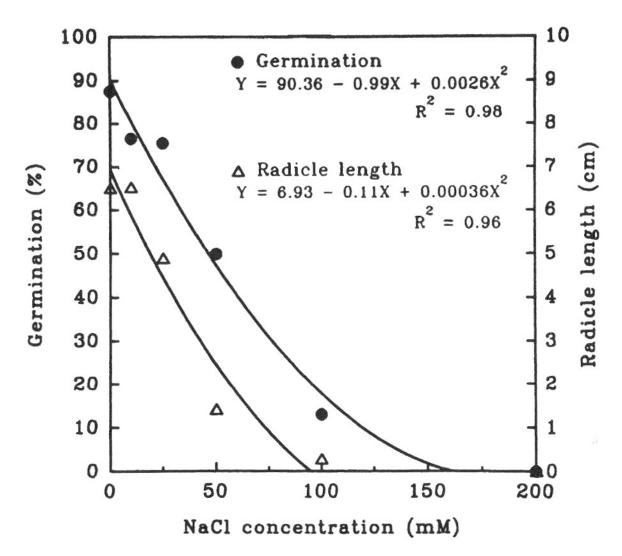


***Figure 3-13.*** *Effect of salinity (NaCl) concentration on germination and radicle growth of hairy beggarticks (*Bidens pilosa*) (from Reddy and Singh, 1992; copyright © by Weed Science by permission).*

nant of vegetation (Kantrud et al., 1989). In coastal wetlands, salinity tolerance can be related to predicted outcomes of saltwater intrusion events (McKee and Mendelssohn, 1989; Pezeshki et al., 1989, 1990b; Pezeshki, 1990, 1992), as well as plant community patterns (Webb and Mendelssohn, 1996).

Substrate type can influence the germination of seeds from seed banks. The seeds of *Boltonia decurrens* germinate more in sand than in clay along the Illinois River (Fig. 3-14; Box 3-1; Smith et al., 1995).

Seed germination constraints can be intuitive for gardeners. Increased depth of burial (planting depth) can decrease the germination of species (Fig. 3-15; Reddy and Singh, 1992), particularly small seeded species (Leck, 1996), which also tend to be more sensitive to soil particle type (Keddy and Constabel, 1986). Burial also influences longevity; long-lived seeds have been buried for hundreds of years in the Okefenokee Swamp (Fig. 3-16; Box 3-2; Gunther et al., 1984).

Temperature is important in the germination of seeds. For some species of cypress swamps, alteration of temperatures greatly increases germination (e.g., *Taxodium ascendens*) (Fig. 3-17). While the seeds of *Taxodium distichum* can be difficult to germinate by the usual routes (e.g., cold stratification, acid treatment), some restorationists have reported success in germinating seeds by unlikely methods. *T. distichum* seeds germinated readily after winter storage in moist plastic bags in a closet in an unheated

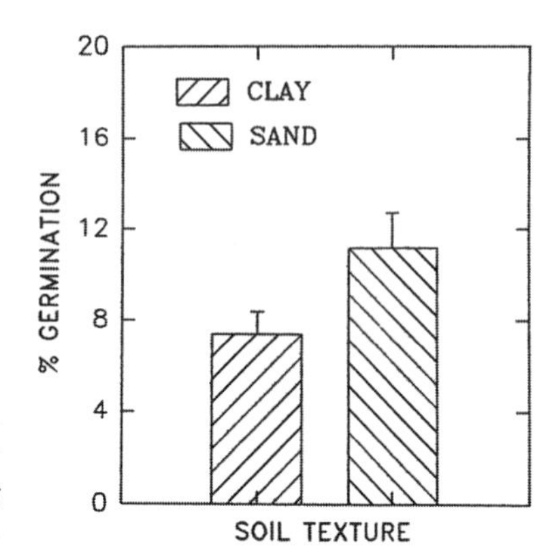


***Figure 3-14.*** *Percent germination of* Boltonia decurrens *after 13 weeks on clay versus sand* ($p < .001$; $df = 58$) *(from Smith et al., 1995; copyright © by Society of Wetland Scientists by permission).*

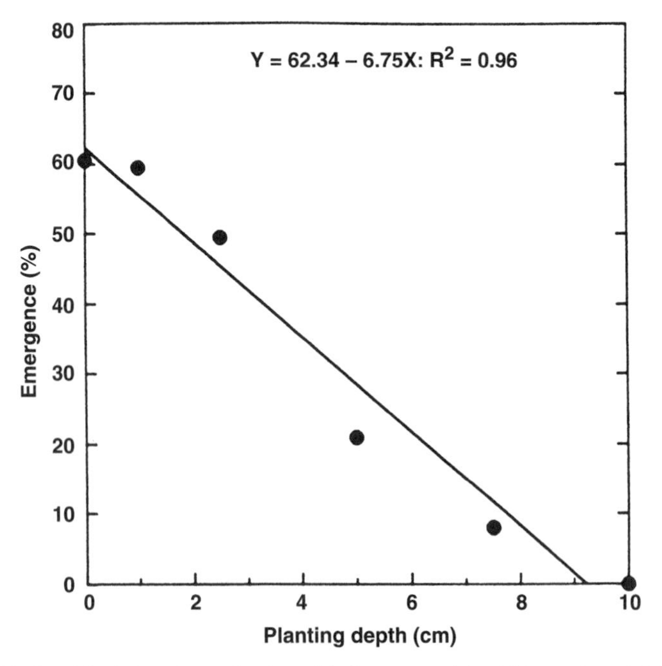


***Figure 3-15.*** *Seed burial and germination of hairy beggarticks (*Bidens pilosa*) 21 days after planting (from Reddy and Singh, 1992; copyright © by Weed Science by permission).*

*BOX 3-1*

## DEPENDENT ON FLOOD PULSING? AN ASTER DISASTER

Some species are declining because their life histories are strategically linked to flood pulsing, a problem in the highly altered riverine landscape of the modern world. *Boltonia decurrens*, the false decurrent aster, is a federally listed threatened perennial in the United States (United States Fish and Wildlife Service, 1988) with characteristics carefully honed to match the primeval environment of the Illinois River. The current state of the Illinois River is degraded by channelization, levees, and water diverted from Lake Michigan via the sewage system of Chicago. *B. decurrens* must cope with both altered flooding and sedimentation regimes (Smith et al., 1993).

Back when the Illinois River still overflowed its banks regularly, the environment of the floodplain was more suitable for *B. decurrens*. The seeds of this species germinate more profusely and its seedlings grow larger on alluvial sand than on other substrates (Smith et al., 1995). While co-occurring species such as *Conyza canadensis* begin to die after a couple of months of flooding, *B. decurrens* survives readily, with its spongy, aerenchymous roots (Stoecker et al., 1995). Even after 10 weeks of saturated soil conditions, *B. decurrens* shows no reduction in stomatal conductance, a measure of the ''running speed'' of plants. Its competitors, *Aster ontarionis, Aster pilosus*, and *C. canadensis*, have decreased stomatal conductance under conditions in which *B. decurrens* thrives. These species eventually die or have lower biomass production (Smith et al., 1993). But flooding is key to *Boltonia*'s survival. Without flooding, other species replace it in three to five years (Smith and Moss, 1998). In the absence of shading competitors, *B. decurrens* has its highest level of photosynthesis and growth in high light conditions (Smith et al., 1993).

*B. decurrens* flourished for the first time in years after the massive floods of 1993–1994 along the Illinois River. Numerous individuals sprang up at sites with the greatest flooding, where water escaped to the floodplain from the channel. Thus, the levees created to dry the floodplain for agricultural and urban usage threaten the extinction of *B. decurrens* (Smith et al., 1998).

*continued*

Boltonia decurrens, *the false decurrent aster, is a federally listed threatened species (photograph by Marion Smith).*

greenhouse (Keith Fessel, personal communication). *T. distichum* seeds also germinated in or under the eavespots of houses (Tim Loftus, personal communication) and on lawns during wet years in southern Illinois (personal observation).

So we revisit the question typically asked by restorationists: How useful is the seed bank in restoring wetland vegetation? The answer is that it depends both on the nature of the wetland and on the severity of the disturbance from which it must recover. In addition, undisturbed wetlands do not always have a high level of correspondence between the species composition of the seed bank, seedlings, and vegetation in the community (Table 3-2). Sometimes fairly rare species in the seed bank are common seedlings in the community and vice versa (Table 3-3). Vegetative propagules are commonly dispersed and can contribute material to the adult stage, but surprisingly little information exists in the literature on this topic.

## Seedling and Sapling Recruitment

To the restorationist, the success of the **seedling** stage, midway in the progression from seed germination to recruitment, is both the most crucial and

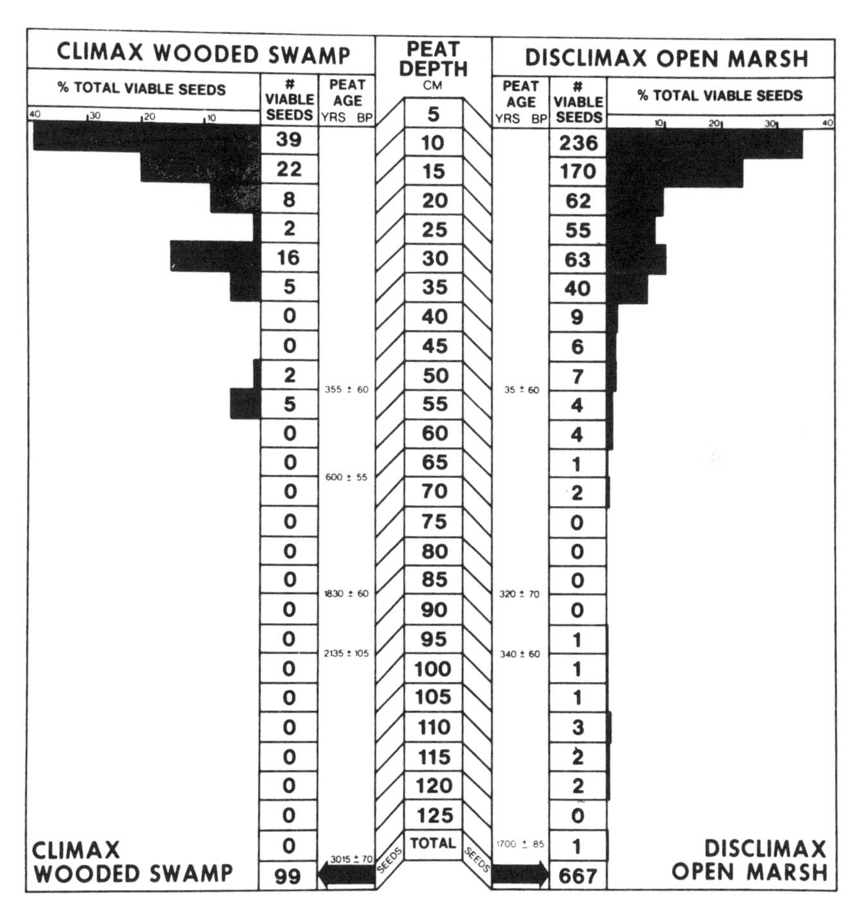


*Figure 3-16. Seed burial and viability at various depths in sediments in Minnies Lake and Chesser Prairie, Okefenokee Swamp (from Gunther et al., 1984; copyright © by Wetland Surveys by permission).*

most tricky part of the restoration process. Too early flooding will cause the death of some species, and species recovery from flooding at the seedling stage is one of the most important factors in the ultimate species composition of bottomland swamp (Bedinger, 1978) and likely other wetland types.

Moving progressively through the life stages of aquatic plants from younger to older, species typically become progressively more flexible in their tolerance for water depth at the seedling (Table 3-4) and sapling (Table 3-5) stages. While seeds of cypress will not germinate underwater, seedlings of cypress will survive extended periods of time under flooded conditions if the flooding does not cover the tips of the plants (Table 3-4). Nevertheless, the seedling stage is particularly vulnerable in that, unlike certain

*BOX 3-2*

**ANCIENT TIME TRAVELERS: SEEDS IN SEED BANKS**

Some seed banks live for a very long time buried in the soil but occasionally are helped to reach the surface via animals, earthquakes, or human interference. In their quest to establish themselves in the evolutionary arms race of the present day, these ancient seeds may be as outmoded as a knight from the Middle Ages jousting with a computerized military tank. Ancient seeds are adapted to the past, but then again, these might recapture their lost genetic variability in populations (Templeton and Levin, 1979).

In the Arctic tundra, seeds can live for centuries frozen in permafrost. In *Eriophorum* tussock tundra, seeds of *Carex bigelowii* and *Luzula parviflora* were among the oldest buried in the soil (197 ± 80 years; McGraw et al. 1991). Viable seeds have been found at depths of 15 cm in the Siberian arctic in the Taymyr Peninsula (Khodachek 1997). Viable seeds of species no longer in the standing vegetation have also been discovered (McGraw et al., 1991).

Some seeds may live even longer. A set of *Nelumbo nucifera* seeds, once claimed to be as old as 50,000 years, were later dated at 466 years (Priestly and Posthumus, 1982). *Lupinus arcticus* seeds in lemming burrows in the Arctic may have been buried for at least 10,000 years (Porsild et al., 1967); however, this claim is unsubstantiated because none of the seeds were radiocarbon dated (Godwin, 1968). In California and northern Mexico, some seeds embedded in adobe bricks were 211 years old (Spira and Wagner, 1983). Obviously, seeds of some species can live for long periods of time buried under the soil.

seeds, if conditions become unfavorable, the seedling cannot regain dormancy.

In nature, seedling mortality is typically very high following flooding episodes. In bald cypress swamps, along the Savannah River of South Carolina, 70-90% of the seedlings died following flooding (Fig. 3-19; Sharitz et al., 1990). Even though both *T. distichum* and *Nyssa aquatica* seedlings can be found on a variety of microsites, including sediment, live wood (knees), and dead wood, few seedlings are recruited into the sapling stage under conditions of flooding (Fig. 3-20; Sharitz et al., 1990). Cypress seed-

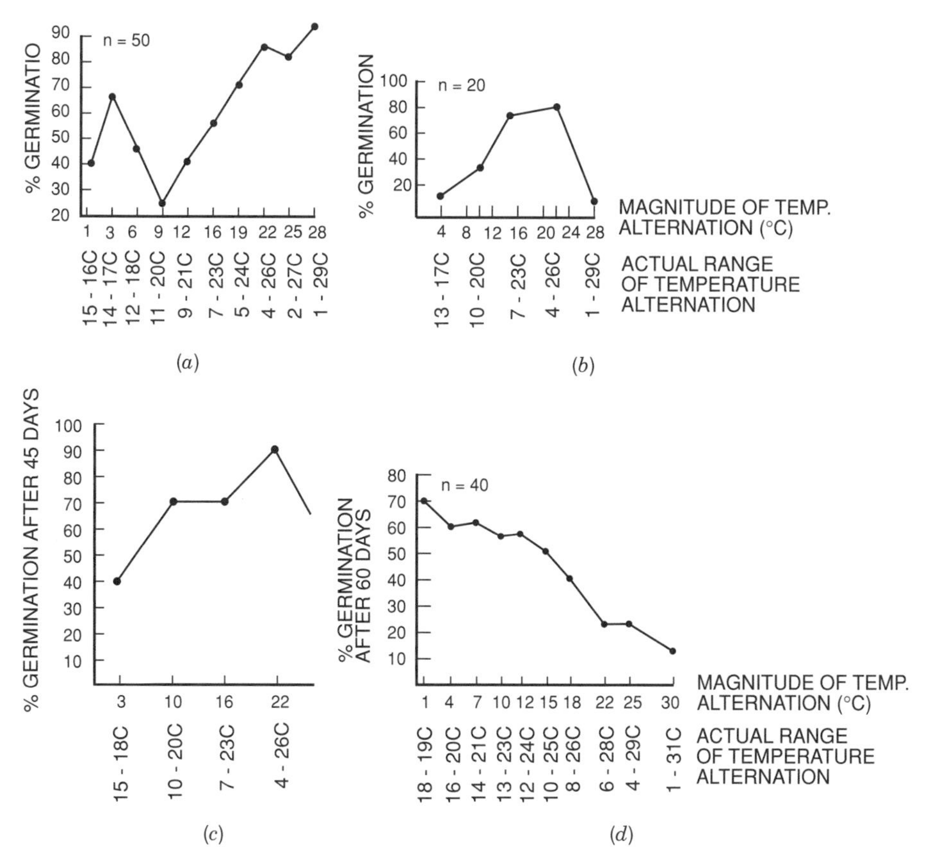


***Figure 3-17.*** *Alternating temperature and germination for (a)* Xyris smalliana*, (b)* Taxodium ascendens*, (c)* Nyssa sylvatica var biflora*, and (d)* Ilex cassine *(from Conti and Gunther, 1984; copyright © by Wetland Surveys by permission).*

lings grow fastest in moist, well-aerated conditions (Fig. 3-18; Mattoon, 1916).

Mortality of first-year tree seedlings is generally high. In southern Louisiana, 35% of the seedlings of *T. distichum* were eaten by nutria between the spring and fall of one growing season (Conner et al., 1986). Typical of species with large, heavy seeds, seedlings of *Quercus nigra* had high seedling survival and were little affected by drought, herbivory, or proximity to a conspecific adult in a floodplain forest of the Big Thicket National Preserve. In contrast, trees with light seeds (*Acer rubrum, Carpinus caroliniana, Liquidambar styraciflua*, and *Ulmus americana*) had low first-year seedling survival due to flooding, drought, damping-off, herbivory, and proximity to conspecific adults. Seedlings of tree species with light seeds

**TABLE 3-2. Comparison of Species Diversity in the Seed Bank, as Field Seedlings, and in Standing Vegetation**

| Wetand Type | Number of Species (Species Richness) | | | | Reference |
|---|---|---|---|---|---|
| | Seed Bank | Seedlings | Vegetation | Total | |
| *Tidal marshes* | | | | | |
| Fresh | 52+ | 19 | — | ~54 | Leck and Graveline (1979) |
| | 35+ | 20 | 24 | 37 | Parker and Leck (1985) |
| | 55+ | 12 | 20 | 58 | Leck and Simpson (1987) |
| | | | | | Leck and Simpson (1995) |
| Salt | 17+ | 5 | 9 | 18 | Hopkins and Parker (1984) |
| *Nontidal marshes* | | | | | |
| Fresh | 45+ | — | 34 | 48 | van der Valk and Davis (1978) |
| | | | | | van der Valk and Pederson (1989) |
| Brackish | 32 | — | 18 | ~35 | Smith and Kadlec (1983) |
| | 6 | — | 11 | 13 | Kautsky (1990) |
| Salt | 9 | — | 14 | 15 | Kadlec and Smith (1984) |
| | 4 | — | 7 | 8 | Ungar and Riehl (1980) |
| *Other freshwater wetlands* | | | | | |
| Fen | 18–26 | — | 10–17 | 24–31 | Meredith (1985) |
| | 48 | — | 59 | 68 | van der Valk and Verhoeven (1988) |
| Lake | 6 | — | 12 | ? | Haag (1983) |
| Lakeshore | 41 | — | 45 | 50 | Keddy and Reznicek (1982) |
| Temporary ponds | 13–18 | 14–22 | 15–24 | 31 | McCarthy (1987) |
| Riverine | 59 | — | 49 | 73 | Schneider and Sharitz (1986) |
| Monsoonal wetland | | | | | |
| | 90 | — | 36+ | 90+ | Middleton (1990); Middleton et al. (1991) |
| | 43 | 13 | — | 48 | Middleton (1990; goose grazing areas only) |
| | 64 | — | 44 | 82 | Mason (1996; 20 wetlands) |
| | 33 | — | 222 | ? | Finlayson et al. (1989, 1990) |

*Source:* Based on Leck (1989) and others.

**TABLE 3-3. Predicted (Seed Density) and Actual Seedling Density, Delta Marsh, Manitoba**

| Species | Seed Density ($m^{-2}$) | Seedling Density ($m^{-2}$) |
|---|---|---|
| | *Emergent Species* | |
| *Scirpus lacustris* | 610 | 400 |
| *Typha glauca* | 220 | 30 |
| *Scolochloa festucacea* | 30 | 100 |
| *Phragmites australis* | 14 | 1 |
| *Puccinellia nuttalliana* | 0 | <1 |
| *Hordeum jubatum* | 1 | 20 |
| | *Mudflat Annuals* | |
| *Atriplex patula* | 16 | 700 |
| *Aster laurentius* | 12 | 20 |
| *Chenopodium rubrum* | 180 | 50 |
| *Rumex maritimus* | 28 | 2 |
| | *Wet Meadow Perennial* | |
| *Stachys palustris* | 3 | 4 |
| *Teucrium occidentale* | 2 | <1 |
| *Lycopus asper* | 10 | <1 |
| *Sonchus arvensis* | 4 | 4 |
| *Cirsium arvense* | 7 | 1 |
| *Urtica dioica* | 3 | 6 |

*Source:* van der Valk and Pederson, (1989).

generally survive in higher numbers if they emerge earlier rather than later in the season (Streng et al., 1989). In laboratory conditions, seedlings of some riparian tree species can withstand at least brief episodes of flooding during the growing season (DeShield et al., 1994) but are more susceptible to these factors in the field, perhaps because of the physical damage of moving debris or sediments in the flood sheet (Mattoon, 1915; Wharton et al., 1982; Kozlowski, 1984; Streng et al., 1989).

Seedlings of halophytes such as mangroves often attain higher total dry weights when grown in saline rather than freshwater conditions (Table 3-6; Greenway and Munns, 1980; Flowers et al., 1986; Pezeshki et al., 1990a). Apparently, the reduced growth of halophytes in fresh water is due in part to the lack of ions for osmoregulation (Greenway and Munns, 1980; Yeo and Flowers, 1980). In one study, *Rhizophora mangle* grew best at a salinity level closest to that of sea water (Stern and Voigt, 1959).

**Saplings** are typically more flexible than seedlings in their tolerance for disturbance (Table 3-5). Both two-year-old *N. aquatica* and *N. sylvatica* var. *biflora* had lower biomasses in stagnant compared to moving water (Harms, 1973). Saplings of *T. distichum* have higher annual growth rates

TABLE 3-4. Seedling Survivorship and Water Depth (Studies Were Conducted in Greenhouse Conditions Unless Otherwise Indicated)

| Species | Survival | | | | Reference | Comments |
|---|---|---|---|---|---|---|
| | Saturated | Flooded Below Tip | Flooded Above Tip | Experiment Length (day) | | |
| *Acer negundo* | — | Yes | No | 32 | Hosner (1958, 1960) | Died if submerged more than 32 days. |
| *Acer rubrum* | — | Yes | — | 21 | DeShield et al. (1994) | |
| | — | — | No | 30 | Hosner (1960) | Died if submerged more than 20 days. |
| | Yes | — | — | 15–60 | Hosner and Boyce (1962) | |
| *Acer saccharinum* | — | — | Yes | 30 | Hosner (1958, 1960) | Seedlings died if submerged for four days. |
| | Yes | — | — | 15–60 | Hosner and Boyce (1962) | |
| *Alnus incana* | Yes | — | — | 10 | Crawford (1972) | Malic acid increases in roots a few hours after flooding |
| *Alnus rugosa* | No | — | — | 6 | McDermott (1954) | 60% mortality in saturated soil for more than eight days. |
| *Avicennia germinans* | Yes | Yes | — | 180 | Pezeshki et al. (1990) | Also survived salinity, water and salinity, and water treatment. |
| | Yes | Yes | — | 240 | McKee (1993) | Constant flooding inhibited root growth. |
| *Baccharis halimfolia* | Yes | — | — | 60 | Tolliver et al. (1997) | 60% mortality at 10 g $L^{-1}$;100% above 20 g $L^{-1}$ salinity |
| *Betula nigra* | — | Yes | — | 21 | De Shield et al. (1994) | |
| *Betula papyrifera* | Yes | No | — | 14 | Tang and Kozolowski (1982) | Stomata lost regulatory function after 14 days. |
| *Betula pubesens* | Yes | — | — | — | Wiegers (1985) | |
| *Celtis occidentalis* | — | — | No | 30 | Hosner (1960) | |
| | Yes | — | — | 15–60 | Hosner and Boyce (1962) | 20% mortality after 60 days. Some mortality of secondary roots. |
| *Celtis laevigata* | Yes | — | — | 15-60 | Hosner and Boyce (1962) | 6.7% mortality after 60 days. All root except primary roots died. |
| *Cephalanthus occidentalis* | — | — | Yes | 30 | Hosner (1960) | |
| *Cladium mariscus* | Yes | Yes/no | No | — | Alexander (1971) | Seedlings killed by both winter drydown and flooding. |
| *Fraxinus caroliniana* | — | Yes | — | — | Putnam et al. (1960) | Seedling established in moist soil from April to July after water recedes. |

| | | | | | | |
|---|---|---|---|---|---|---|
| *Fraxinus pennsylvanica* | Yes | — | No | 15–60 | Hosner (1958); Hosner and Boyce (1962) | Seedling died if completely submerged for 32 days. |
| | Yes | Yes | — | 285 | Good and Patrick (1987) | |
| | — | — | Yes | 30 | Baker (1977) | 91% survival one week after flooding. |
| *Fraxinus profunda* | Yes | — | — | 15–60 | Hosner and Boyce (1962) | |
| *Iva frutescens* | Yes | — | — | 60 | Tolliver et al. (1997) | 60% mortality at 10 g $L^{-1}$;100% mortality above 20 g $L^{-1}$ salinity. |
| *Juniperus virginiana* | Yes | — | — | 60 | Tolliver et al. (1997) | 20% mortality at 10 g $L^{-1}$; 100% mortality above 20 g $L^{-1}$ salinity. |
| *Laguncularia racemosa* | Yes | Yes | — | 180 | Pezeshki et al. (1990) | Also survived salinity, water and salinity, and water treatment. |
| *Larix laricina* | — | Yes | — | 10 | Crawford (1972) | All survived if flooded for eight days. |
| *Liquidambar styraciflua* | — | — | No | 2–32 | Hosner (1958) | |
| | Yes | — | — | 15–60 | Hosner and Boyce (1962) | Secondary roots died. |
| *Liriodendron tulipifera* | — | No | — | 21 | DeShield et al. (1994) | Died after more than seven days of flooding. |
| | — | — | Yes | 30 | Baker (1977) | 68% survival one week after flooding. |
| *Myrica cerifera* | Yes | — | — | 60 | Tolliver et al. (1997) | 60% mortality at 10g $L^{-1}$;100% mortality above 20 g $L^{-1}$ salinity. |
| *Nymphoides indica* | — | Yes | Yes | 52 | Mason (1996) | |
| *Nyssa aquatica* | Yes | — | — | 84 | Dickson et al. (1965) | |
| | Yes | Yes | Yes | 120–180 | Kennedy (1970) | Lost height in treatments with tips held under water. |
| | — | — | Yes | 30 | Baker (1977) | 95% survival one week after flooding. |
| | Yes | — | — | 15–60 | Hosner and Boyce (1962) | |
| *Nyssa sylvatica* var. *biflora* | Yes | Yes? | — | 210 | Hook et al. (1970) | Flooded treatment was 20 cm with "seedlings" one year old. |
| *Phragmites australis* | Yes | Yes | No | >365 | Weisner and Ekstam (1993) | Early germination in year 1 produces taller plants with less second year mortality under flooded conditions. Plant length must exceed water depth. |
| *Pinus echinata* | — | — | No | 180 | Williston (1962) | Seedlings submerged in a pool died after six to seven months. |
| *Pinus taeda* | — | Yes | — | 21 | DeShield et al. (1994) | Some reduction in survival if flooded late in growing season. |

**TABLE 3-4. (Continued)**

| Species | Survival | | | | Reference | Comments |
|---|---|---|---|---|---|---|
| | Saturated | Flooded Below Tip | Flooded Above Tip | Experiment Length (day) | | |
| | Yes | — | — | 60 | Tolliver et al. (1997) | 40% mortality at 10 g $L^{-1}$; mortality above 20 g $L^{-1}$ salinity. |
| | — | — | No | 180 | Williston (1962) | Seedlings submerged in a pool died after six to seven months. |
| *Platanus occidentalis* | — | — | No | 30 | Hosner (1960) | |
| | — | — | No | 30 | Baker (1977) | 35% survival one week after flooding |
| | Yes | — | — | 15–60 | Hosner and Boyce (1962) | 26.7% died after 60 days in saturated soil. |
| *Populus deltoides* | — | — | No | 2–32 | Hosner (1958) | Seedling died if flooded for longer than 16 days. |
| | — | — | No | 30 | Baker (1977) | Cuttings used; 24% survival one week after flooding. Recruitment less successful where groundwater lower than root level (Scott et al., 1993; Segelquist et al., 1993). |
| *Populus deltoides* var. *monilifera* | Yes | No | No | — | Scott et al. (1997) | Seedling regeneration in the field only if sites are bare and moist from scouring flow. |
| *Quercus falcata* | — | — | No | 16 | Hosner (1960) | var. pagodaefolia used in study. |
| | No | — | — | 15–60 | Hosner and Boyce (1962) | 86.7% mortality after 60 days in saturated soil. |
| *Quercus nigra* | Yes | Yes | — | 285 | Good and Patrick (1987) | |
| | — | Yes | — | 21 | DeShield et al. (1994) | |
| *Quercus nuttallii* | — | No | — | 21 | DeShield et al. (1994) | Died after more than seven days of flooding. |
| *Quercus palustris* | — | — | No | 30 | Hosner (1960) | |
| | Yes | — | — | 15–60 | Hosner and Boyce (1962) | Some secondary shoots died. |
| *Quercus phellos* | — | Yes | — | 21 | DeShield et al. (1994) | Some reduction in survival after 21 days of flooding, particularly late in the growing season. |
| | Yes | — | — | 15–60 | Hosner and Boyce (1962) | Some mortality of secondary roots. |

| | | | | | | |
|---|---|---|---|---|---|---|
| *Quercus shumardii* | — | — | No | 30 | Hosner (1960) | |
| | No | — | — | 15–60 | Hosner and Boyce (1962) | 33.3% mortality after 60 days in saturated soil. Some mortality of secondary roots. |
| *Rhizophora mangle* | Yes | Yes | — | 180 | Pezeshki et al. (1990) | Also survived salinity, water and salinity, and water treatment. Saturated seedlings generally survived at higher levels in saline water (Stern and Voigt, 1959). |
| *Rhizophora mangle* | Yes | Yes | — | 240 | McKee (1993) | Root growth stimulated by flooding. |
| *Salix nigra* | Yes | — | Yes | 15–60 | Hosner (1958, 1960); Hosner and Boyce (1962) | Seedlings chlorotic but lived if submerged for 32 days. |
| *Scirpus cyperinus* | — | Yes | — | 21 | DeShield et al. (1994) | Some reduction in survival if flooded, especially late in growing season. |
| *Taxodium distichum* | — | Yes | — | 21 | DeShield et al. (1994) | Died if submerged during growing season. Requires a two-year drawdown for establishment to occur (Shelford, 1954). Recruitment less successful in shade (Shankman 1991). Seedlings successful at the high water mark (Eggler and Moore, 1961). |
| | Yes | No | No | 14 | Demaree (1932) | |
| | Yes | Yes | No | ~30 | Fessel and Middleton (unpublished) | Seedlings have limited ability to survive if submerged during the senescent season. |
| | Yes | No | — | 44 | Mattoon (1916) | Two-week-old seedlings die within two weeks in an inch of water. |
| | Yes | Yes | No | ~10 | Conner and Flynn (1989) | Field study: in deepest water treatment, water level from June to September not above 10 cm. |
| *Taxodium distichum* var. *nutans* | — | — | — | — | McKnight (1980) | Similar to bald cypress. |
| *Ulmus alata* | Yes | — | — | 0–32 | McDermott (1954) | |
| *Ulmus americana* | Yes | — | No | 30 | Hosner (1960); Guilkey (1965) | |
| | Yes | — | — | 15-60 | Hosner and Boyce, (1962) | 6.7% mortality after 60 days. |
| *Ulmus crassifolia* | Yes | — | — | — | McKnight (1980) | |

*Source:* Loosely based on Whitlow and Harris (1979) and others.

**TABLE 3-5. Sapling Survivorship and Inundation**

| | Survival | | Flood | | |
|---|---|---|---|---|---|
| Species | Flood Depth (cm) Above Root Crown | Inundation Period (days) | Survive | Reference | Comment |
| *Acer negundo* | 240–480 | 105 | Yes | Noble and Murphy (1975) | |
| *Acer saccharinum* | >61–122 | 9 | Yes | Harris et al. (1975) | Trees cut to 61–122 cm; complete inundation. |
| *Eucalyptus camaldulensis* | >61–122 | 121 | Yes | Harris et al. (1975) | Trees cut to 61–122 cm; complete inundation. |
| *Fraxinus pennsylvanica* | 90 | 210 | Yes | Broadfoot (1967) | Flooded trees grew 17% more than control trees. |
| | >61–122 | 83 | Yes | Harris et al. (1975) | Trees cut to 61–122 cm; complete inundation. |
| *Gleditsia triacanthos* | 240–480 | 105 | No | Noble and Murphy (1975) | |
| | >61–122 | 53 | Yes | Harris et al. (1975) | Trees cut to 61–122 cm; complete inundation. |
| *Ilex decidua* | 240–480 | 105 | Yes | Noble and Murphy (1975) | |
| *Liquidambar styraciflua* | 90 | 210 | Yes | Broadfoot (1967) | Flooded trees grew 82% more than control trees. |
| | 240–480 | 105 | No | Noble and Murphy (1975) | |
| *Platanus occidentalis* | >61–122 | 44–56 | No | Harris et al. (1975) | Trees cut to 61–122 cm; complete inundation. |
| *Populus x canadensis* | >61–122 | 55–102 | No | Harris et al. (1975) | Trees cut to 61–122 cm; complete inundation. |
| *Populus deltoides* var. *monilifera* | — | — | — | Tolliver et al. (1997) | Do not establish on sites with flood flow >31 $m^3$ $sec^{-1}$. |
| *Populus fremontii* | >61–122 | 55–102 | No | Harris et al. (1975) | Trees cut to 61–122 cm; complete inundation. |
| *Salix* sp. | >61–122 | 114-155 | Yes | Harris et al. (1975) | Trees cut to 61–122 cm; complete inundation. |
| *Salix alba* 'Tristis' | >61–122 | 125 | Yes | Harris et al. (1975) | Trees cut to 61–122 cm; complete inundation. |
| *Salix matusudana* | >61–122 | 69 | Yes | Harris et al. (1975) | Trees cut to 61–122 cm; complete inundation. |
| *Salix terminis* | >61–122 | 125 | Yes | Harris et al. (1975) | Trees cut to 61–122 cm; complete inundation |
| *Taxodium distichum* | 10 | ~730 | Yes | Young et al. (1993) | Periodically flooded saplings developed false growth rings; continuously flooded ones had higher radial growth. |
| | >61–122 | 171 | No | Harris et al. (1975) | Trees cut to 61–122 cm; complete inundation. |

*Source:* Whitlow and Harris (1979) and others.

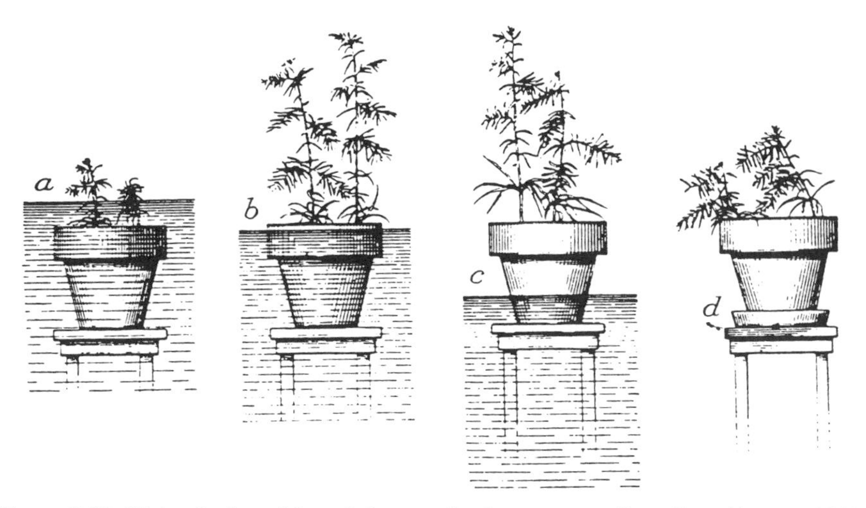


***Figure 3-18.*** *Water depth and the relative growth of cypress seedlings (from Mattoon, 1916; copyright © by Society of American Foresters by permission).*

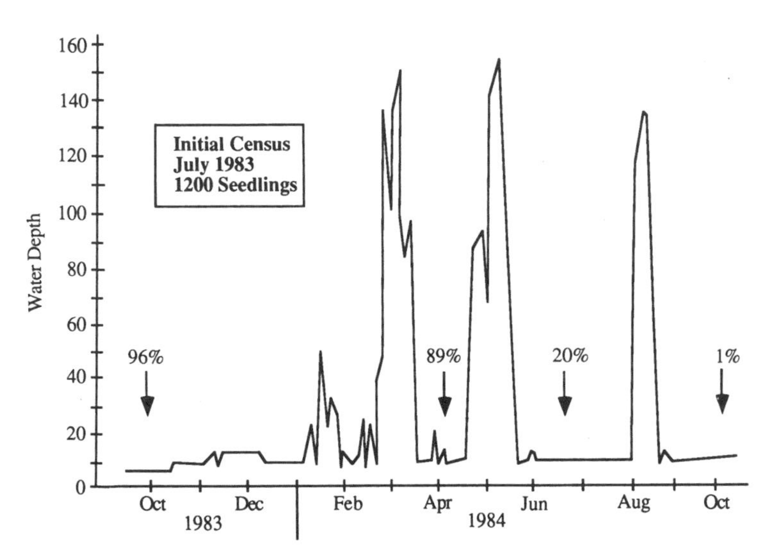


***Figure 3-19.*** *Percent seedling survival (shown by numbers and arrow) of bald cypress and water tupelo in relation to flood events, Savannah River floodplain, South Carolina (from Sharitz et al., 1990; copyright © by Lewis Publishers by permission).*

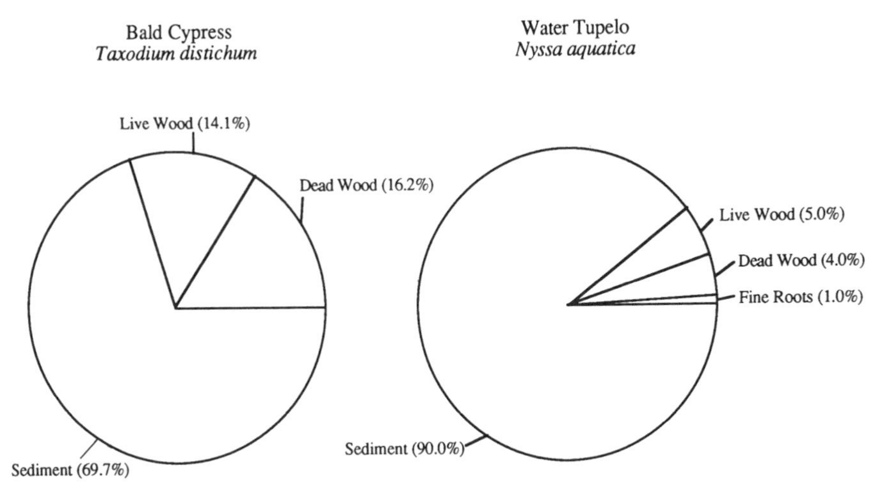


***Figure 3-20.*** *Distribution of bald cypress and water tupelo seedlings on microsites in a deciduous swamp forest, Savannah River floodplain, South Carolina (from Sharitz et al., 1990; copyright © by Lewis Publishers by permission).*

for a few years in continuously flooded conditions versus freely drained conditions (Green, 1947; Eggler and Moore, 1961; Young et al., 1993). However, in long-term flooding, a growth surge is typically followed by a sharp decline. Upstream of a road that impounded water along a floodplain wetland of the Savannah River, in South Carolina, growth of *T. distichum* (trees) accelerated for a few years and then declined (Young et al., 1995). Similarly, after the New Madrid Earthquake lowered the elevation of Reelfoot Lake and permanently flooded some adult trees of *T. distichum*, annual growth of that species increased and then subsequently decreased (Stahle and Cleaveland, 1992). While Stahle and Cleaveland (1992) concluded that the decline in growth was an artifact of the techniques used, Young et al.

**TABLE 3-6. Seedling Total Dry Weight in Response to Treatments of Soil Flooding (5 cm Above Soil) and Salinity (Soil Salinity Maintained at 340 mol m$^{-3}$ NaCl) in Greenhouse Study of *Avicennia germinans, Laguncularia racemosa,* and *Rhizophora mangle* (mean heights of 12.2, 37.0, and 26.6 cm, Respectively)**

| | Total Dry Weight | | | |
|---|---|---|---|---|
| Species | No flood or Salt | Salt | Flood | Flood and Salt |
| *Avicennia germinans* | 36.5b | 40.2a | 29.1b | 40.6a |
| *Laguncularia racemosa* | 31.9a | 37.7a | 26.5a | 36.0a |
| *Rhizophora mangle* | 28.1a | 31.1a | 22.6b | 31.3a |

*Note:* Total dry weight followed by different letters indicates significant differences.
*Source:* Pezeshki et al. (1990).

(1995) argued that the growth may have actually decline at Reelfoot Lake in view of their results recorded via dendrometer bands in South Carolina.

## Adult Water Tolerance, Survivorship, and Root Metabolism

The inherent ability of the adult life stages of species to survive under various water regimes varies widely among herbaceous (Figs. 3-21, 3-22, and 3-23; Table 3-7) and woody (Fig. 3-24; Table 3-8) species and is a major determinant of vegetative composition in wetlands (Bedinger, 1978; van der Valk and Welling, 1988). In bottomland forests of the southeastern United States, forest zones are a product of progressively less flooded conditions from permanently flooded open water to the intermittently flooded bottomland/upland transition (Fig. 3-25; Clark and Benforado, 1981; Ther-

*Figure 3-21.* *Flooding depth (low, medium, high) influences the growth and survival of aquatic plants (pictured, Prakash Mudgal; photograph by Beth Middleton).*

***Figure 3-22.*** Nymphoides indica *(photograph by Beth Middleton).*

***Figure 3-23.*** Ipomoea aquatica *(photograph by Beth Middleton).*

***Figure 3-24.*** Avicennia marina *(photograph by Beth Middleton).*

iot, 1993). In the Pacific Northwest, while the species found in bogs (*Ledum groenlandicum, Vaccinium uliginosum, Kalmia polifolia, Sphagnum* sp.) were little affected by short-term flooding, tree species such as *Tsuga heterophylla* and *Picea sitchensis* died back as a result of a flood in June and July 1948 (Brink, 1954).

The majority of aquatic species have higher rates of production in less flooded conditions. In salt marshes, streamside production of *Spartina alterniflora* on natural levees is higher than it is farther from the stream (Figs. 3-26a, b; DeLaune et al., 1979). However, certain species, such as *Cephalanthus occidentalis* in ponds in Canada, actually increase in abundance with water depths of up to 130 cm (Fig. 3-27; Faber-Langendoen and Maycock, 1989). In peatlands in Norway, oscillations in groundwater level alter the growth of various species of *Sphagnum* (Fig. 3-28; Pedersen, 1975).

Water regime is an environmental factor to which restorationists must

**TABLE 3-7. Herbaceous Species Survivorship and Water Depth**

| Species | Survival | | | | Reference | Comments |
|---|---|---|---|---|---|---|
| | Saturated | Flooded Below Tip | Flooded Above Tip | Experiment Length (day) | | |
| *Aster ontarionis* | Yes | — | — | 70 | Smith and Moss (1998) | 94% survival. |
| *Aster pilosus* | Yes/no | — | — | 70 | Smith and Moss (1998) | 61% survival. |
| *Boltonia decurrens* | Yes | — | — | 56 | Stoecker et al. (1995) | |
| | Yes | — | — | 70 | Smith and Moss (1998) | |
| *Carex praegracilis* | Yes | Yes | — | 50 | Rumburg and Sawyer (1965) | Treatments with water over 12.7 cm for more than 50 days showed decreases in yield; dominant in hay mixture. |
| *Carex rostrata* | Yes | Yes | — | 50 | Rumburg and Sawyer (1965) | Treatments with water over 12.7 cm for more than 50 days showed decreases in yield; dominant in "hay" mixture. |
| *Cladium jamaicense* | — | Yes | — | 730 | Newman et al. (1996) | Water levels drawn down to 5 cm in dry season; otherwise, held at 60 cm. |
| *Conyza canadensis* | Yes/no | — | — | 56 | Stoecker et al. (1995) | About 50% survival after 56 days of surface saturation. |
| | No | — | — | 70 | Smith and Moss (1998) | 28% survivorship. |
| *Eleocharis interstincta* | — | Yes | — | 730 | Newman et al. (1996) | Water levels drawn down to 5 cm in dry season; otherwise, held at 60 cm. |
| *Ipomoea aquatica* | Yes | Yes | — | 120 | Middleton (1990) | |
| *Juncus balticus* | Yes | Yes | — | 50 | Rumburg and Sawyer (1965) | Treatments with water over 12.7 cm for more than 75 days showed decreases in yield; dominant in "hay" mixture. |
| *Ludwiga peploides* | Yes | Yes | — | 120 | Mathis (1996) | Plants died within one week when clipped at ground level in freely drained conditions. |

| | | | | | | |
|---|---|---|---|---|---|---|
| *Nymphoides cristatum* | Yes | Yes | — | 120 | Middleton (1990) | |
| *Panicum hemitomon* | — | Yes | — | 58 | Lessman et al. (1997) | Water levels raised slowly in the experiment to a maximum of 39 cm. |
| *Paspalidium flavidum* | Yes | Yes | — | 120 | Middleton (1990) | |
| *Paspalum distichum* | Yes | Yes | — | 120 | Middleton (1990) | |
| *Phragmites australis* | Yes | Yes | Yes | 455 | Hellings and Gallagher (1992) | Cutting and salinity reduced biomass. |
| | Yes | Yes | — | 120 | Mathis (1996) | Plants died within one week when clipped at ground level in 15 cm of water. |
| *Ranunculus peltatus* | Yes | Yes | ? (27 cm) | ~90 | Volder et al. 1997 | |
| *Scirpus olneyi* | Yes | Yes | — | 37 | Broome et al. (1995) | Salinity above 10 ppt reduced growth. |
| *Spartina alterniflora* | — | Yes | — | 67 | Lessman et al. (1997) | Water levels raised slowly in the experiment to a maximum of 39 cm; at the end of the experiment, some of the roots had died. |
| *Spartina patens* | Yes | Yes | — | 37 | Broome et al. (1995) | Salinity above 10 ppt reduced growth somewhat; increased water depth decreased growth. |
| *Spartina patens* | — | Yes | — | 57 | Lessman et al. (1997) | Water levels raised slowly in the experiment to a maximum of 39 cm. |
| *Typha domingensis* | Yes | Yes | — | 68 | Glenn et al. (1995) | Compared to freshwater treatment; dry mass reduction of 50% at 3.5 ppt salinity 90% at 6 ppt salinity; at 9 ppt salinity growth negligible, though plants survived. |
| | — | Yes | — | 730 | Newman et al. (1996) | Water levels drawn down to 5 cm in dry season; otherwise, held at 60 cm; higher biomass in higher water and nutrient levels. |
| *Typha latifolia* | Yes | Yes | — | 120 | Mathis (1996) | Plants died within one week when clipped at ground level in 15 cm of water. |

*Note:* Studies were conducted in greenhouse conditions unless otherwise indicated.

*Source:* Loosely based on Whitlow and Harris (1979).

**TABLE 3-8. Adult Tree and Shrub Survivorship and Inundation**

| Species | Survival | | | Flood Inundation Length (day) | Reference | Comments |
|---|---|---|---|---|---|---|
| | Saturated Soil | Flooded to Root Crown | Flood Above Root Crown (depth) | | | |
| *Acer negundo* | Yes | Yes | Yes (? cm) | 1460 | Yeager (1949) | |
| | ? | ? | ? | 189 | Bell and Johnson (1974) | |
| | — | — | Yes (? cm) | 73 | Harris et al. (1975) | 99% survival after 73 days. |
| *Acer rubrum* | — | — | Yes (? cm) | ? | Hall and Smith (1955) | Survived if flooded for less than 41% of the growing season. |
| *Acer saccharinum* | Yes | Yes/no | No (? cm) | 730 | Yeager (1949) | 64% survival after 730 days. |
| *Alnus rugosa* | — | — | No (? cm) | ? | Hall and Smith (1955) | Survive if flooded for less than 24% of the growing season. |
| | Yes | — | — | ? | McDermott (1954) | Trees little affected by 16–32 days of soil saturation. |
| *Betula nigra* | Yes | Yes | no (? cm) | 730 | Yeager (1949) | |
| *Carya cordiformis* | — | — | no (? cm) | 148 | Brunk et al. (1975) | |
| *Carya illinoinensis* | Yes | Yes | no (? cm) | 1460 | Yeager (1949) | 11% survival after 730 days. |
| | — | — | no (? cm) | 73 | Harris et al. (1975) | 55% survival after 73 days of flooding. |
| *Carya ovata* | — | — | no (? cm) | 114 | Bell and Johnson (1974) | 38% survival after 114 days. |
| *Carya tomentosa* | — | — | no (? cm) | 149 | Bell and Johnson (1974) | 35% survival after 109 days. |
| *Catalpa speciosa* | — | — | no (? cm) | 67 | Harris et al. (1975) | 60% survival after 67 days. |
| | — | — | no (? cm) | ? | Hall and Smith (1955) | Survive if flooded for less than 18% of the growing season. |
| *Celtis occidentalis* | Yes | No | No (? cm) | 730 | Yeager (1949) | In wet but not flooded conditions, 66% survival after 730 days. |
| | — | — | No (? cm) | 189 | Bell and Johnson (1974) | |
| *Cephalanthus occidentalis* | — | — | No (? cm) | 1460 | Yeager (1949) | 40% survival. |
| *Cercis canadensis* | Yes | No | No (? cm) | 240 | Yeager (1949) | |
| *Cornus florida* | — | — | No (? cm) | ? | Hall and Smith (1955) | Survive if flooded for less than 2% of the growing season. |
| *Diosporos virginiana* | Yes | No | No (? cm) | 1460 | Yeager (1949) | 20% survival if wet but not flooded after 1460 days. |

| | | | | | | |
|---|---|---|---|---|---|---|
| *Fagus grandifolia* | — | — | No (? cm) | ? | Hall and Smith (1955) | Survive if flooded for less than 4% of the growing season. |
| *Forestiera acuminata* | Yes | Yes | No (? cm) | 730 | Yeager (1949) | 30% survival after 730 days. |
| *Fraxinus americana* | Yes | Yes | No (51 cm) | 2555 | Yeager (1949) | 7% survival. |
| *Gleditsia aquatica* | Yes | No | No | 1460 | Yeager (1949) | 8% survival in flooded condition after 730 days. |
| *Gleditsia triacanthos* | — | — | Yes | 189 | Bell and Johnson (1974) | |
| | — | — | No | 730 | Broadfoot and Williston (1973) | None survived after two years of inundation. |
| *Gynocladus dioicus* | Yes | Yes | Yes | 1460 | Yeager (1949) | |
| *Ilex decidua* | Yes | Yes | No | 1460 | Yeager (1949) | |
| *Juglans nigra* | — | — | No | 149 | Bell and Johnson (1974) | |
| | — | — | No | 180 | Kennedy and Krinnard (1974) | |
| *Juniperus virginiana* | — | — | No | 365 | Hall and Smith (1955) | All died in one year. |
| | — | Yes/no | — | 73 | Harris (1975) | 47% survival after 73 days of flooding. |
| *Larix laricina* | — | Yes | — | 2190 | Denyer and Riley (1964) | Survived six years of flooding. |
| *Liquidambar styraciflua* | — | — | No (<30 cm) | 1460 | Broadfoot and Williston (1973) | |
| *Liriodendron tulipifera* | — | — | No | 60 | Kennedy and Krinard (1974) | 11-year-old plantation. |
| *Morus rubra* | Yes | No | No | 240 | Yeager (1949) | |
| *Nyssa aquatica* | — | Yes | Yes (50–86) | 365 | Keeland and Sharitz (1995) | Maximum growth in deep, periodically flooded riverine sites. |
| *Nyssa sylvatica* var. *biflora* | — | Yes | Yes (50–86) | 365 | Keeland and Sharitz (1995) | Maximum growth in deep, periodically flooded riverine sites. |
| *Pinus echinata* | — | — | No (46 cm) | 105 | Williston (1962) | |
| *Pinus resinosa* | — | — | Yes (? cm) | 48 | Ahlgren and Hansen (1957) | 90% survival after 48 days of flooding. |
| *Pinus strobus* | — | — | Yes (? cm) | 48 | Ahlgren and Hansen (1957) | 90% survival after 48 days of flooding. |
| *Pinus taeda* | — | Yes | No (? cm) | 180 | Williston (1962) | 79% of trees survived after 30 days of submergence followed by 90 days of root submergence. |
| | — | — | Yes | 60 | Kennedy and Krinard (1974) | |

**TABLE 3-8. (Continued)**

| Species | Survival | | | Flood Inundation Length (day) | Reference | Comments |
|---|---|---|---|---|---|---|
| | Saturated Soil | Flooded to Root Crown | Flood Above Root Crown (depth) | | | |
| *Platanus occidentalis* | Yes | No | No | 240 | Yeager (1949) | |
| | — | — | Yes | 169 | Bell and Johnson (1974) | |
| *Populus deltoides* | Yes | Yes | No | 730 | Yeager (1949) | |
| | — | — | Yes | 210 | Broadfoot (1967) | |
| | — | Yes | ? | 189 | Bell and Johnson (1974) | |
| *Prunus serotina* | — | — | No | ? | Green (1947) | Dies if flooded for more than 2% of the growing season. |
| *Quercus alba* | — | — | No | 149 | Bell and Johnson (1974) | |
| *Quercus bicolor* | — | — | Yes | 149 | Bell and Johnson (1974) | |
| | — | — | Yes | 60 | Kennedy and Krinard (1974) | |
| *Quercus falcata* | — | — | No | 1095 | Green (1947) | Dies if flooded for more than three years. |
| *Quercus imbricaria* | — | — | No | 149 | Bell and Johnson (1974) | |
| | — | — | No | 129 | Brunk et al. (1975) | Dies if flooded more than 129 days. |
| *Quercus macrocarpa* | Yes | — | No | 730 | Yeager (1949) | |
| | — | — | No (<30 cm) | 1460 | Broadfoot and Williston (1973) | |
| | — | — | Yes | 189 | Bell and Johnson (1974) | |

| | | | | | | |
|---|---|---|---|---|---|---|
| *Quercus muehlenbergii* | — | — | No | 67 | Harris (1975) | 10% survival after 67 days of flooding. |
| *Quercus nigra* | — | — | No (<30 cm) | 1460 | Broadfoot and Williston (1973) | |
| | — | — | Yes | 60 | Kennedy and Krinard (1974) | |
| *Quercus nuttalli* | — | — | Yes | 60 | Kennedy and Krinard (1974) | |
| *Quercus palustris* | Yes | No | No (<76.2 cm) | 730 | Yeager (1949) | |
| | — | — | Yes | 109 | Bell and Johnson (1974) | |
| *Quercus rubra* | — | — | Yes | 50 | Bell and Johnson (1974) | |
| *Quercus stellata* | — | — | No (? cm) | — | Harris (1975) | 45% survival after 73 days. |
| *Quercus velutina* | — | — | No | 109 | Bell and Johnson (1974) | |
| *Robinia pseudoacacia* | — | — | No | 73 | Harris (1975) | 5% survival after 73 days of flooding. |
| *Salix nigra* | — | — | Yes | 1460 | Yeager (1949) | 56% survival. |
| | — | — | No | 1095 | Green (1947) | All died after three years of flooding. |
| *Sassafras albidum* | — | — | No | 149 | Bell and Johnson (1974) | |
| *Taxodium distichum* | — | Yes | Yes (50–86) | 365 | Keeland and Sharitz (1995) | Maximum growth with shallow, permanent flooding (mean, 39 cm). |
| *Ulmus americana* | Yes | No | No | 730 | Yeager (1949) | 49% survival in wet soil after 1460 days. |
| | — | — | Yes | 189 | Bell and Johnson (1974) | 91% survival after 189 days. |

*Source:* Whitlow and Harris (1979) and others.

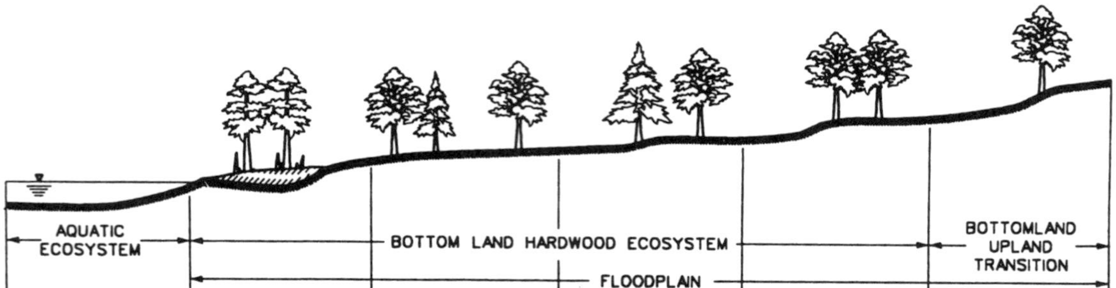

| | | | | | | |
|---|---|---|---|---|---|---|
| | AQUATIC ECOSYSTEM | BOTTOM LAND HARDWOOD ECOSYSTEM | | | | BOTTOMLAND UPLAND TRANSITION |
| | | FLOODPLAIN | | | | |
| ZONE | 1 | 2 | 3 | 4 | 5 | 6 |
| NAME | OPEN WATER | SWAMP | LOWER HARDWOOD WETLANDS | MEDIUM HARDWOOD WETLANDS | HIGHER HARDWOOD WETLANDS | TRANSITION TO UPLANDS |
| WATER MODIFIER | CONTINUOUSLY FLOODED | INTERMITTENTLY EXPOSED | SEMIPERMANENTLY FLOODED | SEASONALLY FLOODED | TEMPORARILY FLOODED | INTERMITTENTLY FLOODED |
| FLOODING FREQUENCY, PERCENT OF YEARS | 100 | 100 | 51-100 | 51-100 | 11-50/100 | 1-10 |
| FLOODING DURATION, PERCENT OF GROWING SEASON | 100 | 100 | >25 | 12.5-25 | 5-12.5 | <5 |

***Figure 3-25.*** *Flooding regime and bottomland hardwood forests in the southeastern United States (from Theriot, 1993, as adapted from Clark and Benforado, 1981; copyright © by U.S. Army Corps of Engineers by permission).*

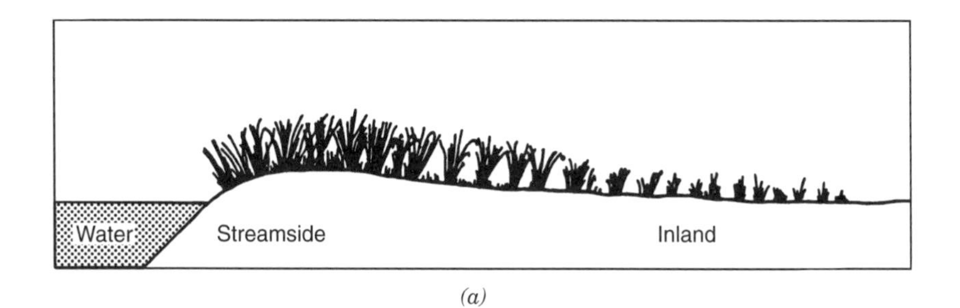


(a)



(b)

***Figure 3-26.*** *(a) Salt marsh of Spartina alterniflora* in cross section and (b) standing crop biomass and height with distance inland from the stream-plant interface, Barataria Bay, Louisiana (from DeLaune et al., 1979; copyright © by Estuaries and Coastal Marine Science by permission).

pay careful attention for successful restoration. While the most water tolerant of the woody species in bottomland swamps of the southeastern United States, such as, *C. occidentalis, N. aquatica*, and *T. distichum* (Table 3-9, can survive for many years under permanent water, they all eventually die (Klimas, 1987).

Many naturally occurring wetlands have been permanently impounded by dams, and this invariably alters the vegetation. Most species flooded above the root collar die in one to six years (Yeager, 1949), although certain

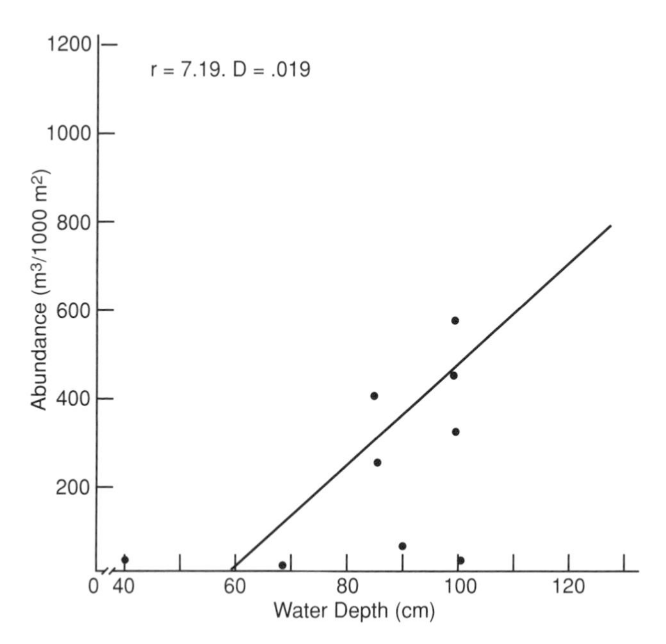


***Figure 3-27.*** *Linear regression of abundance of* Cephalanthus occidentalis *(buttonbush) and water depth, with Spearman correlation (r) (from Faber-Langendoen and Maycock, 1989; copyright © by Canadian Field-Naturalist by permission).*

species are more tolerant. At Caddo Lake, Louisiana/Texas, Klimas (1987) predicts that permanently flooded stands of bald cypress (those that are not exposed for brief periods in late summer; Types C and D; Fig. 3-29) will die over a 50-year interval. In the impounded floodplain at Ocklawaha, Florida, tree mortality has accumulated slowly over an eight-year interval (Conner and Day, 1988). Eighteen years after the impoundment (2.1-3.4 m) of Lake Chicot, Louisiana, all of the tupelo and many of the crowns of the cypress died (Eggler and Moore, 1961). If drawdown is reestablished within one year of impoundment, *T. distichum* can recover quickly, but not following more than five years of constant inundation (Duever and McCollom, 1986).

In cypress swamps of the southeastern United States, even water-tolerant species die in deep, permanent flooding. In Lake Ocklawaha, Florida, no trees of any species survived for more than three years in water over 1.3 m deep (Fig. 3-30). Bald cypress, one of the most water-tolerant species in this wetland type, began to succumb at lower water depths after four years of inundation (0.8-1.2 m; 16% and 50% mortality, respectively). Mortality declined progressively with water depths. The lowest mortality occurred in 0-0.8 m of water (ca. 3%; Harms et al., 1980). Bald cypress can withstand deep, permanent impoundment for several years but die after 15-25 years,

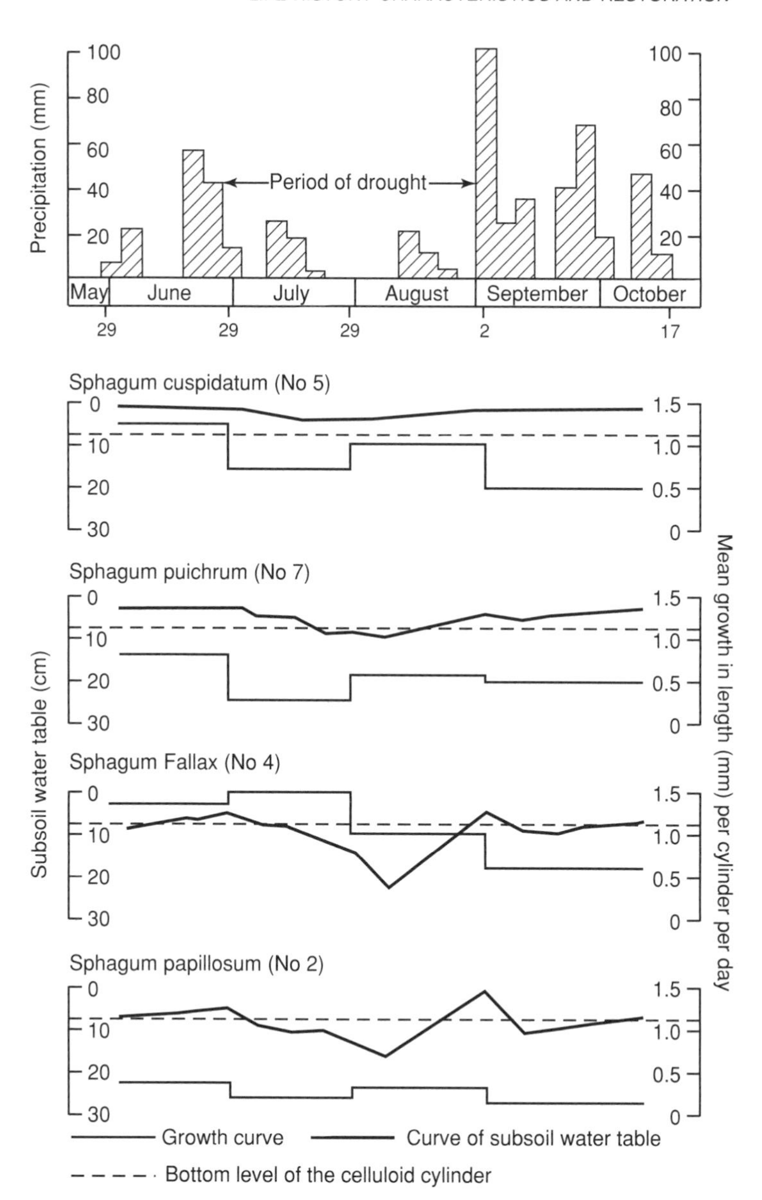


***Figure 3-28.*** Sphagnum *abundance and groundwater regime (from Pedersen, 1975; copyright © by Norwegian Journal of Botany by permission).*

particularly following a sudden flood (Hook, 1984b). After the large 1993 flood along the Mississippi River, many cypress and especially tupelo died in Horseshoe Lake, Illinois (Loftus, 1994), which was impounded with dams after 1929 (Robertson, 1992). Tree death after impoundment has been observed in cypress and other types of hardwood forests in Iowa, Illinois,

**TABLE 3-9. North American Water Tolerance Ratings for Tree Species from Most to Least Tolerant**

| Species | Common Name | Water Tolerance Rating |
|---|---|---|
| *Cephalanthus occidentalis* | Buttonbush | Most tolerant |
| *Forestiera acuminata* | Swamp-privet | Most tolerant |
| *Fraxinus caroliniana* | Carolina ash | Most tolerant |
| *Fraxinus profunda* | Pumpkin ash | Most tolerant |
| *Nyssa aquatica* | Water tupelo | Most tolerant |
| *Nyssa sylvatica* var. *biflora* | Swamp tupelo | Most tolerant |
| *Planera aquatica* | Water elm | Most tolerant |
| *Salix nigra* | Black willow | Most tolerant |
| *Taxodium distichum* | Bald cypress | Most tolerant |
| *Taxodium distichum* var. *nutans* | Pond cypress | Most tolerant |
| *Carya aquatica* | Water hickory | Highly tolerant |
| *Gleditsia aquatica* | Waterlocust | Highly tolerant |
| *Quercus lyrata* | Overcup oak | Highly tolerant |
| *Acer negundo* | Boxelder | Moderately tolerant |
| *Acer rubrum* | Red maple | Moderately tolerant |
| *Acer saccharinum* | Silver maple | Moderately tolerant |
| *Betula nigra* | River birch | Moderately tolerant |
| *Chamaecyparis thyoides* | Atlantic white-cedar | Moderately tolerant |
| *Crataegus* spp. | Hawthorn | Moderately tolerant |
| *Diospyros virginiana* | Persimmon | Moderately tolerant |
| *Fraxinus pennsylvanica* | Green ash | Moderately tolerant |
| *Gleditsia triacanthos* | Honeylocust | Moderately tolerant |
| *Gordonia lasianthus* | Loblolly-bay | Moderately tolerant |
| *Ilex decidua* | Possumhaw | Moderately tolerant |
| *Liquidambar styraciflua* | Sweetgum | Moderately tolerant |
| *Magnolia virginiana* | Sweetbay | Moderately tolerant |
| *Persea borbonia* | Redbay | Moderately tolerant |
| *Pinus elliottii* | Slash pine | Moderately tolerant |
| *Pinus serotina* | Pond pine | Moderately tolerant |
| *Pinus taeda* | Loblolly pine | Moderately tolerant |
| *Platanus occidentalis* | American sycamore | Moderately tolerant |
| *Populus deltoides* | Eastern cottonwood | Moderately tolerant |
| *Quercus nuttalli* | Nuttall oak | Moderately tolerant |
| *Quercus palustris* | Pin oak | Moderately tolerant |
| *Quercus phellos* | Willow oak | Moderately tolerant |
| *Ulmus americana* | American elm | Moderately tolerant |
| *Ulmus crassifolia* | Cedar elm | Moderately tolerant |
| *Carpinus caroliniana* | American hornbeam | Weakly tolerant |
| *Carya illinoinensis* | Pecan | Weakly tolerant |
| *Carya laciniosa* | Shellbark hickory | Weakly tolerant |
| *Celtis laevigata* | Sugarberry | Weakly tolerant |
| *Celtis occidentalis* | Hackberry | Weakly tolerant |
| *Ilex opaca* | American holly | Weakly tolerant |
| *Juglans nigra* | Black walnut | Weakly tolerant |
| *Magnolia grandiflora* | Southern magnolia | Weakly tolerant |
| *Morus rubra* | Red mulberry | Weakly tolerant |
| *Nyssa sylvatica* | Blackgum | Weakly tolerant |
| *Pinus glabra* | Spruce pine | Weakly tolerant |
| *Quercus falcata* var. *pagodifolia* | Cherrybark oak | Weakly tolerant |
| *Quercus laurifolia* | Laurel oak | Weakly tolerant |
| *Quercus michauxii* | Swamp chestnut oak | Weakly tolerant |

**TABLE 3-9. (Continued)**

| Species | Common Name | Water Tolerance Rating |
|---|---|---|
| *Quercus shumardii* | Shumard oak | Weakly tolerant |
| *Quercus virginiana* | Live oak | Weakly tolerant |
| *Ulmus alata* | Winged elm | Weakly tolerant |
| *Asimina triloba* | Pawpaw | Least tolerant |
| *Cornus florida* | Flowering dogwood | Least tolerant |
| *Fagus grandifolia* | American beech | Least tolerant |
| *Juniperus virginiana* | Eastern redcedar | Least tolerant |
| *Liriodendron tulipifera* | Yellow poplar | Least tolerant |
| *Ostrya virginiana* | Eastern hophornbeam | Least tolerant |
| *Pinus echinata* | Shortleaf pine | Least tolerant |
| *Prunus serotina* | Black cherry | Least tolerant |
| *Quercus alba* | White oak | Least tolerant |
| *Sassifras albidum* | Sassafras | Least tolerant |
| *Ulmus rubra* | Slippery elm | Least tolerant |

*Source:* Hook (1984b).

Tennessee, Mississippi, and Louisiana (Hall et al., 1946; Green, 1947; Penfound, 1949; Broadfoot and Williston, 1973; Bell and Johnson, 1974; Dellinger et al., 1976).

Another aspect of adult production/mortality patterns in relation to water depth has to do with the latitudinal position of species. Ecoclinal populations typically produce less than others elsewhere in the range of the species (Turner, 1976; Westlake, 1981; Middleton, 1994b). One would expect patterns of production to reflect environmental gradients across biomes or regions as well (Gosz and Sharpe, 1989). Moving from south to north starting near the equator, standing crops of *Phragmites australis* decrease with latitude (Figs. 3-31 and 3-32; Westlake, 1981). Similarly, salt marshes of *Spartina alterniflora* decrease in annual production from about 27° to 44° N latitude (Fig. 3-33; Turner, 1976). Cypress swamps decrease in production, as measured by litterfall, with increases in latitude across North America (Table 3-10; ca. 29° to 37° N latitude; Middleton, 1995b). Other environmental factors also influence production, including the level of nutrients (Brinson et al., 1981a; Mitsch et al., 1991). The functional differences of the boundary with the main portion of the biome is a relatively new area of study (Hansen et al., 1992; Risser, 1995), but it is likely an important constraining factor in restoration.

From a restoration perspective, it is likely that species comprised of populations that vary along their geographical ranges **(ecocline)** may be more difficult to reestablish at the edge of the range. Regarding biome boundaries, the **resilience**, or relative ability of a site to return to its initial condition, could be lower than at sites in other landscape positions because

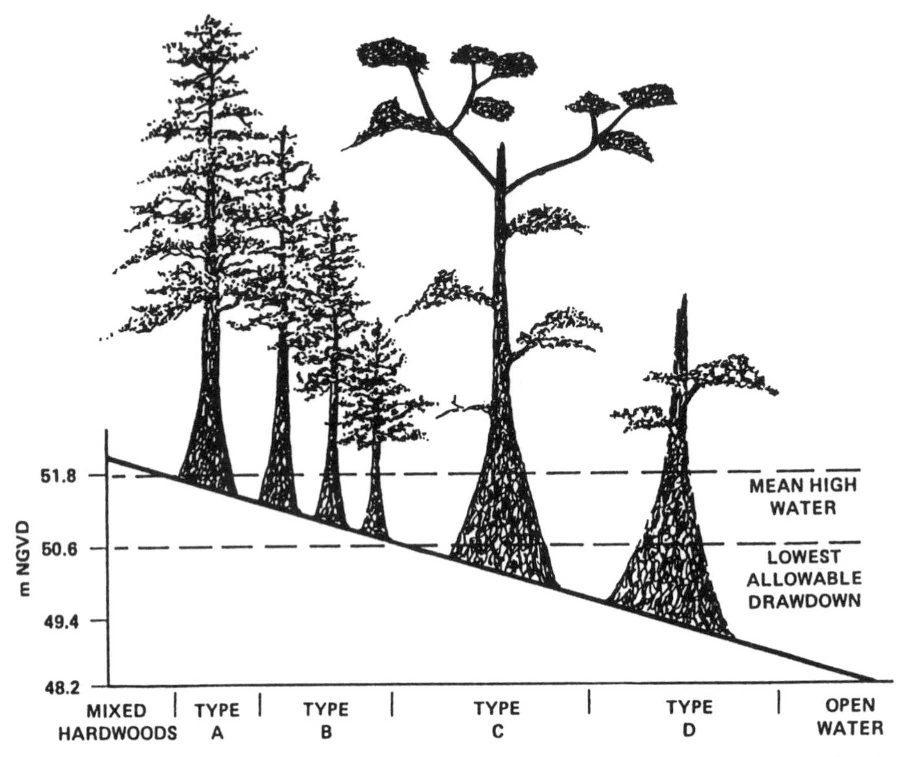


***Figure 3-29.*** *Flooding regime and bald cypress type in the permanently impounded Caddo Lake, Louisiana/Texas. Type A has the best growth, with annually exposed soil and healthy trees. Type B is inundated most of the time but is exposed briefly 1 year in 2 to 1 year in 40. With Type C, stand vigor declines with increasing water depth to Type D, with individual trees showing extreme stress. Klimas (1987) projects that after 50 years of permanent flooding, Types C and D would be eliminated (from Klimas, 1987; copyright © by Wetlands by permission).*

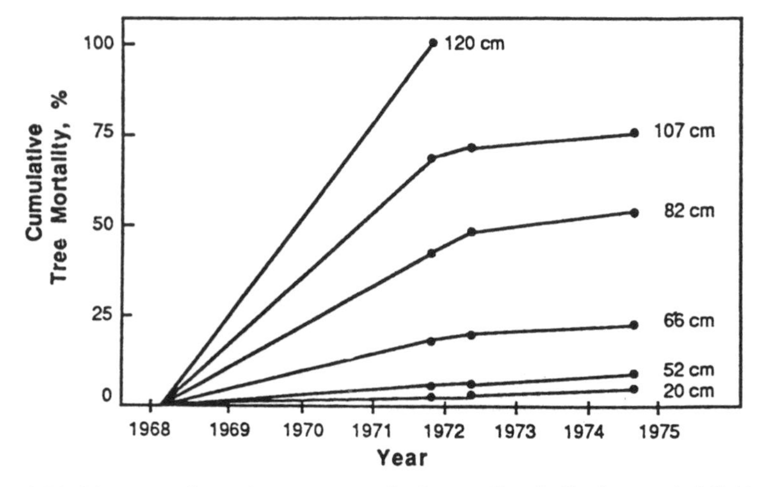


***Figure 3-30.*** *Tree mortality and mean water depth over time in the impounded Ocklawaha, Florida, floodplain (from Conner and Day, 1988, as adapted from Lugo and Brown, 1984; copyright © by North Carolina State University by permission).*

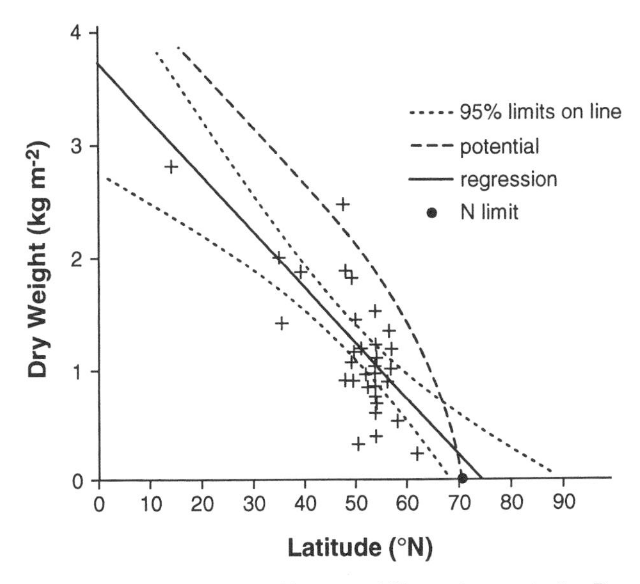


***Figure 3-31.*** *Latitude and standing crop biomass of* Phragmites australis. *Regression with S = 3.683 − 4.952 L,* r = 0.705, $r^2$ = 0.497, n = 35, p << .01 (from Westlake, 1981; copyright © by Royal Botanical Society of Belgium by permission).

***Figure 3-32.*** *Production (standing crop biomass) sampling in a monsoonal wetland, Keoladeo National Park, India (pictured, Soren Singh and Laxmi Kant Mudgal; photograph by Beth Middleton).*

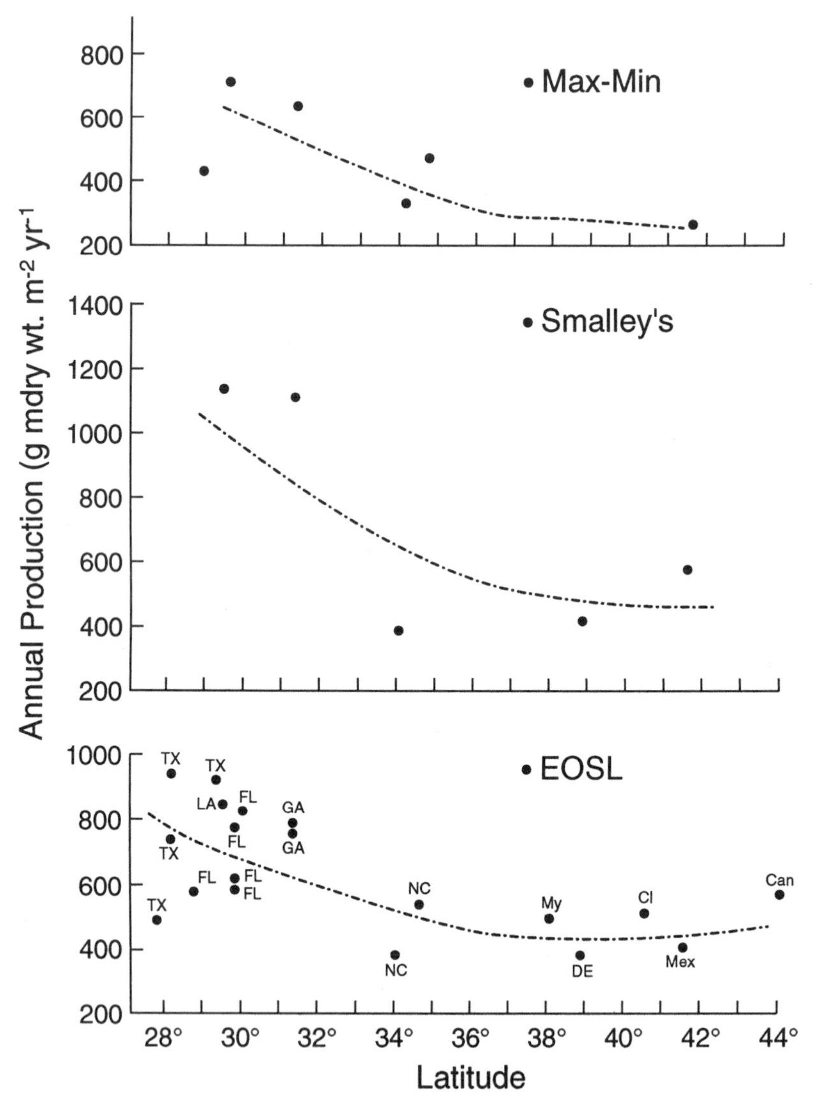


***Figure 3-33.*** *Latitude and annual production of* Spartina alterniflora *by three methods: top max-min, middle Smalley's method, and bottom EOSL estimate. Curves fit by eye (from Turner, 1976; copyright © by Contributions in Marine Science by permission).*

**TABLE 3-10. Litter Fall Totals of Biomass in kg ha$^{-1}$yr$^{-1}$ Phosphorus (mg/L) and Nitrate (mg/L) in Cypress Swamps Arranged by Latitude**

| Location | Forest Type | Latitude | Total Biomass | Phosphorus (mg/L) | Nitrate (mg/L) | Source |
|---|---|---|---|---|---|---|
| Henderson Slough, KY | Mixed cypress<br>1. Slow flow, semi-permanent flooding | ~37°50' | 1360 | 0.20 | — | Mitsch et al. (1991) |
| | 2. Permanent flooding | ~37°50' | 2530 | 0.08 | — | Mitsch et al. (1991) |
| Cypress Creek, KY | Cypress, impounded by levee, stagnant | ~37°25' | 630 | 0.12 | | |
| Heron Pond, IL | Cypress-tupelo; semi-permanent/intermittent | 37°21' | 3480 | 0.06–0.28 | 0.01 | Dorge et al. (1984) |
| Lower Cache, IL | Cypress-tupelo impounded by dam | 37°17' | 2346 | 0.04$^{-1}$ 0.28 | 0.05–0.78 | Middleton (1994a) |
| Pitt Co., NC | Cypress tupelo | 35°35' | 6428 | na | na | Brinson et al. (1980) |
| Okefenokee, GA | Cypress | 30°30' | 5950 | 0.10 | na | Schlesinger (1978) |
| Lac des Allemands, LA | Red maple-cypress: | 29°59' | | | | Conner and Day (1992) |
| | 1. Natural | | 4740 | na | na | |
| | 2. Impounded | | 3580 | na | na | |
| Alachua and Collier Co., FL | River floodplain cypress | 29°15' | 5970 | 0.24 | na | Brown (1981) |

na = not available.

*Source:* Adapted from Middleton (1994a) and Mitsch et al. (1991).

species at the edge of their range can be stressed (e.g., lower production of biomass and seeds, less successful regeneration, lower water level tolerances). Hence, sites at the boundaries of biomes may be more difficult to restore.

## Flood Tolerance in Plants

The widely variable tolerance of aquatic species for water depths is the key to understanding the potential of specific species for restoring vegetation in wetlands. The ability of aquatic plants to survive flooded conditions at all has to do with their anatomical, morphological, and metabolic adaptations to flooding (Pezeshki, 1994). An elementary understanding of the root physiology associated with **anaerobic** conditions will greatly help the restorationist appreciate the critical importance of drawdown for emergent species. Also, it is important to remember that even flood-tolerant species cannot live under flooded conditions without oxygen forever; oxygen deprivation eventually results in mortality (Crawford, 1983; Armstrong et al., 1994).

***Figure 3-34.*** *Water lily (*Nymphaea *sp.), a species with pressurized ventilation (photograph by Beth Middleton).*

Under normal, well-oxygenated conditions, from 1 molecule of sugar, 38 molecules of ATP are produced, along with $CO_2$ and water. If oxygen is not present, however, the glycolytic pathway ultimately produces only two molecules of ATP, along with $CO_2$ and ethyl alcohol. Ethanol is toxic to plants (Raven et al., 1992; Pezeshki, 1994). For a plant to get enough ATP to grow, it must get oxygen, either by making contact with the atmosphere or by photosynthesis (Armstrong et al., 1994).

Tolerance to flooding is due primarily to physical mechanisms involving gas transport (Armstrong et al., 1994). One anatomical feature that promotes aeration of roots is **aerenchyma** or cavities in the stems and roots of aquatic plants (Arber, 1920; Sculthorpe, 1967 McKevlin et al. 1998), as, for example, in *Carex rostrata* (Fagerstedt, 1992). Oxygen can move along narrow films along the surface layer of rice (Raskin and Kende, 1985). Analogs of pressurized ventilation, first discovered in water lilies (Fig. 3-34; Dacey, 1980, 1981), have now been discovered in other species, including *P. australis* (Armstrong and Armstrong, 1991), *Taxodium distichum* (Grosse et al., 1992), *Typha domingensis*, and *Typha orientalis* (Brix et al., 1992).

It is crucial for some part of the plant to protrude, like a snorkel, from the water for any of these aeration mechanisms to function. This provides another key for the restorationist in understanding herbivory as a distur-

**TABLE 3-11. Comparison of Percent (%) Survivorship for Various Species After Experimental Cutting Underwater and Length of the Experiment (Days)**

| Species | Not Cut | Cut | Days | Source | Study Location |
|---|---|---|---|---|---|
| *Eleocharis acuta* | 100 | 0 | 75 | Blanch and Brock (1994) | Llangothlin Lagoon, NSW, Australia |
| *Ipomoea aquatica* | 100 | 9–16 | 80 | Middleton (1990) | Keoladeo National Park, Rajasthan, India |
| *Ludwigia peploides* | 100 | 100 | 45 | Mathis (1996) | Carbondale, Illinois, USA |
| *Myriophyllum variifolium* | — | No regrowth | 75 | Blanch and Brock (1994) | Llangothlin Lagoon, NSW, Austrailia |
| *Najas marina* | — | Regrew | >44 | Agami and Waisel (1986) | Tel Aviv, Israel |
| *Nymphoides cristatum* | 100 | 100 | 80 | Middleton (1990) | Keoladeo National Park, Rajasthan, India |
| *Paspalidium punctatum* | 100 | 1–16 | 80 | Middleton (1990) | Keoladeo National Park, Rajasthan, India |
| *Paspalum distichum* | 100 | 1–49 | 80 | Middleton (1990) | Keoladeo National Park, Rajasthan, India |
| *Phragmites australis* | — | 0 | 30 | Weisner and Granélli (1989) | Krankesjön and Häljasjön Lakes, Sweden |
| *Phragmites australis* | 100 | 25 | 45 | Mathis (1996) | Carbondale, Illinois, USA |
| *Phragmites australis* | 100? | 0 | 540 | Hellings and Gallagher (1992) | Lewes, Delaware, USA |
| *Potamogeton lucens* | — | Regrew | >44 | Agami and Waisel (1986) | Tel Aviv, Israel |
| *Typha angustata* | — | No regrowth | 30 | Singh et al. (1976) | Hissar, Haryana, INDIA |
| *Typha angustifolia* | 80–85 | 0–5 | 21 | Sale and Wetzel (1983) | southern Michigan, USA |
| *Typha latifolia* | 100 | 42 | 38 | Mathis (1996) | Carbondale, Illinois, USA |
| *Typha latifolia* | 80–85 | 0–5 | 21 | Sale and Wetzel (1983) | southern Michigan, USA |
| *Typha* sp. | — | No regrowth | — | Weller (1975) | Rush Lake, Iowa, USA |

*Note:* In cases where several treatments were applied to experimental plants, the results of the most severe treatment in the experiment (e.g., clipping every week at ground level) are given. — indicates that the information is not available.

*Source:* Middleton, accepted.

bance. Herbivory can kill emergent species in a restored wetland. While this is not necessarily a problem if the species have become fully established and incorporated into the seed and propagule bank (see the discussion of succession in Chapter 2), it can be a problem if it occurs too early in the vegetative development of the stand, that is before planted species have set seed. Because of the critical role of stem protrusion from the water column in aerating the roots, many emergent species die if cut underwater (Table 3-11).

Other species, such as *Nymphoides cristatum* in India, regrow quickly to the surface of the water after cutting and become very common in grazed locations (Middleton, 1990). Some aquatic species, such as *Nymphoides peltata*, induce rapid stem elongation via ethylene production, and perhaps this contributes to the successful regrowth of some aquatic species (Osborne, 1982). *Callitriche platycarpa* also grows more quickly when submerged because it produces ethylene (Jackson, 1982).

After stems are cut underwater, oxygen levels in roots drop (Fig. 3-35; Sale and Wetzel, 1983; Weisner and Granélli, 1989; Mathis, 1996). In the field, herbivory or hand cutting of many plant species underwater can either

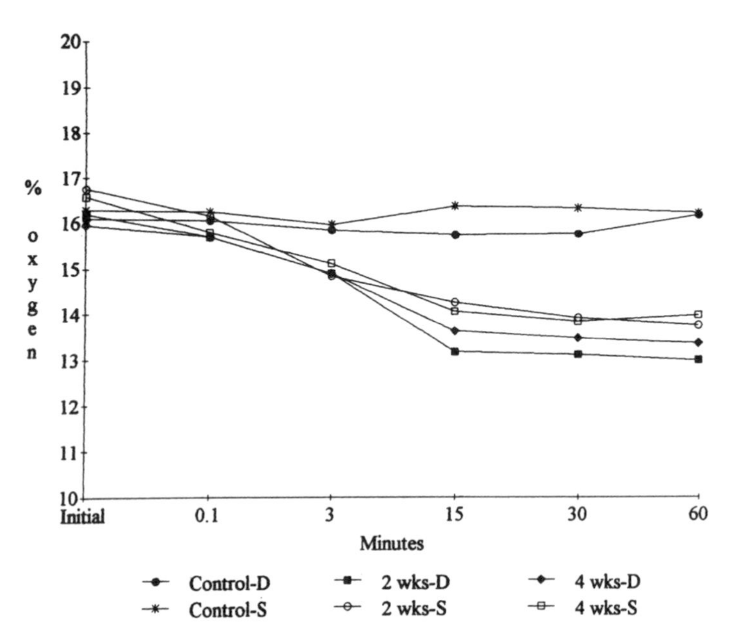


*Figure 3-35. Oxygen levels (percent) in rhizomes of* Phragmites australis *in clipped and unclipped plants immediately before (initial) and 60 minutes following clipping. Treatments included plants growing in submerged (S) and drawndown (D) conditions and not clipped (control), clipped every two weeks (2 wks) and clipped every four weeks (4 wks) (from Mathis, 1996; copyright © by Marilyn Mathis by permission).*

***Figure 3-36.*** *Clipped (left) and unclipped (right) individuals of* Cyperus alopecuroides. *In the underwater treatment, the clipped plants die (photograph by Beth Middleton).*

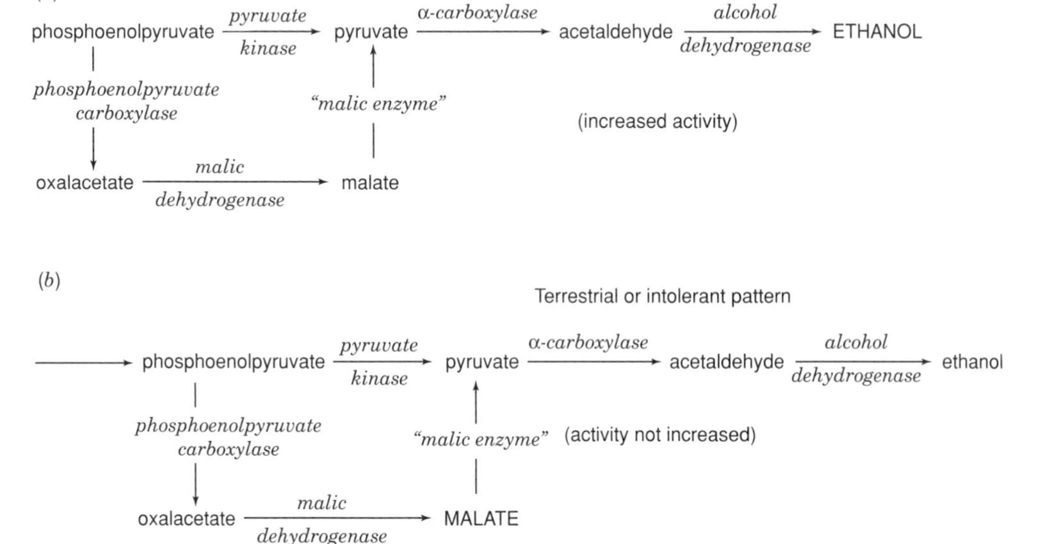


**Facultive aquatic or tolerant pattern**

***Figure 3-37.*** *Pathways of anaerobic metabolism in flooded roots of (a) terrestrial (flood-intolerant) versus (b) facultative aquatic (flood-tolerant) species. Flood-tolerant species accumulate malate in anaerobic conditions rather than produce ethanol, as do flood-intolerant species. Enzymes are in italics; substrates are in roman (from McMannon and Crawford, 1971; copyright © by New Phytologist by permission).*

kill them or cause them not to regrow (Fig. 3-36). Species that do survive cutting often have reduced levels of biomass, particularly below ground biomass, and this is also true in aerated conditions (Middleton 1990; Dumotier 1996).

Metabolic adaptations also confer tolerance to flooding, as aquatic plants may have to undergo anoxia for periods of up to several months (Armstrong et al., 1994). One theory of flood tolerance has to do with avoidance of the production of ethanol, which is toxic to plants (Fig. 3-37a). Species that are flood-tolerant store malate in their roots until flooding ceases (Fig. 3-37b) and then use this stored material in plant activities (McMannon and Crawford, 1971). Nevertheless, anaerobic conditions lead to low energy yield for plants, and their activities cannot be maintained indefinitely (Crawford, 1978; Armstrong et al., 1994).

Flood intolerance can be detected by the presence of alcohol dehydrogenase (ADH) during flooded conditions because this enzyme is involved in the production of ethanol. During flooding, flood-intolerant species such as *Pisum sativum* increase their ADH activity, whereas flood-tolerant species such as *Phalaris arundinacea* decrease it. Similarly in the seeds of flood-tolerant species (e.g., rice), ethanol remains low after soaking, while germination rates remain high. In contrast, flood-intolerant peas greatly increase their ethanol after soaking and germinate poorly (Hook, 1984a). These metabolic processes explain why water level fluctuation is essential in most wetland types, a fact that must not be ignored by restorationists.

# *Part III*

# *Restoring Disturbance Dynamics in Wetlands*

# 4

# *Restoration and Disturbance Dynamics in Restored Landscapes*

*The ultimate success of a restored wetland as a functional ecosystem will depend on the level to which its reengineered or passively rejuvenated environment can endure natural disturbances. Restored wetlands that will persist are likely to resemble natural wetlands in the region.*

*Each restoration project is different, and its success is largely dependent upon the skills and knowledge of the restoration team. It is not the type of reengineering approach to hydrology that makes a particular project work. Rather, it is the ability of the team to visualize how the site was disturbed and continues to be disturbed, and what measures are necessary to remedy this situation in the context of the landscape position of the site. How closely the site resembles the target type in the end is dependent on the collective vision of the engineers, hydrologists, botanists, geologists, and others working on the reconstruction of the ecosystem.*

## GOALS OF RESTORATION

Long-term resilience and persistence in restored wetlands come with the reinstatement of the biophysical processes and interactions that are a part of natural wetlands. Some wetland restorations fall short of this goal by their design; a wetland totally bounded by dikes and filled by pumps will never experience the stochastic changes in water level in response to climate variability that are assumed by an unbounded wetland (Box 4-1). A large number of wetland restoration projects lead to the creation of novel systems

*BOX 4-1*

**STEPS IN THE PROJECT FRAMEWORK**

| Step | Procedure |
|---|---|
| Project objectives | Determine desired outcome from environmental perspective, that is, restore site to specific preexisting conditions and ameliorate environmental problems. Summarize in a formal manner. |
| Feasibility/planning study | Define existing habitat condition, (e.g., channel type, old field in pothole). Evaluate cause of existing condition. Survey specific problems within the proposed restoration site. Elicit public opinion and involvement of landowners and general public. |
| Design | Survey land elevations across restoration site and/or map cross-sectional measurement of stream channel. Determine natural state of the habitat type of the proposed restoration site by observing the dimensions of the unaltered stream channel in the same river or in adjacent catchments or the landscape position of other similar regional wetlands. Examine historical documents. Base project methodology on criteria to create a resilient system capable of existing without constant human intervention. Where possible, project timing of natural disturbance reintroduction, that is, flood pulsing, fire, natural herbivore grazing (necessary to remove fences for natural herbivores at some point). Formalize plans for revegetation (planting, natural regeneration), channel/landform shaping, bank protection in streams. Prepare detailed plan with documentation of information sources. |
| Construction | Choose equipment for various operations. Discuss philosophy of restoration with site workers. Designer ususally needs to supervise construction work. |

| Step | Procedure |
|---|---|
| Clean-up phase | Revegetate site through replanting, reseeding (donor seed bank or seed sewing), or natural regeneration. Stockpiles of stones for stream bank protection retained. |
| Maintenance plan | Written plan to detail future maintenance permitting and its timing. Acknowledgment of natural disturbance, such as flood pulsing, fire, natural herbivore grazing. |
| Postproject appraisal | Monitor the extent of environmental (hydraulic/morphological) and biological success. Evaluate views of the public. |

*Source:* Adapted from Brookes (1990).

*BOX 4-2*

## SHORT OF THE IDEAL: THE LEITBILD CONCEPT

The Leitbild Concept is an approach to stream restoration developed in Germany to guide decisions in restoration planning. The idea is to first determine the ideal state of the specific river or stream—in other words, what it would look like in a relatively undisturbed state independent of economic and political constraints (Kern, 1992b). This ideal state is determined by the study of undisturbed streams to arrive at an understanding of the dynamics of the channel, sediments, nutrients, and biota. A grasp of the of nature of anthropogenic disturbance in the watershed is also important to determine the difference between the current situation and the ideal (Larson, 1995).

Due to economic, political, and social constraints, as well as irreversible changes to the abiotic and biotic systems, it is usually impossible to restore the ideal state of the system. The optimal design is likely to be quite different from the ideal one (Brookes and Shields, 1996).

on sites, either because the engineering practices involved have no hope of recapturing the original condition (e.g., releasing a small amount of water from a dam to restore some flow) or because the site has been permanently impaired (Box 4-2; Ebersole et al., 1997). An understanding of the historical changes leading to the degradation of a riverine system should precede project planning in restoration projects (Kondolf and Larson, 1995).

Wetland restoration should focus on regaining ecosystems with an environment resembling that under which the biota evolved (Ebersole et al., 1997), that is, self-perpetuating ecosystems with water regimes characteristic of the region. This means reversing any human-created drainage system by removing/blocking drainage tiles, setting back levees, and/or removing dams or embankments. Particularly in Europe, downcut rivers and streams are restored by reestablishing the connection between the channel and floodplain through stream remeandering or dechannelization. The historical original is at best poorly known, so relatively undisturbed wetlands in the region should be used as targets. In this chapter, a variety of engineering approaches are compared that vary widely in their potential to restore the original hydrologic regime of wetlands.

## FLOOD PULSING AND DISTURBANCE IN RESTORED WETLANDS

Because of the major role of water level fluctuation and other disturbances in the dynamics of natural wetlands, it is critical to allow these forces to resume in restored wetlands. **Persistence** should be the restoration goal of wetland managers, not **constancy**. Regulatory standards should require restored wetlands to have the long-term ability to persist despite water level changes (Erwin, 1991), and/or other natural disturbances (Willard and Hiller, 1990). In mitigation, the Army Corps of Engineers monitors the site at the end of five years to determine if the mitigation succeeded and is in compliance with the permit. The current goal of **mitigation** is very short term (Nancy Rorick, personal communication). Despite a boom in activity, concern is growing over the functional equivalence of created and restored wetlands (Malakoff 1998).

Of these disturbances, water regime needs the most emphasis in the reengineering of wetlands. Part of the evaluation of restored and created wetlands should be based on their ability to function amid the flood pulses associated with extreme events in natural systems. It is also important to recognize that even most natural riverine systems have increased hydrologic extremes compared to their original state because of wetland losses in watersheds (Loucks, 1990). Guidelines in planning self-maintaining wetlands in a landscape context are critical if wetland restoration is to be a truly successful part of mitigation. The ultimate test of the success of a wetland restoration/creation project should be based on the ability of the site to withstand the natural disturbances characteristic of the region (Fig. 4-1).

A restoration project that does not succeed reflects a lack of understanding of the system (Bradshaw, 1987). Usually this is due to a failure to

1. understand the target natural wetland, particularly regarding landscape function,
2. reverse the anthropogenic perturbations underlying the current system,
3. recognize the role of natural disturbance events in shaping and maintaining wetlands (Fig. 4-1),
4. properly assess the pre-or postproject hydrology and stratigraphy,
5. connect the site to any other wetlands,
6. be managed by an agency, and/or
7. construct the proper wetland type for the site, such as a floodplain forest in an area that is protected from flooding by levees.

## Information Gaps in Restoration

There are many uncertainties that need to be addressed in wetland restoration (Table 4-1). For stream and river channels, there is a serious lack of knowledge concerning the accommodation of spatial and temporal changes in the amounts of sediment and water discharge along the course of a river channel due to changes in land use across the watershed (Brookes and Shields, 1996). In the restored wetlands of small watersheds, it is likely that the frequency of extreme flood pulse events is increased over that of similar wetlands in undisturbed watersheds because of regional loss of hydrologic retention (Fig. 4-2; Loucks, 1990).

In the past, wetland restoration projects were smaller than 1 ha (Morlan and Frenkel, 1992), and this may be problematic for restoration success. It is not clear what minimum size is required for any wetland type or what level of connectivity it must have to be functional within its landscape setting. These size/position relationships are likely highly dependent on the position of the wetland in the landscape, but at this point there are no written guidelines.

While restoring a more natural channel morphology can potentially improve a riparian habitat for certain species, there has been little study of the effects on the biota of specific engineering practices (Table 4-1). Little is known about the maintenance requirements of a reengineered stream (Brookes and Shields, 1996). For example, if dams or weir structures are

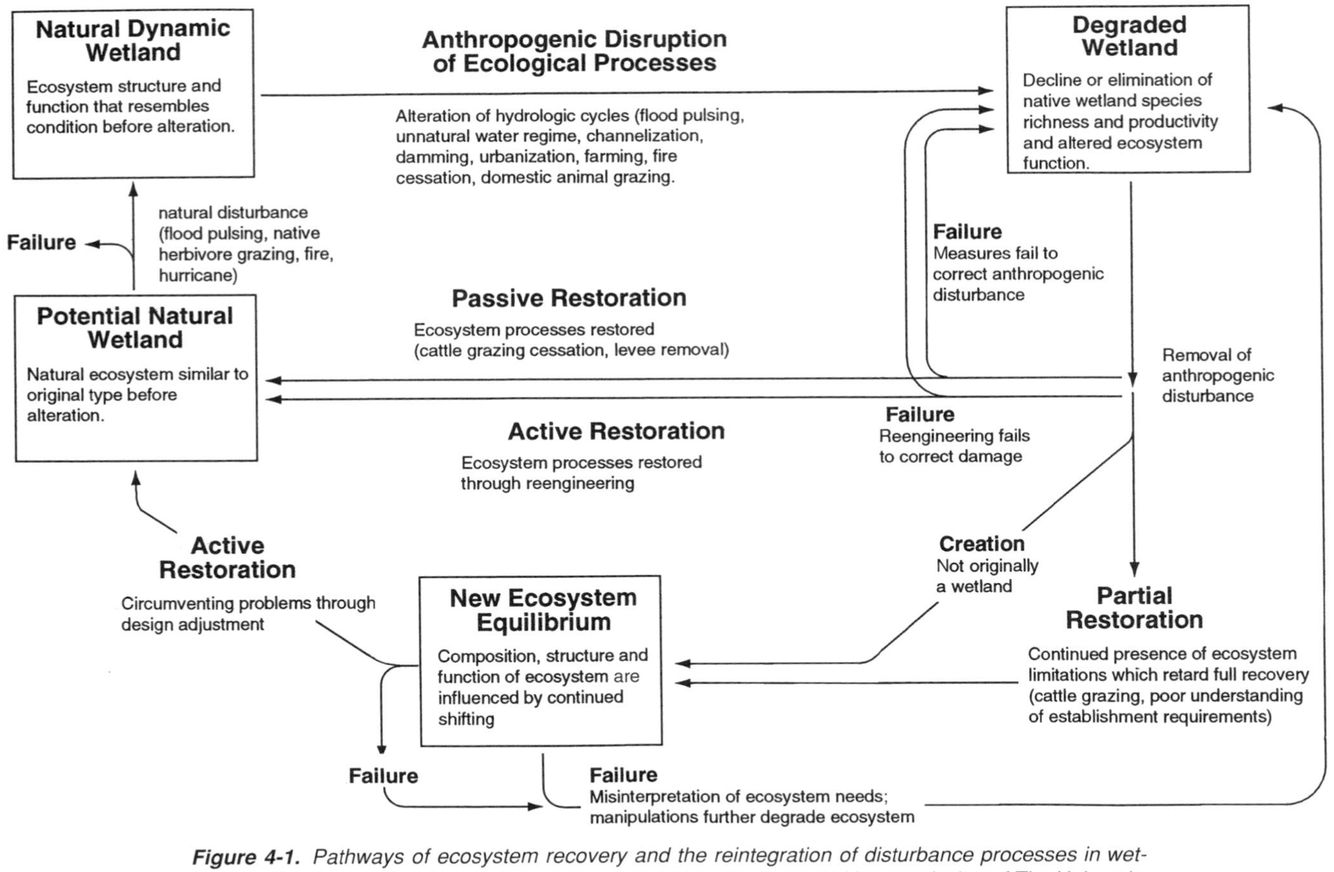


***Figure 4-1.*** *Pathways of ecosystem recovery and the reintegration of disturbance processes in wetlands (adapted from Kauffman et al., 1995; copyright 1995. Reprinted by permission of The University of Wisconsin Press).*

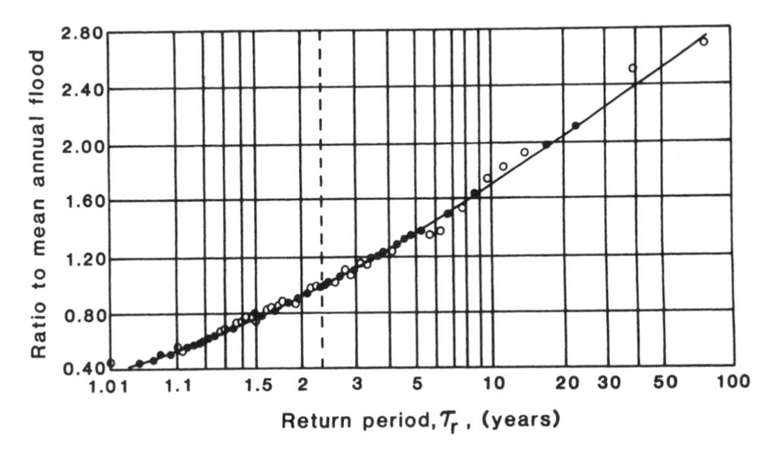


*Figure 4-2. Regional-flood-frequency curve for selected stations in the Youghiogheny and Kiskinetas River basins (Pennsylvania and Maryland), showing the return time for pulsed events expressed as a ratio to mean annual flooding. A reduction in the hydrologic detention of wetlands raises the slope of this curve (from Linsley et al., 1975, as presented in Loucks, 1990; courtesy of the U.S. Geological Survey).*

placed across streams to increase water levels upstream of the site to deepen pools for fishing, how often will these need to be removed to deal with the siltation that builds up behind them? And how often and during which seasons will the water need to be drawn down to avoid the massive die-off of trees and emergent species that plagues permanent impoundments?

While peat subsidence may occur during the drainage of various types of wetlands, little is known about the negative impacts of artificially filling a wetland basin with either peat or soil to enhance its elevation level (Table 4-1). While fire and herbivory are important parts of the vegetation dynamics of natural wetlands, when is it appropriate to reintroduce these disturbances to the system?

In addition, while the goal of wetland restoration could be to restore a wetland closely resembling the original on the site (i.e., similar species, elevation), that may be impossible. Certain sawgrass prairies in Florida exist on irregularly exposed limestone rock. These areas are "rock plowed" before farming; in other words, the limestone rock and marl are broken up to form a soil. To restore these wetlands, materials must be removed, which lowers the elevation of the site. Also, water management practices have altered the groundwater relationships of the sawgrass prairie region in Florida (Dalrymple et al., 1993). These sites are very difficult, if not impossible, to restore to their original quality. Many questions regarding the reengineering and maintenance of restored and created wetlands can be answered only through continued research.

**TABLE 4-1. Examples of Uncertainty and Certainty in River Channel and Wetland Restoration**

| Issue | Certain Knowledge | Uncertainties |
|---|---|---|
| | *River Channel* | |
| Regulating catchment land use | Sediment and water discharges are likely to alter as a consequence of land use change, and potentially there will be adjustments in downstream river morphology and aquatic habitats. | In what direction do sediment load and instream flow patterns change? By how much? Over what period of time? How does this affect potential plant reintroduction and composition? |
| Improving water quality | Riparian buffer strips reduce nutrient leakage to the watercourse. | How wide should a buffer strip be? How should it be maintained? What species should be planted? |
| Restoring a more natural channel morphology | Alters and potentially improves the aquatic riparian habitats. | What is the optimal channel size? What size of substrate? In what ways does the biota recover following modification of a particular channel type? |
| Assessing the effect of restoration measures on flood conveyance | The addition of roughness elements (e.g., a more natural, sinuous channel, and vegetation) increases the hydraulic resistance, thereby reducing flood conveyance. | How to measure roughness? What value to use for computation? Seasonal variation caused by the growth and dieback of vegetation. |
| Determining the subsequent maintenance requirements | There may be a need to intervene if the project does not function as anticipated (e.g., if the flood conveyance is jeopardized). | How often? In what way? What are the potential negative impacts? |
| | *Freshwater Wetland, Salt Marsh, Mangrove, and Northern Peatland* | |
| Functional attributes of reestablishment such as biomass production and species richness levels | Organic matter and fertilizer levels often low or limiting in restored wetlands. Dispersal-limited species do not readily reestablish themselves in restored wetlands. | Does organic matter level increase adequately over time to support levels of biomass production equivalent to natural wetlands? Does current level of habitat fragmentation make it unlikely that dispersal limited species will reestablish eventually? |
| Peat subsidence during drainage | The original elevation of the wetland may be much higher than the current level, particularly in the case of diked mangrove swamp and salt marsh and ditched or tiled freshwater marshes and bogs. | What are the negative impacts of artificially filling a wetland basin to enhance the elevation level? |

**TABLE 4-1.** *(Continued)*

| Issue | Certain Knowledge | Uncertainties |
|---|---|---|
| Water fluctuation return interval | Water level fluctuations are necessary to maintain emergent plant species. | How often does water drawdown need to occur to mimic natural systems? In systems where water is maintained artificially (damming, pumping), how often should drawdown occur? |
| Fire return interval | Fire typically increases biomass production and species richness in natural wetlands. | How often should fires occur? In the absence of natural lightning strike information, should fire be set? In situations where the peat has been drained, what are the consequences of fire? |
| Natural herbivory regime | Herbivores such as geese and ducks can destroy replanted stands of vegetation in restored wetlands. | How long should a replanted stand of wetland vegetation be protected from grazers? |

*Source:* Adapted from Brookes and Shields (1996).

## Site Selection

Site selection is specific to particular types of wetlands. In tidal wetlands, a site should be selected where a surface is either available or can be recontoured to support the desired vegetation from the perspective of inundation period and salinity level. A site that is accessible only by boat adds to the cost of the restoration. It is important to select a site that is not very exposed to waves. While these ideas provide some guidelines for site selection in tidal situations, better methods for predicting site success are required (Broome, 1990).

## Passive versus Active Restoration

***Passive Restoration*** Often, the most successful form of restoration is **passive** restoration if that approach results in the resumption of ecosystem processes (Bradshaw, 1987). This depends entirely on the nature of prior anthropogenic disturbances, but often the damaged wetland is resilient, so that passive restoration leads to rapid recovery (Fig. 4-1).

A variety of anthropogenic influences have led to the degradation of riparian corridors, particularly agriculture and grazing (Sedell et al., 1991), but passive approaches can provide a remedy (Elmore and Beschta, 1987). In the Hart Mountain National Antelope Refuge, northern Great Basin, the key objectives of the restoration plan are to attempt to reintroduce natural cycles of fire and flooding, as well as to limit cattle grazing (Pyle, 1995).

In eastern Oregon, the reduction of cattle grazing led to the redevelopment of perennial water flow and riparian vegetation along degraded intermittent streams (Fig. 4-3); Elmore and Beschta, 1987). In riparian floodplains along the lower Snake River in Oregon, the removal of cattle via fencing (or reduced utilization) resulted in the rapid reinvasion and establishment of riparian species without further remediation (Kauffman et al., 1995). Improved cattle management lessened grazing intensity, which resulted in streambank stabilization and increased production of riparian vegetation along Alder Creek, Parlin, Colorado. Grazing was limited along riparian zones by watering cattle on ridges and by placing half salt blocks away from watering areas (Cairns and Pratt, 1995).

In another example of passively restored wetlands, large tracts of riverine rainforests along the Kinabatangan River in North Borneo regenerated. Until the 1850s, the Dutch used these as tobacco plantations. Now the disturbed plantation sites, along with contiguous undisturbed rainforest, have revegetated into one of the most important biospheres in Southeast Asia, harboring numerous rare fauna (Spencer, 1995). Natural revegetation is a component of passive restoration, which will be discussed in Chapter 5.

***Active Restoration*** **Active restoration** involves the removal of an anthropogenic disturbance, often through reengineering the restoration site (Fig. 4-1), but this should be done only following and in conjunction with passive restoration (or the cessation of land use activities that halt degradation) (Kauffman et al., 1995). Techniques to restore wetlands involve reestablishing flood regimes by reestablishing river flow, removing water control structures, and halting drainage through tiles, as well as managing topography and biota. It may also be necessary to control contaminants in the system (National Research Council 1992).

Neither creation nor partial restoration (in which not all degradation factors are addressed) results in the total resumption of the natural processes of the original community (Fig. 4-1). Partial restoration is the most commonly used approach in riverine settings (Brookes et al., 1996).

Active restoration is necessary in situations where the hydrology of a wetland has been altered but abusive land practices cannot be offset by passive approaches and can result in several alternative outcomes, including

1. the restoration of the ecosystem,
2. failure of the project with little change, or
3. failure of the project with additional degradation to the site.

Active restoration must be approached cautiously. The best projects attempt to reengineer geomorphological processes, so that water release and

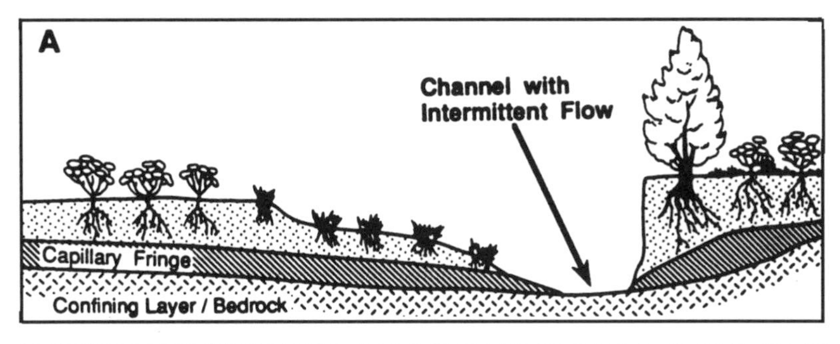


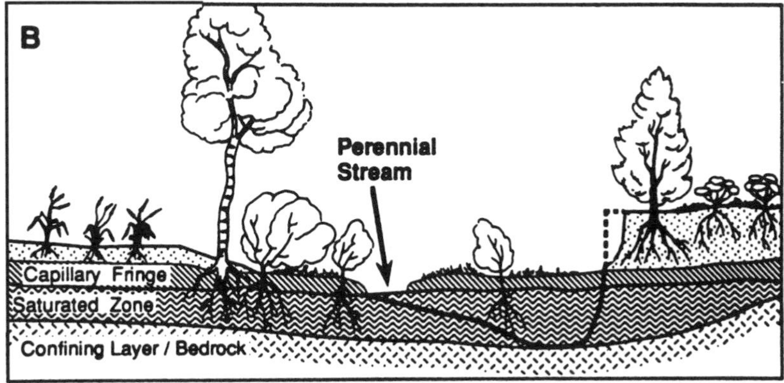


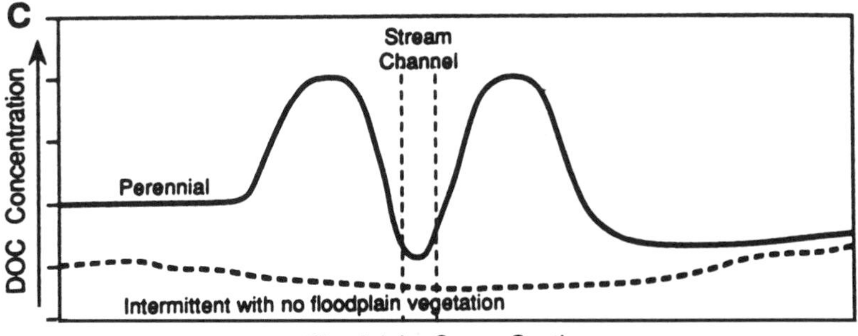


***Figure 4-3.*** *General characteristics and function of riparian areas.*

*(A)* Degraded riparian area*: little vegetation to protect and stabilize banks, little shading; lowered saturated zone, reduced subsurface storage of water; warm water in summer and icing in winter; poor habitat for fish and other aquatic organisms in summer or winter; low forage production and quality; low diversity of wildlife habitat. (B)* Recovered riparian area*: vegetation and roots protect and stabilize banks, improve shading; elevated saturated zone, increased subsurface storage of water; increased summer streamflow; cooler water in summer, reduced ice effects in winter; improved habitat for fish and other aquatic organisms; high forage production and quality; high deversity of wildlife habitat (C) Dissolved organic carbon (DOC) concentrations in flood plain ground water, along transects for perennial and intermittent streams (from Elmore and Beschta, 1989; copyright © by Rangelands by permission).*

sediment management from a dam may be emphasized (Ligon et al., 1995). In streams, structures meant to restore the system can actually result in further degradation (Fig. 4-1). While instream structures such as log weirs, gabbions, or rip-rap may ameliorate certain problems, they decrease channel sinuosity and thus reduce the interchange between the riparian ecosystem and hydrologic processes. Active restoration has a high failure rate because the reengineering fails to correct the problems created by anthropogenic disturbance. This approach should never be used unless there is a clear understanding of how the system will respond (Kauffman et al., 1995).

## REENGINEERING WET ECOSYSTEMS

### Restoring the Original

Even optimal engineering structures and approaches designed to restore wetlands do not in themselves create a natural condition. Instead, they work in concert with all other factors constraining the site, including past and present anthropogenic disturbances, irreversible changes, and position in the watershed. The best engineering projects revitalize systems that have been altered by past anthropogenic modifications and require little human intervention in the future.

Active restoration approaches that have the potential to re-create natural wetlands include, in riverine situations, levee removal, artificial channel backfilling, and dechannelization and, in coastal situations, embankment removal and the filling of canals (Box 4-3). These are fairly new procedures, and almost no research exists to document their success. According to Zedler (1986), restoration technology is still experimental, and testing procedures should be incorporated into the restoration plan to gain more information regarding the process. Also, the success of projects may be higher if interactive adjustments are made over a period of years to attain the proper hydrologic conditions required by the species involved.

Dechannelization offers the best hope of restoring the hydrologic conditions originally present in floodplains along rivers and streams, but it is extremely difficult and expensive. A channelized stream downcuts its banks so that the water in the channel flows below its original elevation (see Chapter 1). In recent efforts to dechannelize the Kissimmee River in Florida (see the Kissimmee River case history in Chapter 6), using a series of weirs, water from the canal was rediverted into parts of the old river channel, and the canal was backfilled with the original canal dredge (Toth, 1996). New channels will be excavated in some areas to connect remnant river channels. At the Kissimmee River test plot, dechannelization of the demonstration

*BOX 4-3*

**GENIUS LOCI**

Every place on earth is different from every other place; this identity is known as the *genius loci* or the *spirit of the place*. The features that make a place distinctive include waterfalls, rock formations, landform, soil color, wind-swept trees, distinctive smells or acoustics, and the growth form, color, and composition of plant species. The knobby "granny" oaks of Sherwood Forest in the United Kingdom symbolize the Robin Hood legend. In a restoration design, particularly if the wetland setting is somewhat distinctive, the genius loci should be enhanced or re-created (Bell, 1995).

Wetland restorationists sometimes comment that no two restoration projects are the same. This is no doubt a reflection of the genius loci phenomenon.

area by backfilling and new channel formation in an area with dredge spoil removed (borrowed) has led to rapid revegetation of areas adjacent to the channel (Fig. 4-4; Toth, 1996). While there is very little research on the success of the dechannelization, early results show promise.

The portion of the River Cole in Oxfordshire, England, above an old mill site has been redirected into the former stream bed. Below the mill the stream was remeandered, which may rejuvenate former wet meadows. Fish such as sea trout and salmon can now move freely along the River Brede in South Jutland, Denmark, because weirs and other obstacles have been eliminated along the channel (Nielsen, 1996).

One very innovative restoration project, a miniature version of the dechannelized Kissimmee River, is planned for property owned by the Wisconsin Department of Natural Resources in the Lodi Wildlife Area (Fig. 4-5a). According to the restoration plan, tile lines into a channelized stream will be broken. The water in the stream will be diverted while the channelized stream bed is dechannelized (Fig. 4-5b). The channel will be filled and grade contour bars (Fig. 4-5c) added to prevent future downcutting. After construction, the water will be placed back into the reconstituted stream course. The meanders will reestablish contact between surface and ground water, as well as create an overbank flood zone.

While this is a small project in comparison to the Kissimmee Project, it

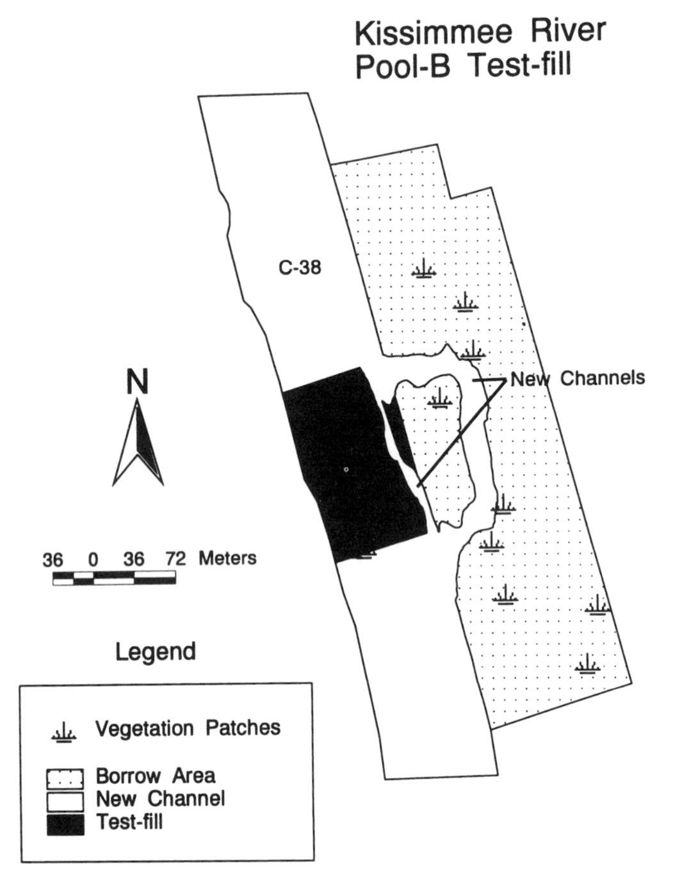


*Figure 4-4. New channel formation formed through backfilling the canal and borrowing the original material dredged from the canal along the Kissimmee River. Patches of vegetation colonizing the area are also shown (from Toth, 1996; copyright © by Wiley, Chichester, by permission).*

has the potential to restore a wetland resembling the original on the upper portion of this small watershed. The original stream had three water sources: rain, substantial meltwater from snow, and groundwater. It is hoped that the restructured floodplain relinked to its stream can support wet prairie and sedge meadow vegetation (Perry Rossa, Mead and Hunt Consulting and Doris Ruesch, Wisconsin Department of Natural Resources, personal communications).

In response to major floods in various locations in the United States from 1993 to 1997 and because of the high cost of levee maintenance, some levees have been removed and/or set back (Trepagnier et al., 1995; Christenson, 1997). Alternatively, a river breach through an existing embankment has the potential of re-creating certain sedimentation and hydrological processes in the outfall area (Good, 1993). Along the Upper Rhône River of France, an accidental break in an embankment in 1919 led to geomorphological changes, which in turn, led to the formation of new islands in the former channel over the course of the next 30 years (Bravard et al., 1986). Levee removal or breakage can reestablish or partially reestablish the linkage between the channel and the floodplain. Little research indicates that this or other related procedures can rejuvenate floodplain function (Brookes et al., 1996), although wetland vegetation has revegetated behind breached embankments in Louisiana (Trepagnier et al., 1995) and France (Bravard et al., 1986).

Attempts to restore coastal features such as fresh water and salt marsh, sea grass bed, and mangrove swamp have been variously successfully. The active methods used to reverse damage to coastal wetlands include removal of embankments and roads, filling of canals, rediversion of freshwater rivers, and manipulation of sediments (dredging or filling).

Coastal salt marshes diked for pasture have a higher potential for restoration than those subsequently filled for development or converted to ag-

(a)

***Figure 4-5.*** *(a) Unrestored prairie wetland near the Driftless Area, Lodi Wildlife Area, Wisconsin; photograph by Beth Middleton. (b) Site plan for a site in the Lodi Wildlife Area. The channelized stream will be dechannelized. The remeandered stream will restore groundwater exchange and flood pulsing dynamics via overbank flow to the floodplain with a drop structure placed at point 500 along the stream (see Fig. 4–5c; blueprint from Mead and Hunt, Madison, Wisconsin, by permission). (c) Grade control bars in the stream bed to prevent future downcutting and drop structure details for point 500 along the stream (blueprint from Mead and Hunt, Madison, Wisconsin by permission).*

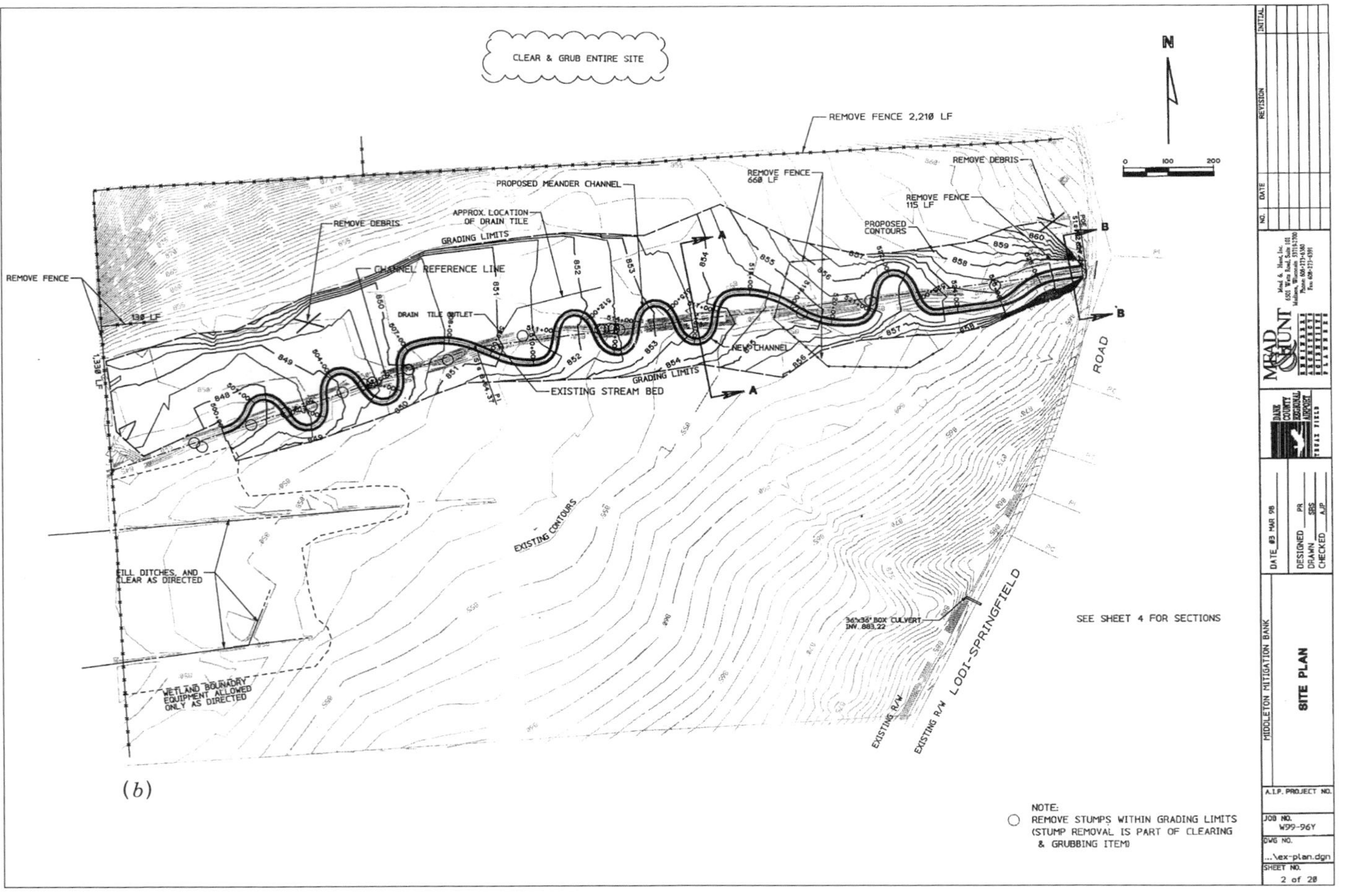


*Figure 4-5.* (*Continued*)

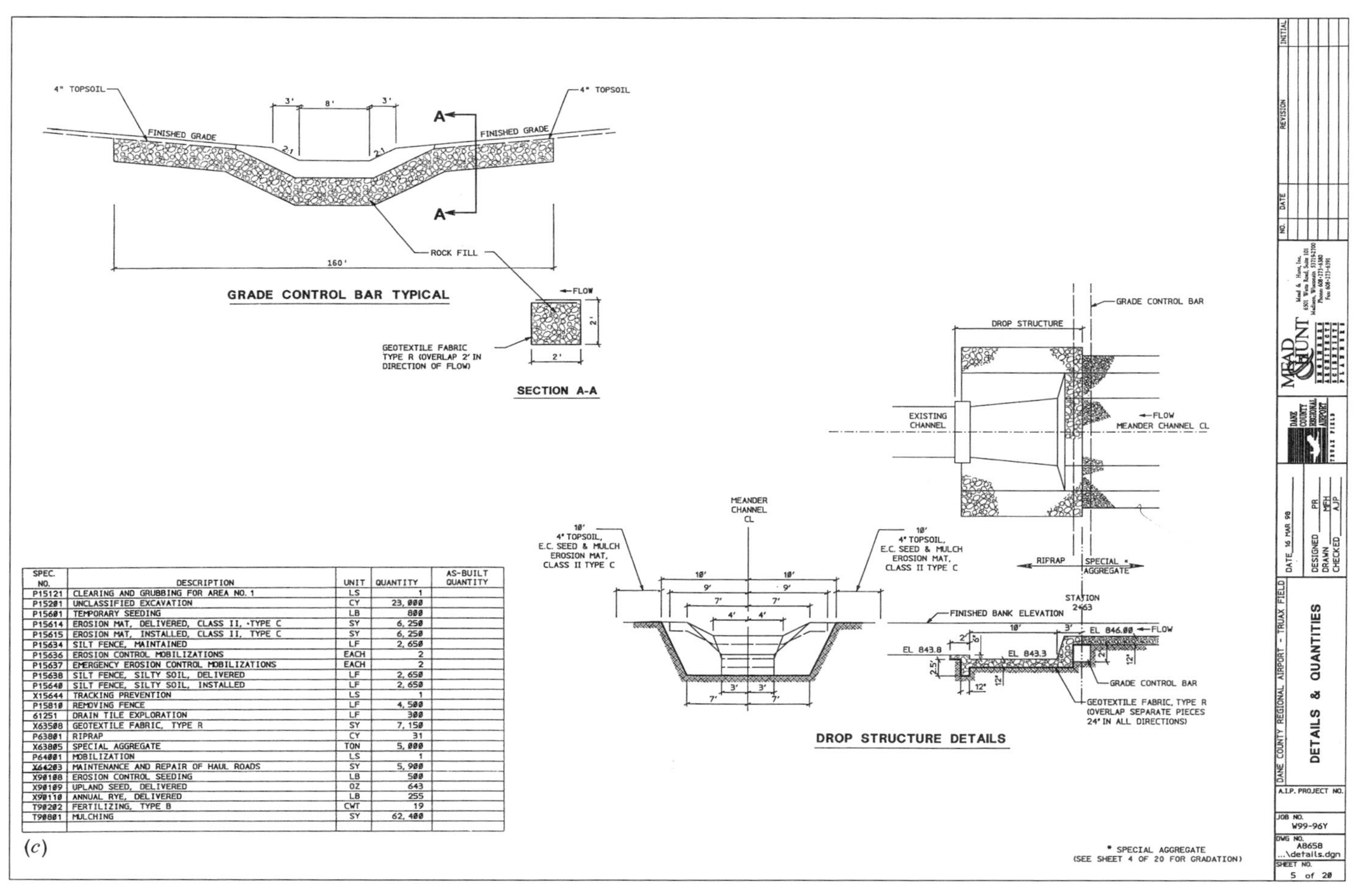

| SPEC. NO. | DESCRIPTION | UNIT | QUANTITY | AS-BUILT QUANTITY |
|---|---|---|---|---|
| P15121 | CLEARING AND GRUBBING FOR AREA NO. 1 | LS | 1 | |
| P15201 | UNCLASSIFIED EXCAVATION | CY | 23,000 | |
| P15601 | TEMPORARY SEEDING | LB | 800 | |
| P15614 | EROSION MAT, DELIVERED, CLASS II, TYPE C | SY | 6,250 | |
| P15615 | EROSION MAT, INSTALLED, CLASS II, TYPE C | SY | 6,250 | |
| P15634 | SILT FENCE, MAINTAINED | LF | 2,650 | |
| P15636 | EROSION CONTROL MOBILIZATIONS | EACH | 2 | |
| P15637 | EMERGENCY EROSION CONTROL MOBILIZATIONS | EACH | 2 | |
| P15638 | SILT FENCE, SILTY SOIL, DELIVERED | LF | 2,650 | |
| P15640 | SILT FENCE, SILTY SOIL, INSTALLED | LF | 2,650 | |
| X15644 | TRACKING PREVENTION | LS | 1 | |
| P15810 | REMOVING FENCE | LF | 4,500 | |
| 61251 | DRAIN TILE EXPLORATION | LF | 300 | |
| X63508 | GEOTEXTILE FABRIC, TYPE R | SY | 7,150 | |
| P63801 | RIPRAP | CY | 31 | |
| X63805 | SPECIAL AGGREGATE | TON | 5,000 | |
| P64001 | MOBILIZATION | LS | 1 | |
| X64203 | MAINTENANCE AND REPAIR OF HAUL ROADS | SY | 5,900 | |
| X90108 | EROSION CONTROL SEEDING | LB | 500 | |
| X90109 | UPLAND SEED, DELIVERED | OZ | 643 | |
| X90110 | ANNUAL RYE, DELIVERED | LB | 255 | |
| T90202 | FERTILIZING, TYPE B | CWT | 19 | |
| T90801 | MULCHING | SY | 62,400 | |



**Figure 4-5.** *(Continued)*

ricultural uses. In the Salmon River Estuary of Oregon, after an embankment was removed, a 22-ha pasture was restored to salt marsh. No grading, planting, or other active restoration measure was necessary. Because the site had subsided due to oxidation of the organic soils, low marsh rather than the original high marsh redeveloped on the site, which was dominated by pasture species during the 17 years following embankment. The reconnected tidal creeks support numerous juvenile fish, including salmonids. Because tidal circulation has been reestablished on this site, the marsh surface is rising toward its preembankment level as organic matter and sediments accumulate. This project suggests that it is sometimes more effective to work with natural processes in restoration rather than to replicate historical conditions that are typically unrealistic and not fully known (Morlon and Frenkel, 1992).

Alternatively, in Phase 1 of the South Slough National Estuarine Research Reserve, the materials from the dike removal will be spread out over the subsided pastureland to raise the elevation of the subsided surface (Rumrill and Cornu, 1995). It can be argued that the simple removal of a tidal barrier is sometimes too simplistic an approach because shoreline erosion, channel migration, or unexpected species invasion can occur (Haltiner et al., 1997; Turner and Lewis, 1997).

In canals dug for transportation through salt marsh (brackish-estuarine marsh) in Louisiana, backfilling following abandonment has had mixed success. These canals are refilled to eliminate the waterway and spoil bank and to reestablish vegetation. After less than ten years, only half of the canals studied had significant emergent reestablishment, often because the depth of the filled canal exceeded 1 m. However, the deeper backfilled canals supported emergent vegetation. The procedure was successful in increasing overland water flow, which could raise the elevation of sites with lowered elevations through the accumulation of land-building sediments and belowground plant materials (Turner et al., 1988).

Mangrove swamps with disrupted hydrology can sometimes be restored. A roadway that had severed a tidal connection from the ocean to mangroves was removed. This tidal disruption had killed over 100 ha of mangrove swamp in Laguna Boca Quebrada, Vieques, Puerto Rico. The vegetation recovered when the hydrology was remedied by the removal of the roadway (Lewis, 1990).

The purpose of many coastal enhancement projects often is not to recreate the original conditions, but rather to enhance water quality or provide a habitat for waterfowl. In freshwater fringe wetland along the Great Lakes, embankments may be created to provide water level control. The purpose of embanked lakeshore is primarily to support waterfowl (Levine and Willard, 1989) and protect shorelines. However, such structures cut

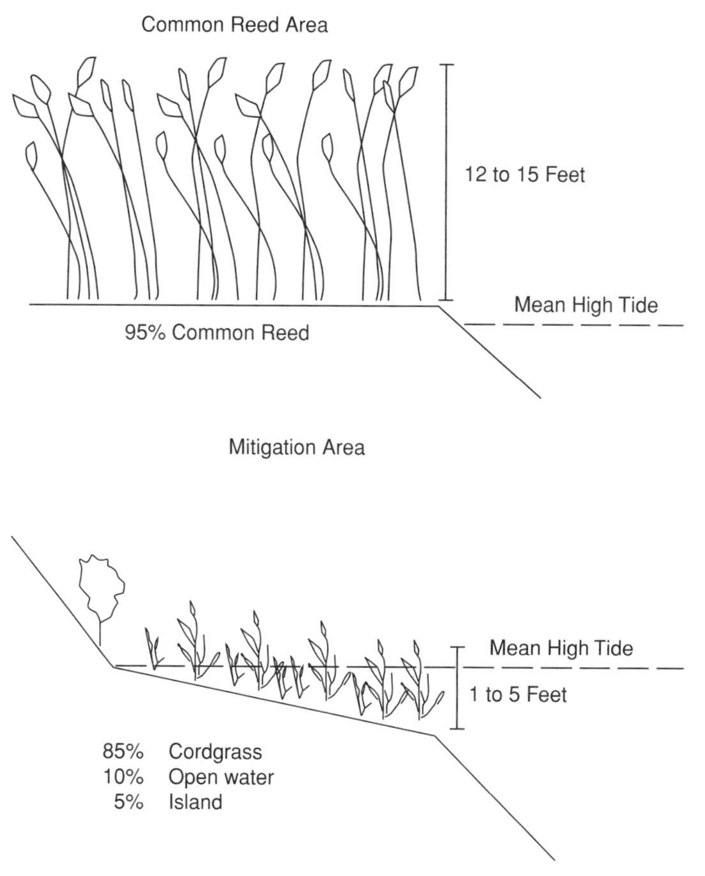


***Figure 4-6.*** *Excavation of coastal freshwater wetland previously disturbed through dredge spoil deposition (from Bontje and Stedman, 1991; copyright © by Hillsborough Community College, Florida by permission).*

off normal hydroperiods along the shoreline wetlands (Kusler and Smardon, 1992).

In a coastal freshwater marsh project in New Jersey, a *Spartina alterniflora* marsh was created by excavation in a *Phragmites australis* stand created by the deposition of dredge spoil. (Fig. 4-6). The site once had been vegetated by *Spartina alterniflora*, sedges and a large stand of white cedar (*Chamaecyparis thyoides*). Because of the reduced freshwater flow due to the Oradell Reservoir and the prior usage of the site as a settling basin for dredge spoil, it was not feasible to restore the white cedar swamp (Bontje, 1988). This project was a habitat enhancement rather than an ecosystem restoration because 1. it did not attempt to restore the original estuarine ecosystem 2. the site was degraded in a regional context and therefore could not be "restored" within the context of current land use, and, 3. the project

***Figure 4-7.*** *Aerial photograph of a mangrove restoration site featuring flushing cuts along Henderson Creek in the Rookery Bay Estuarine Research Reserve. The site was originally destroyed by leveling and filling in 1972. The mitigation procedure included removing exotics, reestablishing the original elevation, and planting red mangroves. The flushing cuts set the stage for the natural regeneration of white and black mangrove because propagules moved freely in the area with the tides (Rookery Bay NERR staff photo).*

created intertidal marsh with mud flats with raised islands of woody vegetation for waterfowl which were unlike the high marsh of *Spartina patens* and *Distichilis spicata* that probably originally existed there. However, the resulting ecosystem had improved water and wildlife quality *(National Research Council*, 1992).

Fill was also removed in a mangrove mitigation project in the Rookery Bay National Estuarine Research Reserve along Henderson Creek, Florida. The site, originally vegetated by red, white, and black mangroves (*Rhizophera mangle, Avicennia marina*, and *Laguncularia racemosa*) (Karen McKee, personal communication), was destroyed by leveling and filling in 1973 (Shirley, 1992). As part of the restoration approach, exotic vegetation was removed and the original elevation reestablished by removing fill. To facilitate water movement (tidal pulsing), flushing cuts were excavated from a tidal creek (Henderson Creek) to a pond (Fig. 4-7). While red mangrove was planted in December 1990 (Shirley, 1992), because propagules were dispersed freely via the flushing cuts, substantial numbers of white and black mangrove established naturally (Mike Shirley and Karen McKee, personal communication).

In North Carolina, a salt marsh was created on a dredge spoil island in the Cape Fear River estuary. Later, this site was planted in *Spartina alterniflora*. After 12 years, the site harbored mixed species of salt marsh plants and macrofauna (Sacco et al., 1988). In California, salt marsh was restored by lowering the level of dredge spoil at the Connect Marsh and the Marisma de Nacion; however, these constructed salt marshes were poorly vegetated at high elevations. While reference sites shared this trait to a lesser extent, the constructed wetlands had especially hypersaline groundwater because of infrequent tidal inundation, high evaporation rates, and coarse dredge substrates (Haltiner et al., 1997)

Dredge spoils also have been used to restore sea grass beds. In general, projects are more successful if the site is somewhat protected. In the Huffco Project, sea grasses (*Syringodium filiforme*, *Thalassia testudium*, and *Halodule wrightti*) were successfully planted and established on the leeward side of a spoil island but not on the high-energy side of Hog Island in Redfish Bay, Texas (Carangelo, 1988).

A variety of approaches to recapturing the original site less satisfactory than levee removal and dechannelization have been used to restore streams. Reducing the side slopes of channelized streams allows the stream to remeander, albeit at a lower elevation than the floodplain (Fig. 4-8; Peterson et al., 1992). In this procedure, either one or both banks are excavated to restore stream sinuosity, as in a project at Elbaek, Denmark (Brookes, 1990). Sometimes the slope is reduced along with channel dredging (Ward et al., 1994).

One of the most important measures in stream restoration in agricultural areas is to set aside space adjacent to the stream. Buffer strips can ameliorate certain stream sediment and agricultural runoff problems (Fig. 4-9; Nutter and Gaskin, 1988; Peterson et al., 1992).

Streams are often endpoints for drainage tiles from wetlands where nutrients are carried directly from fields into streams. Used particularly in Europe, one system to offset this problem is to create ''horseshoe'' wetlands within the buffer strip by excavating into the tile area away from the stream (Fig. 4-10; Peterson et al. 1992).

Although sediments accumulate behind weirs (Ward et al., 1994), they can be placed in the channels of streams to reverse some of the adverse impacts of downcutting, including increased flow and downstream sedimentation. Stone weirs were added to the channel of an incised stream in northwest Mississippi in conjunction with existing gabions to stabilize the channel (Fig. 4-11). This resulted in a modest benefit to the fishery (Shields et al., 1995). In a downcut channel of East Fork Lobster Creek, Oregon, instream rock-filled gabions and boulders improved the habitat for juvenile coho salmon, but only for about ten years. Interim placement of instream structures can be of value in fish habitat improvement until longer-term watershed restoration occurs (House, 1996).

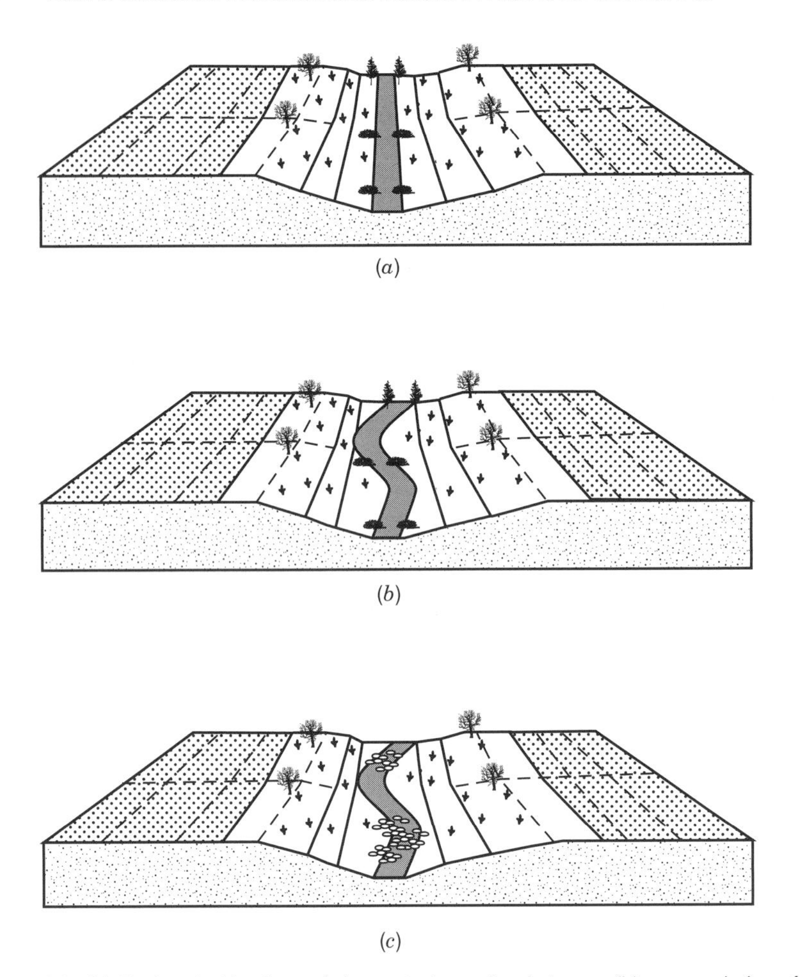


***Figure 4-8.*** *(a) Reduced side-slope of downcut channelized stream, (b) remeandering of the channel, and (c) riffles and pools encouraged by placement of rocks and gravel in the channel (from Peterson et al, 1992; copyright © by John Wiley & Sons, Chichester, U.K., by permission).*

Weirs have been placed instream in the Peace-Athabasca Delta of northern Canada to attempt to raise water levels which dropped precipitously after the construction of the W. A. C. Bennett Dam in 1968. While the large lakes have filled with water, high-elevation and perched lakes have not (Prowse et al., 1996).

The focus of restoration efforts in riverine systems should be the reintegration of the floodplain with the channel so that fluvial disturbance and connectivity are reestablished (Ward and Stanford, 1995a). While instream structures may deepen pools and sometimes improve the stream habitat for game fish, they do not re-create original flow conditions. None of these approaches have the potential to restore the original connection to the flood-

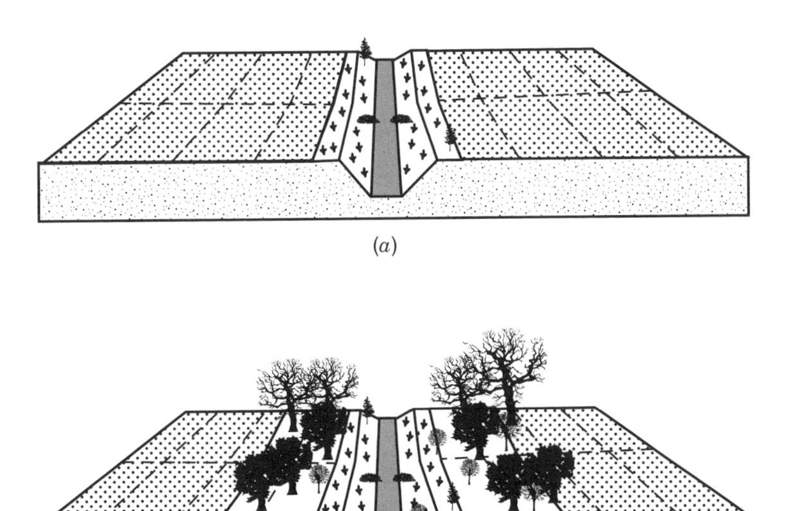


***Figure 4-9.*** *(a) Channelized stream in an agricultural area and (b) a buffer strip planted with trees for nutrient retention and wildlife (from Peterson et al, 1992; copyright © by Wiley, Chichester, by permission).*

***Figure 4-10.*** *Horseshoe wetland excavated in a buffer strip at the end of a tile line terminating in a stream (from Peterson et al, 1992; copyright © by John Wiley & Sons, Chichester, U.K., by permission).*

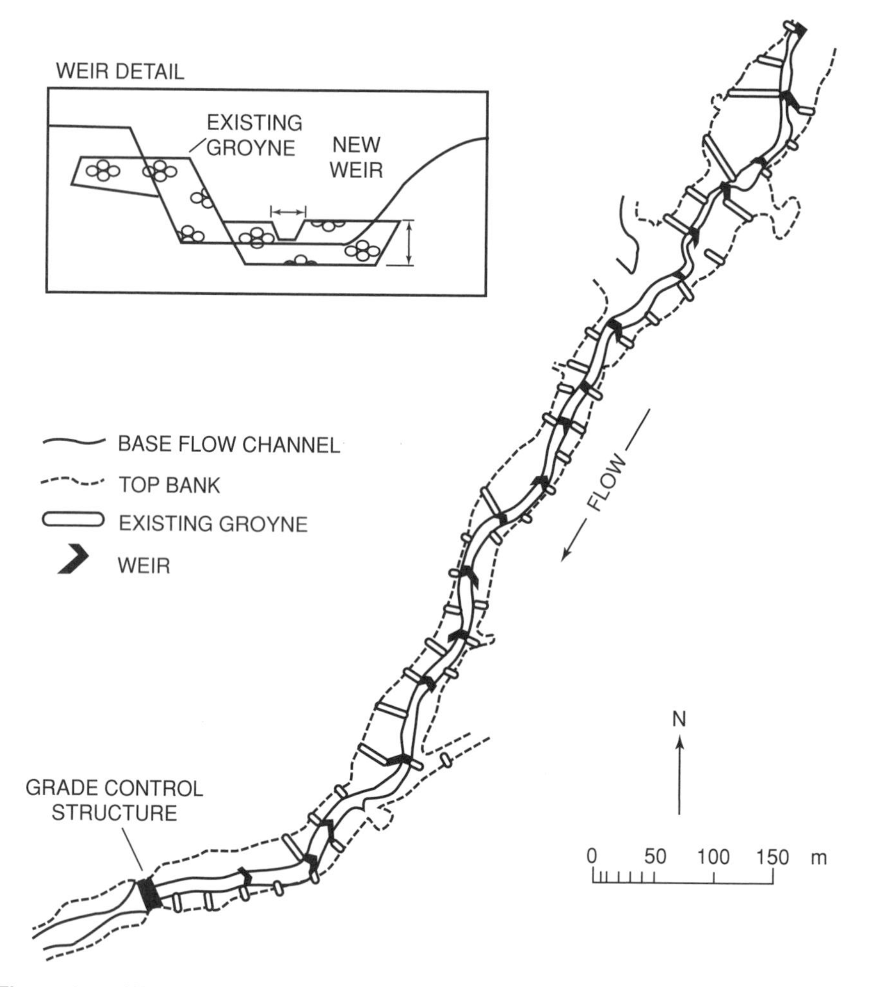


*Figure 4-11. Weir structure in an incised channel to alleviate the sedimentation problem and improve the fish habitat (from Shields et al., 1995; copyright © by Regulated Rivers: Research and Management by permission).*

plain via the flood pulse, but they, can still be useful in bandaging particular problems, often these associated with channelization and subsequent down-cutting.

## Rewatering the Wetland

In freshwater wetlands, a number of techniques have been developed to put water back on wetlands, including breaking the tile line and interrupting the flow of water through a stream or ditch. While the techniques discussed in this section seldom re-create anything resembling the original hydrology of the site, the resulting wetlands often have tremendous value for improving water quality and wildlife habitat. In agricultural areas, most streams are reduced to drainage ditches, and much can be done to re-create some aspects of their original function (Peterson et al., 1992).

Depending on the situation, most of these techniques only approximate the original hydrologic setting of the wetland and are dependent on some form of maintenance to remain functional. In many cases, the water table has dropped or so much water has been diverted from the region that wetland restoration is possible only through artificial measures. Thus, many projects in wetland restoration do not attempt a full restoration or **rehabilitation** of a wetland but are simply enhancements or creations (Brookes and Shields, 1996). In most cases, the techniques described below create pools by obstructing the flow of water, but they do not create a natural hydrologic regime.

Where drainage is by tile, lines can be broken in creative ways to allow water to flow into a former marsh. If the tile line in the watershed is high and if its interruption will not interfere with neighboring property, it is possible in some cases to restore the natural hydrology. Tile lines are incapacitated either by breaking them up or by removing a segment of the tile line and replacing it with an impermeable tile (Galatowitsch and van der Valk, 1994). Old tile lines can be difficult to find.

Where the water can be interrupted only temporarily, or where water must not exceed certain elevations, the flow through the tile line into a basin can be controlled in a variety of ways. Embankments can be created around the restoration site, and the drainage tile can be brought to the surface with an outlet to a depression and then water levels controlled by utilizing preexisting ditches or tiles (Fig. 4-12; Galatowitsch and van der Valk, 1994). To deepen and/or create islands in the basin in the Prairie Pothole Region of the United States, project sites are often excavated first (Galatowitsch and van der Valk, 1994). These sites can also be dynamited, but the subsequent wetlands lose this increased depth in approximately 20 years due to sedimentation (Strohmeyer and Fredrickson, 1967).

Wetland restoration in agricultural or urban areas usually requires water control to avoid economic damage by flooding. Also, water control structures may be helpful to manipulate water levels, particularly early in the plant reestablishment phase (Harker et al., 1993). Emergency spillways can be used to control water, usually to a maximum level (Fig. 4-13). Outflowing water passes over the weir opening onto an apron and then passes into the downstream channel. Emergency spillways can discharge large amounts of water, are very stable, and typically require little maintenance (Wenzel, 1992). However, the water level control achieved by these struc-

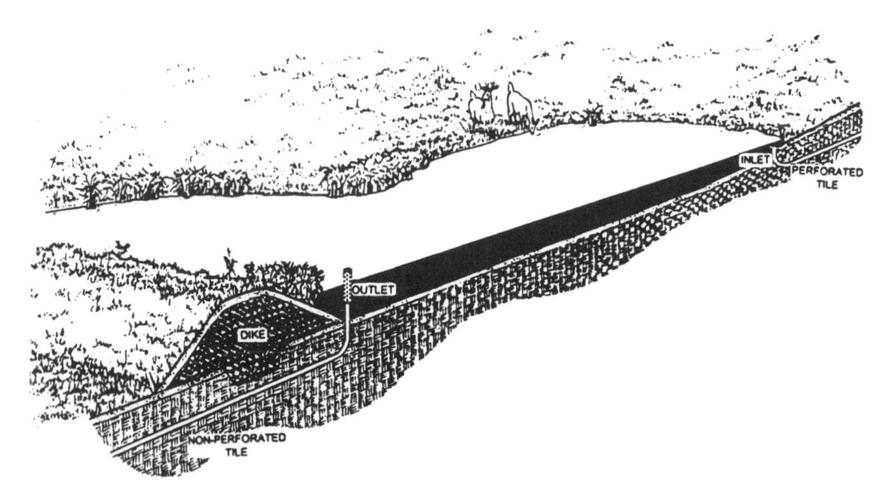


*Figure 4-12.* Wetland restored between embankments and supplied with water from a tile line to an outflow pipe. The overflow pipe to the tile line is pictured at a lower elevation (from Galatowitsch and van der Valk, 1994; copyright © by Iowa State University Press, Ames, Iowa, by permission).

tures is not very flexible and therefore cannot recapture the original hydrologic setting.

Stop log structures (Fig. 4-14) also allow water control, but with the added advantage that they can be adjusted to maintain various water depths in the restored wetland (Hammer, 1997). However, stop log structures require constant maintenance, so they cannot be used in unmanaged sites. A related structure is the "Wisconsin tube" (Fig. 4-15; Iowa Department of Conservation, 1989) and the "locking type control structure" (U.S. Army Corps of Engineers Waterways Experiment Station, 1993). Using these structures, water levels are maintained in restored wetlands by removing or adding boards with a winch (Lee Gladfelter, personal communication). These are relatively labor intensive to build and maintain, though less likely to become blocked than simpler pipe structures. Beaver are particularly adept at conquering these structures, and if they are not kept clear, after a period of time extraction of the accumulated deposits may require explosives or heavy equipment (Hammer, 1997).

Relatively simple pipe structures can control water levels in restoration projects. At least one project in Iowa employs a pipe overflow system from an excavated wetland basin that interrupts a **bull ditch** originally dug to drain surface water from cropland into a stream (Fig. 4-16). In function, this approach is related to the *horseshoe* wetland idea used in Europe in buffer strips, as both have the potential to improve the quality of water discharged from agricultural fields into streams.

One very simple way to increase water retention time in areas of poor

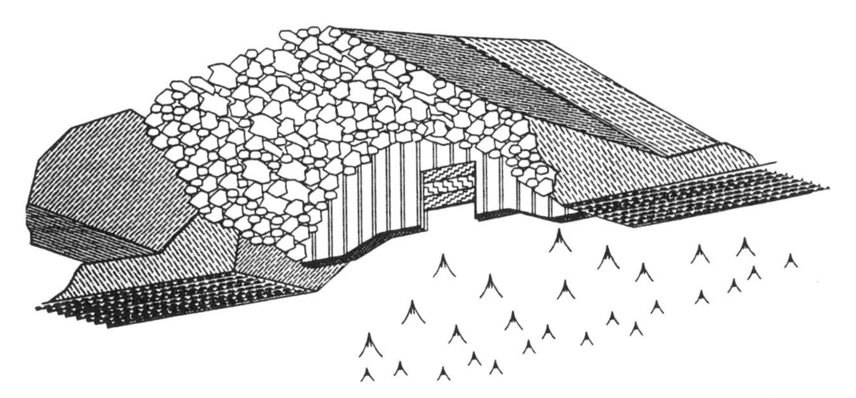

*Figure 4-13.* *Wetland restoration constrained by embankments, with the weir acting as an overflow emergency spillway (from Wenzel, 1992; copyright © by Minnesota Board of Water and Soil Resources, St. Paul, Minnesota, by permission).*

water quality is to create **ditch plugs** (earthen dams) across ravines, ditches, or low-flow streams without any additional water control structure (Fig. 4-17). This method is more widely used in somewhat arid regions (e.g., western Iowa more commonly than central Iowa: personal observation). The drawback with this method is that the hydrologic conditions created typically do not resemble any long-term natural disturbance.

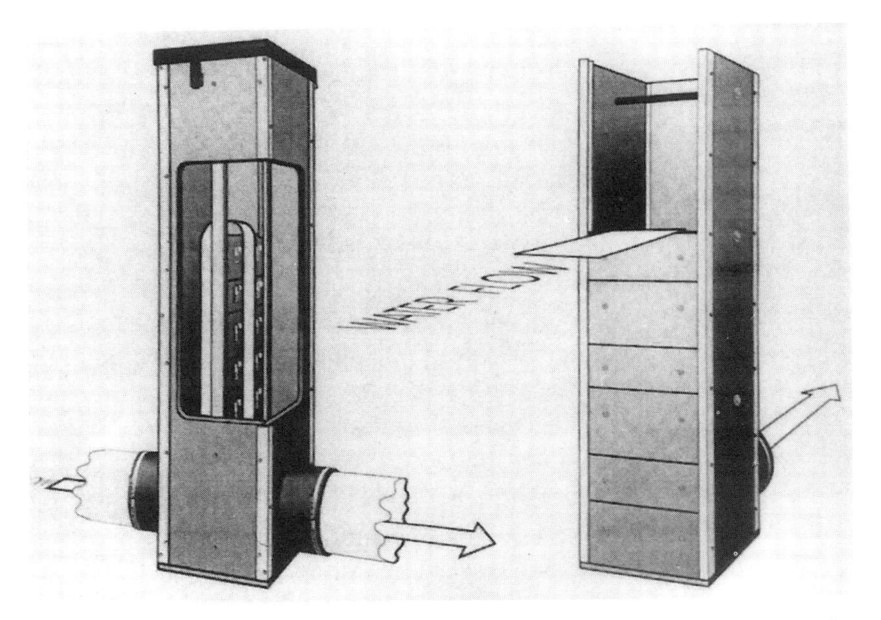


*Figure 4-14.* *Stop log structure to control water levels in medium-sized to small systems (from Hammer, 1997; copyright © by Lewis Publishers, Boca Raton, Florida, by permission).*

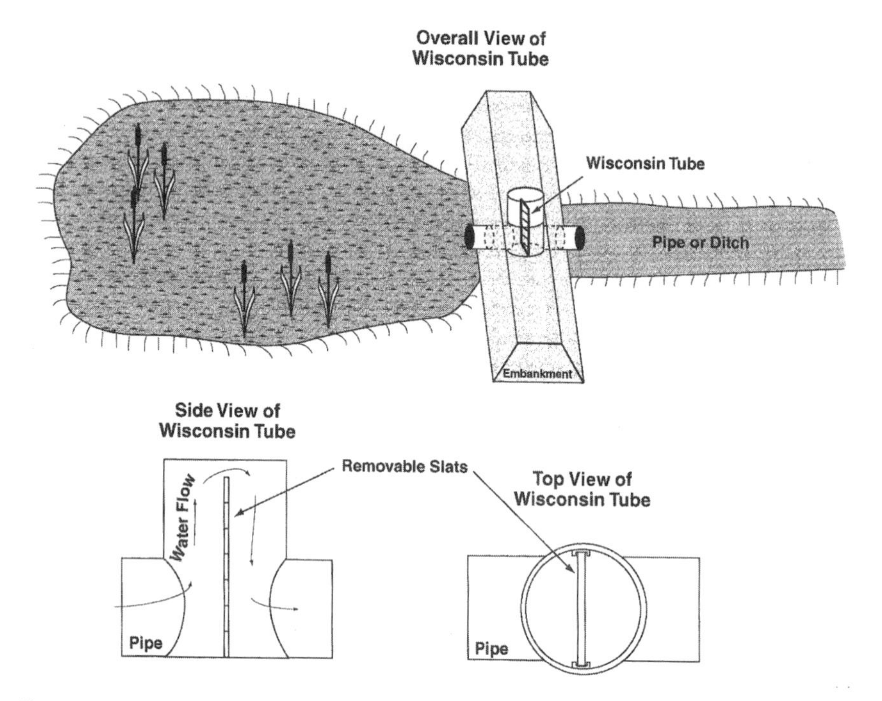


*Figure 4-15. Wisconsin tube variation of a stop log structure to control water levels in a wetland fed by drainage tiles. (Wetland restoration design from personal notes taken during work for the Research Unit in Landscape Ecology, Iowa State University, Ames, Iowa; details of the "Wisconsin tube" from blueprint plans of the Iowa Department of Natural Resources, 1989; copyright © by Iowa Department of Natural Resources, Des Moines, Iowa, by permission).*

Water pumps to divert water from rivers, lakes, or groundwater also can rewater wetlands. Along the shores of Lake Erie, there are a number of projects that embank the lakeside edge of the wetland and then pump water into the enclosed area (Mitsch, 1989). At the Fern Forest Nature Center in Pompano Beach, Florida, a southern forested wetland was rewatered by pumping water from the Cypress Creek canal to the park. The pumping compensated for the hydrology of the site, which had been altered by canals and water retention ponds. The water was directed through the park, utilizing elevation and stop log structures, with an attempt to simulate water flux during the wet and dry seasons of south Florida (Weller, 1995). It goes without saying that approaches such as these require high maintenance and are not self-sustaining in the long term, although they can have beneficial wildlife and/or water quality effects.

The techniques described thus far in this chapter have been mainly concerned with the restoration of wetlands that are not primarily groundwater fed. To restore groundwater-fed wetlands, excavating below the level of

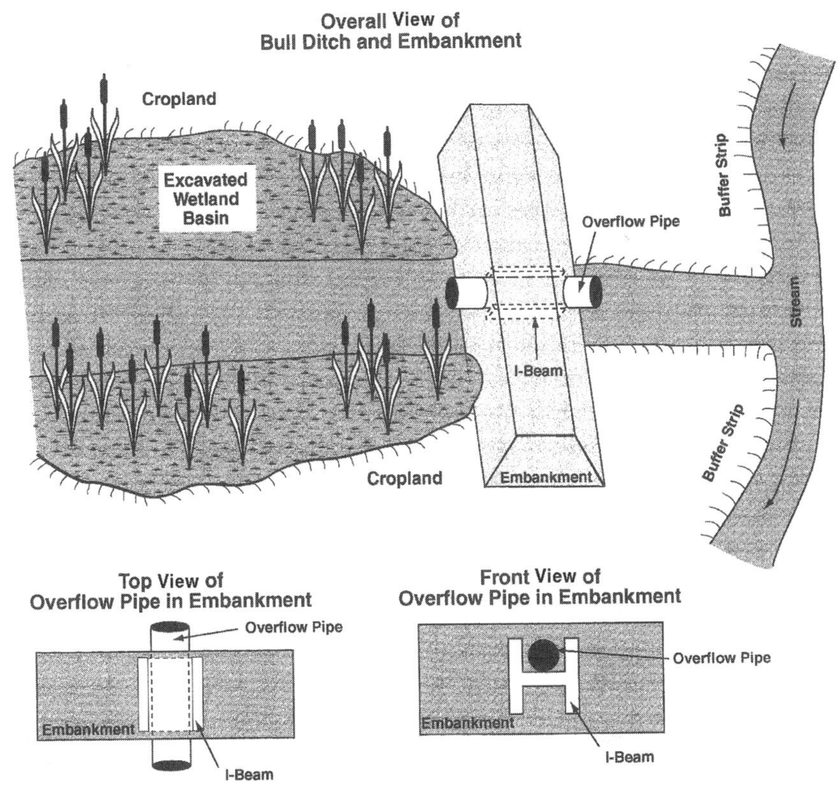


***Figure 4-16.*** *Related to the horseshoe wetlands of Europe in design and function, this wetland design discovered in Iowa interrupts water flow through a bull ditch into an excavated basin before the water continues into the ditch, leading to a stream. In this design, water level is controlled by an overflow pipe held by an I-beam (wetland restoration design from personal notes taken during work for the Research Unit in Landscape Ecology, Iowa State University, Ames, Iowa; details of I-beam construction from the Iowa Department of Natural Resources; copyright © by Iowa Department of Natural Resources, Des Moines, Iowa, by permission).*

the preconstruction elevation of the water table may not produce the desired hydrology. The rate of water loss can be increased, causing greater drainage of the uplands and increased evapotranspiration in the excavated depression. If the wetland is excavated to the annual low water level, the wetland may not be saturated even in the wettest part of the year. In part this is because the rooting level of the plants is lowered to a depth at which the plants were previously unable to reach the water. Evapotranspiration increases to a point where the water table cannot recover, even during the nongrowing season. Mathematical models can be used to aid the restorationist in making predictions about the behavior of groundwater given a particular wetland design (Winston, 1997). In addition, groundwater wells should be installed

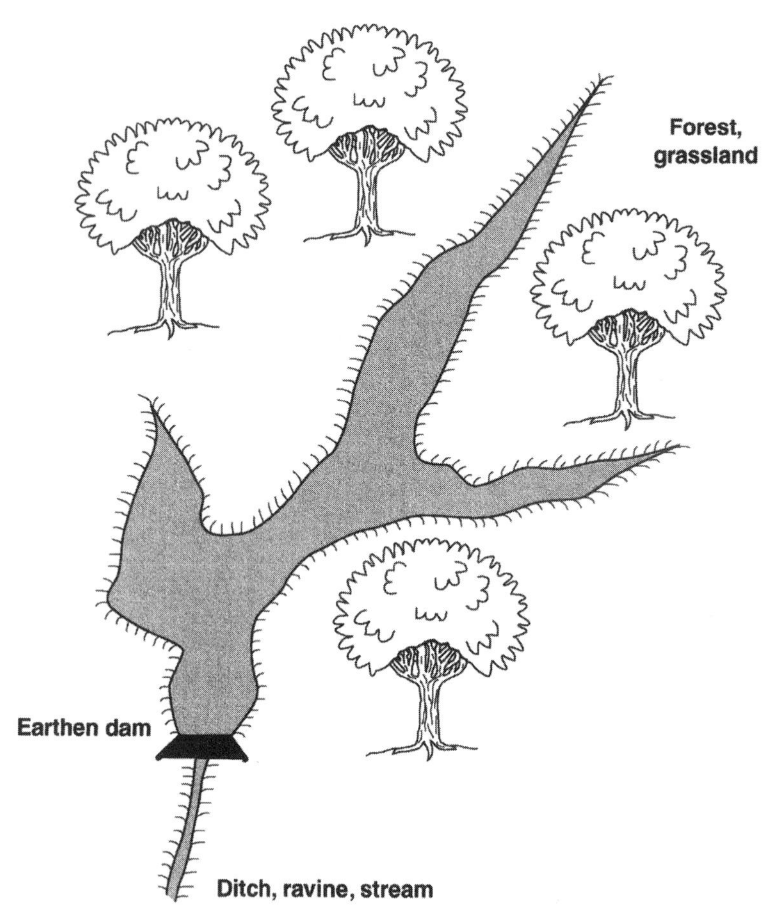


***Figure 4-17.*** *A ditch plug used to obstruct water flow through low-flow streams, ravines, or ditches (wetland restoration design from personal notes taken during work for the Research Unit in Landscape Ecology, Iowa State University, Ames, Iowa).*

at restoration sites and monitored for one year. The water-level data can be compared to rainfall data to determine if the year of monitoring is representative (Nancy Rorick, personal communication).

Many groundwater-fed wetlands in the Netherlands are losing their unique vegetation. *Ericetum tetralis* heathland can be restored by sod-cutting, but Cirsio-Molinietum communities cannot. At Staverden, barrages (dams) were placed into drainage ditches to stop the artificial drainage of the wetlands; at Stroothuizen, the sod was cut (10–20 cm) to mineral soil. Both of these treatments increased the desired species at these sites (Jansen et al., 1996). Along the River Vecht, near Utrecht, fen vegetation might be established by cutting floating forests and seeding the sites with fen species (van der Valk and Verhoeven, 1988).

Peat extraction turns bogs into fens by altering their hydrology and water chemistry (Wind-Mulder et al., 1996). To restore peatland in North America after peat mining, mined bogs are regraded and the discarded vegetation is placed in islands over the surface. The drainage ditches are also plugged (Carpenter and Farmer, 1981; Pakarinen, 1994).

Restoration alternatives can be quite limited and natural restoration targets arguable. In the brook valleys of the Gorecht area of the Netherlands, restoring the low-productivity, species-rich meadows that existed there around 1900 is problematic because of the large amount of nutrients stored in the soil. To restore the mesotraphent fen communities that were present there until the Middle Ages, mineral-rich, nutrient-poor groundwater is required. Abstraction at the present level does not allow groundwater to fill ponds. The area could be restored to surface water wetlands, except that it would require that groundwater abstraction cease and that agricultural drainage be greatly reduced (van Diggelen et al., 1994). Very often, wetland restoration options are limited and totally dictated by current conditions and long-term changes in the regional landscape.

# 5

# *Revegetation Alternatives*

*An essential but sometimes difficult step in restoration is the revegetation of the site. The conditions which exist at the restoration site after its reengineering will determine which species can persist. Allowances should be made for inevitable natural disturbances such as flood pulsing, water level fluctuation, fire, and/or herbivory by native fauna. Especially in very disturbed sites, e.g., those that have been excavated or farmed for many decades, the reestablishment of vegetation may be facilitated by human intervention. Natural revegetation can be used to wholly or partially revegetate sites, depending on the species richness of the seed/propagule bank and the level of interconnectedness of intact sites of similar wetland type. This chapter will explore approaches in natural and active revegetation to reestablish vegetation in a variety of world wetland scenarios.*

The revegetation of some wetlands following their restoration can occur spontaneously but in others may require massive intervention. In all cases, the success of this venture will depend on the fit of the species to the environmental setting. Most species have different tolerances for various water levels in their seed, seedling, and adult stages and tend to become more tolerant of extremes as mature individuals (see Chapter 3). According to the Initial Floristic Composition Concept, the first species to establish are likely to dominate the site (Egler, 1977). Following Island Biogeography Theory, species are likely to become more diverse over time due to dispersal and invasion (MacArthur and Wilson, 1967); and disturbance allowing opportunities for establishment (Johnstone, 1986) nonetheless, it is clear that the initial mix of species at a site can determine the species composition at the restoration site for a long time. Decisions by restorationists regarding the appropriate technologies for revegetation are critical for the ultimate success of the restored wetland, at least regarding its functional attributes associated with the composition of species and their value to wildlife.

## NATURAL REVEGETATION

One route to the revegetation of restoration sites is to mimic the natural process through natural revegetation (Primack, 1996). The ability of sites to revegetate themselves after restoration without further intervention is very situation dependent. This approach is inexpensive (Trepagnier et al., 1995), and when it can be used successfully, it ensures that there will not be an artificial quality to the composition or spatial configuration of the vegetation. In many cases, the natural seed bank along with dispersal can be at least partially successful in regenerating appropriate species given adequate attention to the initial design of the site regarding hydrologic and biotic connectivity to the landscape.

Natural revegetation is itself a passive approach, but it may be used in combination with either the passive removal of anthropogenic disturbances such as cattle grazing along streams (Sedell et al., 1991; Kauffman et al., 1995) or active measures to restore flood pulsing to the floodplain.

In the world as a whole, vast tracts of land are abandoned to agriculture, mining, or other man-made disturbances. In Central Europe, little active reclamation is done on abandoned old-field and mining lands in the Czech Republic and Hungary. No woody species appear in wet old-field sites there. On other types of abandoned sites with moist soil, the wind-dispersed *Betula pendula* commonly establishes spontaneously (Prach and Pyšek, 1994); In England, both *Betula pendula* and *Fraxinus excelsior* naturally revegetate (Harmer et al., 1997).

Bottomland hardwood forests in the United States revegetate naturally if there is a seed source and water to transport the seeds of these species (Clewell and Lea, 1990). Because of the patterns of seed deposition due to wind dispersal, red maples and sweetgum form a "seed wall" at the edge of the nearest mature forest facing the devegetated site (Clewell, 1986, in Clewell and Lea, 1990).

Bottomland hardwood forest along the Lower Mississippi Valley were largely undisturbed until the 1930s, when they were massively altered by water control, logging, and agriculture. Many of the species of these forests establish spontaneously, including the wind-dispersed *Fraxinus pennsylvanica* var. *subintegerrima, Ulmus americana, Acer rubrum*, and *Salix nigra* (Newling, 1993). The heavier-seeded species of oaks and hickories regenerate poorly after disturbance without human intervention. These species include *Carya aquatica, Carya illinoinensis, Quercus falcata var. pagodaefolia, Quercus lyrata, Quercus michauxii, Quercus nigra, Quercus nuttallii, Quercus palustris*, and *Quercus phellos.*

In cypress swamps of the southeastern United States, very few seeds of the dominant species may be viable after one to three years of farming (Newling, 1993; Middleton, 1995a), but this does not mean that natural regeneration cannot be successful. Seed banks that stay viable for long periods of time are certainly useful in restoration (van der Valk and Pederson, 1989). Without long-lived seed banks, natural revegetation can still be successful if dispersal can supply seeds to restoration areas. Where the flood pulse is still operational or has been reconnected in cypress swamps, dominants such as *Nyssa aquatica* and *Taxodium distichum* regenerate naturally (Newling, 1993; Middleton, 1995a). Many other characteristic swamp species disperse in the water (Schneider and Sharitz, 1988; Middleton, 1995b), so that the natural regeneration of swamps may be possible where flood pulsing has been restored (Middleton, 1995a).

In salt marshes, establishment by seed is very dependent on wave energy and light availability. On the Bolivar Peninsula of Texas, sandy material dredged from the Gulf Intracoastal Waterway accumulates in Galveston Bay after disposal. On this site, seeds of *Spartina patens* and *Spartina alterniflora* established naturally above the high tide zone (Webb et al., 1988).

In mangrove restoration, the number of floating propagules from nearby sources can be large enough to revegetate sites without human intervention (Buchanan, 1989). Propagules are often as successful as or more successful than transplanted saplings. Where large numbers of floating propagules are available, the use of natural revegetation can be effective and save huge amounts of money (Lewis, 1990).

## ACTIVE REVEGETATION TECHNOLOGY

Natural revegetation has limitations. Isolated sites may have few parent trees and thus too little seed available. Seed predation, herbivory, or competition levels may be very high at the site (Harmer and Kerr, 1995). The agent of natural seed dispersal may no longer be present (e.g., granivores, flood pulsing). The habitat may be fragmented to the extent that seed dispersal from a source site cannot be transferred to the restoration site. Also, the restoration site may be suitable for particular species, but safe sites are not available; for example, certain species may be present, preempting the invasion of other species. Because of these hindrances to natural revegetation, it may be necessary to reintroduce some or even most species at a restoration site (Primack, 1996).

Even so, a perceived lack of natural restoration potential indicates that further thought should be given to the restoration plan. Some thought should be given to the following questions:

1. Can the landscape linkages for hydrology and dispersal be reconnected at this site?
2. Can the natural hydrologic regime be restored, including seasonal and interannual water fluctuations?

If adjustments can be made to the plan to improve the natural restoration potential via dispersal and/or the extent seed bank, then these should be incorporated into the plan. This will help ensure the long term self-sufficiency and resilience of restoration sites.

While active revegetation is often necessary, these practices need to be used judiciously to avoid turning wetland restoration into a gardening exercise. Cypress trees grow nicely on lawns in the southeastern United States, but most lawns are not wetlands.

## Site Preparation

Because final water depths are dependent on elevation, restoration sites are often adjusted by grading. The site should be carefully surveyed and mapped before, during, and after construction. To dry the area sufficiently for machinery, a temporary berm can be constructed, the water piped around the area, or construction postponed until the dry season. It is sometimes necessary to put machinery on pontoons or floats (Miller, 1987).

In restoration projects where a wetland seed bank is on site before construction, after the site is excavated the topsoil is reapplied to the depression. This approach was used along with replanting in a mitigation site restored by the Illinois Department of Transportation along Highway 13, Saline County, Illinois (Nancy Rorick, personal communication).

Nurse species can be planted on sites to prevent the invasion of undesirable species. These provide shade, promote organic matter, and discourage highly competitive grasses. In bottomland hardwood forests, cottonwood or volunteer willow are sometimes used this way (Clewell and Lea, 1990).

## Direct Seeding

The success of revegetation by direct seeding can either yield a very natural-looking restoration site (Buchanan, 1989) or be a complete failure (Buchanan, 1989; Harmer and Kerr, 1995). Tree species can be difficult to revegetate from seed because typically these germinate only after some time, giving granivores opportunities to destroy seeds (Evans, 1988). Also, the weather may become too dry for germination or a storm may wash away the seeds (Buchanan, 1989). Seed germination is at best unpredictable; viable seeds may not germinate under field conditions (Gosling, 1987), so it is important to be aware of the specific germination conditions of the species involved (see Chapter 4). Furthermore, after germination, trees may grow slowly compared to other species and therefore may not be able to compete initially with annual and perennial species (Davies, 1987)

Seeds of species from nearby sources can be collected by hand or with mechanical seed collectors (Mahler, 1988). Seeds should be collected just as they mature but before they shatter (Knutson and Woodhouse, 1982; Woodhouse and Knutson, 1982). For salt marshes in North America, the time of collection varies from late September in the north to as late as November on the south Atlantic coast (Woodhouse and Knutson, 1982). Seeds heads can be clipped from plants with scissors, clippers, or knives (Broome, 1990). Alternatively, a garden tractor with a cutting blade and attachment bag can be used to collect seed (Broome et al., 1988). Salt marsh seeds should be stored in sea water to inhibit germination (Broome, 1990).

Collected seeds can be either propagated to produce plants or spread on sites (Broome, 1990). Alternatively, seeds and/or plantings can be purchased from specialty nurseries, lists of which are provided in various books (Thompson, 1992; Harker et al., 1993; Galatowitsch and van der Valk, 1994) and on the Internet. Surveys of local flora and community types abound worldwide, and all of these should be consulted for planning the revegetation of sites. Lists of species that occur in various community types exist by region for the United States (Harker et al., 1993). While there is no consensus about what constitutes a local ecotype, it is logical that plants collected from a local source would have a better chance of success at a restoration site than those collected at some distance from the site (Marburger, 1992).

After collection, close attention must be paid to the requirements of specific seeds for germination. In temperate regions, most seeds germinate better after temperature **stratification** (periods of cold from 0° to 5°C varying in length from approximately one day to five months; Marburgher, 1992). Seeds can be stored at ambient temperatures or in a refrigerator for lengths of time varying according to species.

The initial seeding of aggressive native species can be followed in later years by slower-growing ones. For prairie **potholes** in the southern part of the region, the initial mix of dominants seeded by mechanical planters (Stage 1) should include late-flowering grasses such as cord grass (*Spartina pectinata*) along with aggressive prairie forbs such as coreopsis (*Coreopsis* spp.) and compass plants (*Silphium* spp.; Table 5-1; Galatowitsch and van der Valk, 1994; Admiraal et al. 1997).

**TABLE 5-1. Species for Seeding/Plugging at Various Stages of Revegetation in Prairie Potholes of the Southern Region**

| Plants Often Recolonizing without Planting | Stage 1 Plants | Stage 2 Plants | Stage 3 Plants | Weedy Plants to Be Avoided |
|---|---|---|---|---|
| | | *Wet Prairie* | | |
| *Aster simplex* | *Andropogon gerardii* | *Anemone canadensis* | *Carex gravida* | *Agropyron repens* |
| *Ambrosia* spp. | *Calamagrostis canadensis* | *Asclepias incarnata* | *Carex stricta* | *Cirsium arvense* |
| *Bidens* spp. | *Desmodium canadense* | *Aster novae-angliae* | *Chelone glabra* | *Helianthus grosseserratus* |
| *Elymus canadensis* | *Elymus canadenssis* | *Aster puniceus* | *Cicuta maculata* | *Lythrum salicaria* |
| *Erigeron* spp. | *Epilobium coloratum* | *Eupatorium perfoliatum* | *Gentiana andrewsii* | *Melilotus alba* |
| *Verbena hastata* | *Helenium autumnale* | *Liatrus pycnostachya* | *Gentiana puberulenta* | *Phalaris arundinacea* |
| | *Panicum virgatum* | *Phlox pilosa* | *Lilium michiganense* | *Phragmites australis* |
| | *Ratibida pinnata* | *Pycnanthemum virginiana* | *Lythrum alatum* | western strains of prairie grasses (e.g. Blackwell switchgrass) |
| | *Silphium perfoliatum* | *Stachys palustris* | *Pedicularis lanceolata* | |
| | *Silphium lanciniatum* | *Teucrium candense* | *Phlox glaberrima* | |
| | *Spartina pectinata* (plugs) | *Veronicastrum virginicum* | *Thelypteris palustris* | |
| | | *Zizia aurea* | *Tradescantia ohiensis* | |
| | | *Sedge Meadow* | | |
| *Bidens* spp. | *Aster* spp. | *Asclepias incarnata* | *Carex lacustris* | *Cirsium arvense* |
| *Carex vulpinoidea* | *Eupatorium perfoliatum* | *Aster novae-angliae* | *Carex languginosa* | *Helianthus grosseserratus* |
| *Cyperus* spp. | *Eupatorium maculatum* | *Aster puniceus* | *Carex stricta* | *Lythrum salicaria* |
| *Juncus dudleyi* | *Glyceria striata* | *Calamagrostis canadensis* | *Gentiana andrewsii* | *Phalaris arundinacea* |
| *Juncus torreyi* | *Mimulus ringens* | *Chelone glabra* | *Gentianopsis crinita* | *Phragmites australis* |
| *Leersia oryzoides* | *Stachys palustris* | *Lycopus* spp. | *Pedicularis lanceolata* | *Typha angustifolia* |
| *Rumex altisimus* | *Verbena hastata* | *Lysimachia* spp. | | *Typha latifolia* |
| | | *Scutellaria* spp. | | |
| | | *Shallow Emergent* | | |
| *Eleocharis erythropoda* | *Alisma* spp. | *Acorus calamus* | *Carex atherodes* | *Lythrum salicaria* |
| *Eleocharis obtusa* | *Carex comosa* | *Iris virginica* | *Carex lacustris* | *Phalaris arundinacea* |

| | | | | |
|---|---|---|---|---|
| Polygonum amphibium<br>Polygonum hydropiper<br>Polygonum pensylvanicum<br>Rumex orbiculatus<br>*Rumex verticillatus*<br>*Scirpus atrovirens*<br>*Scirpus fluviatilus*<br>*Typha latifolia* | Carex lupulina<br>Carex lupuliformis<br>Carex squarrosa<br>Carex spp.<br>*Eleocharis palustris*<br>*Eleocharis* spp.<br>*Sagittaria* sp.<br>*Sparganium eurycarpum* | Lysimachia thrysiflora<br>Sium suave | | Phragmites australis |
| | | *Deep Emergent* | | |
| *Scirpus acutus*<br>*Scirpus tabernaemontanii*<br>(S. *validus)* | None | None | *Ludwigia palustris* | *Phragmites australis*<br>*Typha angustifolia* |
| | | *Submerged Aquatic* | | |
| *Ceratophyllum* spp.<br>*Najas* spp.<br>*Potamogeton foliosus*<br>*Potamogeton nodosus*<br>*Potamogeton pectinatus*<br>*Potamogeton zosteriformis*<br>*Utricularia vulgaris* | None | *Potamogeton* spp.<br>*Ranunculus flabbelaris*<br>*Ranunculus* spp. | *Vallisneria americana* | *Myriophyllum spp.*<br>*Potamogeton crispus* |
| | | *Floating Aquatic* | | |
| *Lemna minor*<br>*Lemna triscula*<br>*Spirodela* spp.<br>*Wolffia* spp.<br>*Wolffiella* spp. | None | None | *Nuphar microphylla*<br>*Nuphar advena*<br>*Nymphaea tuberosa* | None |

*Note:* During Stage 1, plants (unless noted) can be mechanically seeded; in Stage 2, plants are hand seeded; in Stage 3, plants are transplanted as seedlings or plugs.

*Source:* From Admiraal et al. (1997) as modified from Galatowitsch and van der Valk (1994) as adapted from Schramm (1992) and Betz (1986).

In Stage 2, seeds can be sewn by hand two to three years after the restoration site is initiated, after the weedy species have begun to decline (Galatowitsch and van der Valk, 1994). To hand sew seeds, it is recommended that seeds be broadcast while following sets of parallel transects (Harker et al. 1993). Species of Stage 3 (e.g., sedges, orchids, gentians) must be planted as plugs at the site because these are slower growing and more sensitive to competition than those of Stages 1 and 2 (Galatowitsch and van der Valk, 1994).

In prairie potholes, seed planting is best done in the spring because fall-planted seeds are more likely to be consumed by rodents. The site should be mown in the late fall of the year preceding planting to facilitate the elimination of weeds. In the spring when the soil dries, several passes with a disk will break up existing roots. Light harrowing until planting time will reduce weeds. In mid-June the site is lightly cultivated, followed by cultipacking to firm the soil. Seeds are then broadcast (approximately 22 kilos $ha^{-1}$ or 20 pounds $acre^{-1}$ using an all-terrain spreader or fertilizer applicator followed by cultipacking. Alternatively, seeds can be spread with hydromulchers that spray seeds onto prepared soil. Seeds drills can be used, though these leave obvious "rows" that are visible for years after planting (Galatowitsch and van der Valk, 1994).

In salt marshes, direct seeding is possible only in the upper 20–30% (Woodhouse and Knutson, 1982) or sometimes in the upper 50% of the tidal range. Establishment must take place during a storm-free period so that seeding is most successful in sheltered sites (Broome et al., 1988).

In bottomland hardwood forests, large-seeded species such as oaks and pecans can be planted with a specially outfitted agricultural planter (e.g., John Deere maxi-merge planter). Lighter-seeded species can be disseminated with spreaders or from airplanes. Large-seeded acorns are planted at the rate of approximately 11 kilos $ha^{-1}$, smaller ones at about half that rate. One person with a tractor-drawn planter can plant approximately 160 ha $day^{-1}$ and an airplane or commercial agricultural spreader (terra-gator) can plant 20 ha $day^{-1}$. On the same day, several tractor operators with disks or other covering devices are needed to cover the acorns (Wilkins, 1994). In riparian areas in the western United States, seeds are generally spread by hydroseeding at a rate of 202 kilos $ha^{-1}$ (180 pounds $acre^{-1}$ Miller, 1987.)

Seeds have certain advantages over saplings in bottomland forest restoration. Direct seeding of oaks is three times less expensive than planting saplings. Two people (tractor driver and planter) working a planter operation to plant saplings can plant only 6 ha in an eight-hour day. Saplings must be planted by hand during the flooded season (January through March in the Lower Mississippi Valley), and they may die if conditions become either too dry or flooded. Saplings also require much more storage space.

Sites replanted with saplings revegetate rapidly but may not differ in the long run from seeded ones (Wilkins, 1994).

Controlled time-release fertilizers are usually applied at planting time (Miller, 1987). Coastal salt marshes, particularly in sediment-rich estuaries, may not require fertilization except perhaps on newly deposited sand (Woodhouse and Knutson, 1982). Fertilization effects were mixed in one sea grass experiment, but no negative effects were observed. Because it is relatively inexpensive, fertilizer (phosphorus and nitrogen) is recommended in sea grass restoration (Fonseca et al., 1994).

Organic matter is generally lower in restored or created wetlands compared to natural wetlands. In Pennsylvania, natural wetlands had 11.1–19.8% more organic matter than created wetlands, and they were typically lower in the percentage of total nitrogen and higher in bulk density (Fig. 5-1). However, the organic matter at a depth of 5 cm depth did not tend to increase over time (Fig. 5-2; Bishel-Machung et al., 1996). The root infection rate of VAM mycorrhizae was higher in the roots of *Typha angustifolia* and *Phragmites australis* in created than in natural palustrine wetlands in Connecticut, most likely because organic matter and fertilizer levels were lower in the created wetlands (Fig. 5-3; Confer and Niering, 1992). Similarly, in coal ponds created as wetlands (Fig. 5-4), organic matter levels were much lower than in natural wetlands. In such situations, fertilization may cause plants to grow larger and subsequently produce more plant bio-

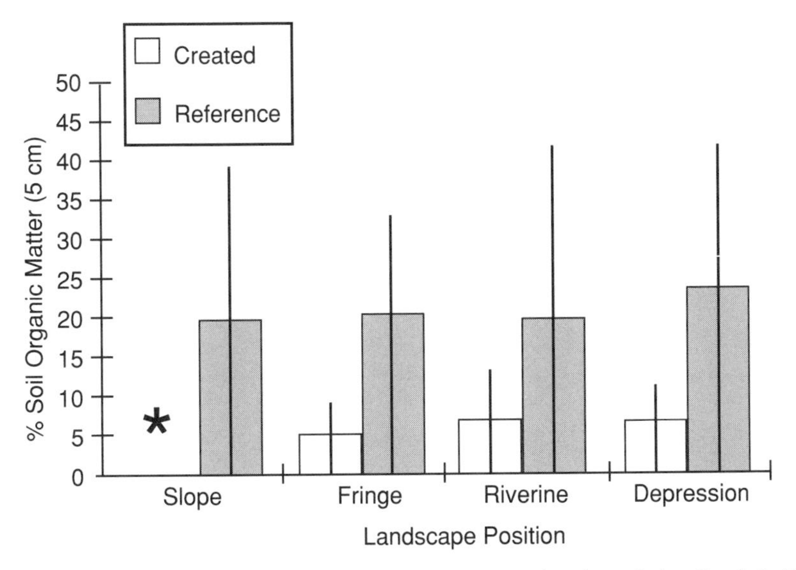


***Figure 5-1.*** *Organic matter in soil compared between natural and created wetlands in Pennsylvania. Missing data are denoted with an asterisk. Error bars indicate one standard deviation (from Bishel-Machung et al., 1996; copyright © by Wetlands by permission).*

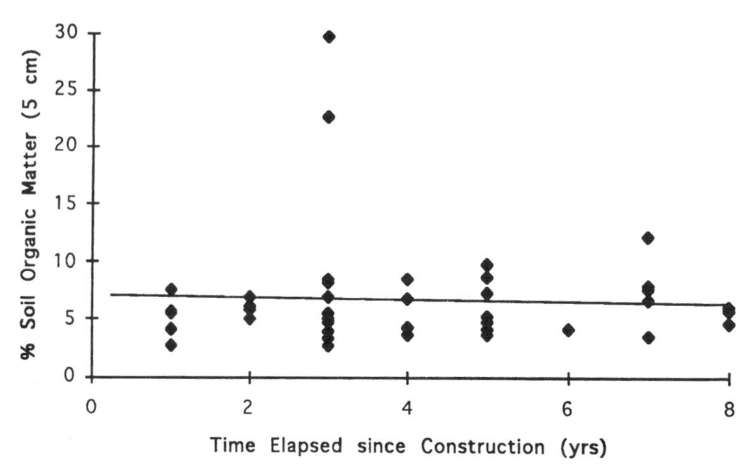


***Figure 5-2.*** *Organic matter in soil at 5 cm and time elapsed since construction in years in created wetlands in Pennsylvania (from Bishel-Machung et al., 1996; copyright © by Wetlands by permission).*

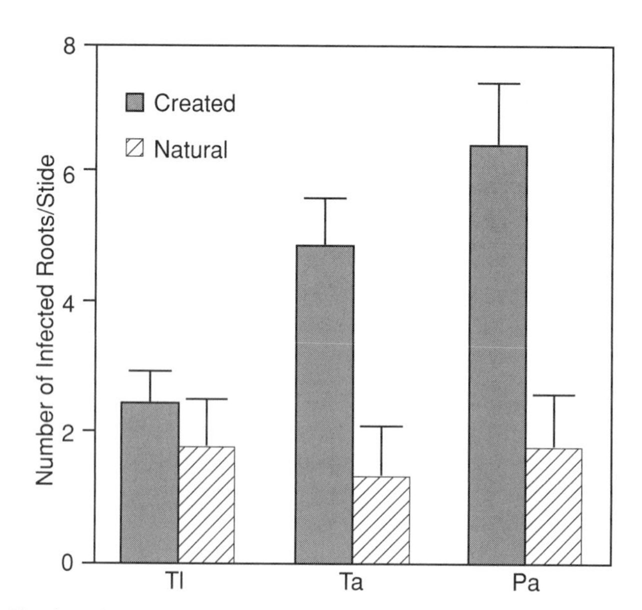


***Figure 5-3.*** *Number of roots infected with VAM mycorrhyzae in created versus natural wetlands in Connecticut. (from Confer and Niering, 1992; copyright © by Wetlands Ecology and Management by permission).*

***Figure 5-4.*** *Coal slurry pond reclaimed as a wetland, Burning Star #5, DeSoto, Illinois (photograph by Beth Middleton).*

mass to contribute to future organic matter levels in the soil. Fertilizer additions also increase seed production of the species involved (Middleton, 1995c).

## Donor Seed Banks

One approach to the revegetation of wetlands is to transfer soil from an existing wetland to a restoration site via a **donor seed bank** (Helliwell, 1989; van der Valk and Pederson, 1989). This approach is useful if a wetland is being destroyed near one that is being restored. Because the majority of the seeds and roots are generally in the upper few centimeters of the soil, the uppermost layer of the wetland soil is scraped from the surface with a front-end loader (Galatowitsch and van der Valk, 1994). This surface soil can be stored at the site while a lower layer is scraped from the donor site, transported to the transfer site, and spread. Following this procedure, the surface soil can be transported to the transfer *site* and then spread over the already transferred soil (Helliwell, 1989). The soil can be transported to the restoration site via dump truck, dumped in situ, and then smoothed with a small bulldozer and scraper. In many wetlands, such as moist grasslands in England (Helliwell, 1989) or sedge meadows in North America (Galatowitsch and van der Valk, 1994), the sod may be useful for restoration. It can be lifted from the donor wetlands, transported, and placed on the surface of the restoration site as if laying sod on a lawn (Helliwell, 1989; Galatowitsch and van der Valk, 1994). The soil, as well as any sod, should be laid out in appropriate zonal positions as based on elevation within the restoration site (Helliwell, 1989; Galatowitsch and van der Valk, 1994).

Alternatively, small amounts of soil from one or even several intact wetlands can be transferred to the restoration site over a period of years. Potentially, these inoculations could increase the number of species in the restored wetland (Galatowitsch and van der Valk, 1994).

The selected donor wetland should be free of troublesome species (e.g., *Lythrum salicaria* in North America). Because the soil sometimes contains seeds of undesirable species, a seed bank study should be conducted before the donor soil is transferred to the new site (Galatowitsch and van der Valk, 1994).

The spreading of "wild hay" collected after seed ripening may also facilitate prairie wetland revegetation and has been successful in prairie restoration (Galatowitsch and van der Valk, 1994). Because many species of sedge meadow set seed in midsummer, cutting and transfer of hay would likely increase species richness at that time of year (Galatowitsch and van der Valk, 1994).

Drift can be collected from swamps and other wetland systems with large amounts of floating seeds. The material tends to collect on curves in streams and at the receding water elevation. Drift can be collected and then spread on the restoration site. The advantage of this method is that it spreads not only seeds and spores, but also insect larvae and possibly other disseminules from the swamp.

## Direct Planting

The planting of propagated nursery plants or vegetative fragments is a common practice at restoration sites. Because transplanting (Fig. 5-5) is time-consuming and because transplants cannot be held for long after removal, it is helpful to construct an on-site nursery to hold the plants. Pits of water 30 to 60 cm deep can be dug, lined with plastic, and filled with 20 to 30 cm of water. This type of bed also can be used to propagate plants on site. Water can be provided through an irrigation system (Marburger, 1992). Small numbers of tubers can be kept moist in an ice chest (Marburger, 1993).

The specifics of transplanting plants in restoration projects are numerous and situation dependent. Plants of tidal wetlands are harder to dig in old marsh than in sites with looser soil, such as dredge disposal sites and those

***Figure 5-5.*** *Hand-planted emergents at an Illinois Department of Transportation mitigation site, Saline County, Illinois (photograph by Paul Hilchen, Illinois State Geological Survey, Carterville, Illinois).*

near inlets and along the edges of marshes. Dug plants should be kept in wet sand until transplanted in the field. Plants grown in peat pots from seed transfer better to the field than field-dug ones because their root systems are not disturbed. A tractor with a tobacco or vegetable planter can aid in rapid planting of the transplants. Hand planting is more often used in intertidal restoration. The plant should be firmly placed in the soil to prevent it from becoming washed out with the tide. Plants can be fertilized individually or the fertilizer can be broadcast. In North America, spring (April 1–June 15) is the ideal planting time for *Spartina alterniflora* (Broome, 1990).

In China, *Spartina anglica* is propagated using rhizomes and roots in the open or in paddies (Chung, 1982, 1989). If it is difficult to transport live materials, seeds are germinated either on moist filter paper or in a refrigerator and then the plants are grown in pots. Planting time affects survivorship, so more plants survive when planted in April and May (98% survival), followed by January and February (90% survival), September and October (80% survival), and July and August (30% survival). Planted sites accrete materials so that elevation increases as much as 80 cm in seven years (Chung, 1989).

Sea grass restoration has been attempted for various species in North

America, Japan, the the United Kingdom, France, subtropical/tropical North America, and Australia (Phillips, 1982). In sea grass restoration, because seeds of some species are almost never found, transplanting is the operative method of restoration (Holtz, 1986). After being gathered from beds, sea grasses can be held in aerated sea water (Eleuterius, 1976). Using a sea witch (scooping bucket with a cutting edge), sea grasses can be lifted from the donor site, cut into smaller pieces, bagged, and then transferred directly by boat to be hand planted at the restoration site (Holtz, 1986).

Sea grasses either can be planted directly in the substrate (nonanchored) or anchored in some way. Large intact mats, individual plants, or plugs of plants can be planted at the restoration site. Alternatively, individual shoots with leaves can be attached to pipes or rods with rubber bands, fixed to concrete rings, and tossed overboard or placed in the sediment with a plastic anchor (Phillips, 1982). Sea grass turions fastened to concrete anchors with rubber bands did not establish well in restoration plots in the Florida Keys (Lewis and Phillips, 1979).

Sea grasses are fragile and must be handled carefully (Fonseca, 1990). While *Thalassia testudinum* is slow-growing and does not recover readily from harvesting, the one-time collection of plants from beds of *Halodule wrightii* and *Syringodium filiforme* at 0.5- to 1.0-m intervals did not have a chronic impact on the beds (Fonseca et al., 1994).

A few studies have compared the results of different planting approaches. For the sea grass *H. wrightii,* the area revegetated by plants after 96 days was highest using a coring method (Fig. 5-6) and for *S. filiforme* using a staple method. In the coring method, a sod plugger extracted the plugs. In the staple method, a 2-mm-diameter metal rod with 20-cm arms anchored these plugs into the substrate (Fonseca et al., 1994). After 16 months, *Zostera marina* shoots woven into synthetic mesh and planted on 1-m centers were equivalent in density to a natural population (ca. 700 shoots $m^{-2}$). Fall transplanting was successful, but spring transplanting was not (Kenworthy et al., 1979).

Transplanted salt marsh species typically have higher survival rates at sites with lower salinity. In one study, some species survived at higher rates if transplanted in peat pots rather than as vegetative fragments—for example, *Distichilis spicata, Panicum repens, Panicum amarum*, and *S. patens* but not *Juncus roemerianus, Phragmites communis*, and *S. alterniflora* (Eleuterius, 1976). In another study, *J. roemerianus* transplantings were successful whether or not rhizomes had a bud (Coultas, 1979).

Mangroves from a wide variety of species have been successfully planted around the world (Lewis, 1982). Propagules can be collected, and should be kept moist and then transported quickly to the restoration sites. These are buried partially in the soil (Buchanan, 1989). Alternatively, plants can

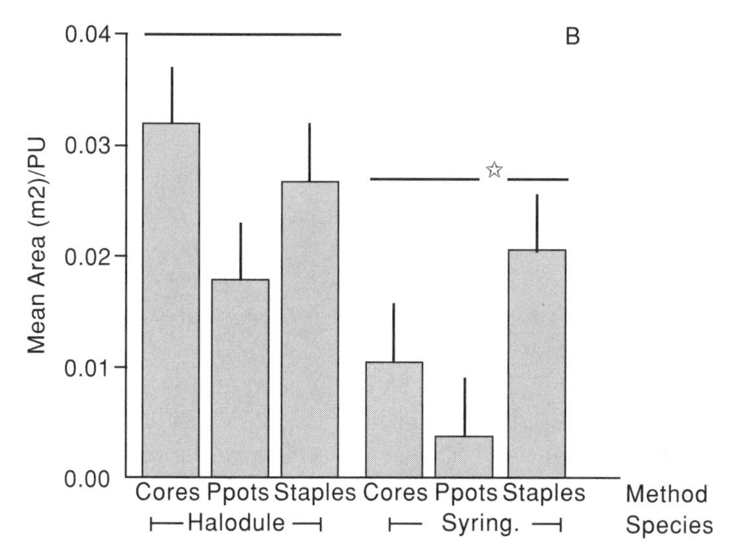


***Figure 5-6.*** *Mean area covered of seagrasses per planting after 96 days (from Fonseca et al., 1994; copyright © by Restoration Ecology by permission).*

be grown in the nursery, but they must be properly acclimated to salt water. The site may need to be protected from wave action for a time so that the propagules can become established (Buchanan, 1989). Both mangrove and salt marsh species are tied to particular tidal elevations somewhere between sea level and mean high water (Lewis, 1982), though mangrove species typically can exist at slightly lower elevations and thus seaward of salt marsh, where both communities exist together (Lewis, 1990).

Mangroves hit the ground running, so that the propagules shed by trees are actually seeds that have already germinated. In Australia, gray mangrove propagules are pushed 1–2 cm into the ground in sheltered locations where they are less likely to wash away. Seedlings can also be lifted with a 100-mm-diameter PVC pipe from under mature stands and transported to the restoration sites in plastic garbage cans. These should have no more than six to ten leaves and no pneumatophores. The seedlings can be slid out of the PVC pipe by pouring water into the pipe. Planting the seedlings in clumps spaced 20–25 cm apart is often successful because the inner plants are protected (Buchanan, 1989).

In bottomland hardwood forests, oaks and pecans are the most difficult species to establish. Saplings of these species are planted by hand with dibble sticks or mechanical planters. The root stocks of *T. distichum* saplings can be placed in socks and then dropped directly into the water for planting (Joy Marburgher, personal communication). Larger seedlings generally fare better than smaller ones (Williston et al., 1980). At least for oaks, taller seedlings survive flooding at higher rates than do shorter ones (Wilkins, 1994).

Most tropical species in northern India, particularly tropical grasses such as *Paspalum distichum*, can be rooted from vegetative sprouts from internodes or rhizome fragments. Plants can be cut into internodal fragments, and the pieces floated in water until sprouting occurs and then replanted (Middleton, personal observation). In Florida, plugs of leather fern and sawgrass transplant well (Lewis, 1990).

Bamboos rarely flower and set seed, so seed for propagation can be difficult to procure (McClure, 1967). Bamboo seeds also lose viability very quickly (Farrelly, 1984). McClure (1967) gives extensive advice and experimental results for propagating bamboo species worldwide. Bamboo rhizomes from the edges of patches are typically of an optimal age for propagation (McClure, 1967). In bamboos with a clumping growth form (typically tropical; Farrelly, 1984), rhizomes should be cut only at the neck (McClure, 1967). In species with a trailing growth form (typically temperate, warm climate; Farrelly, 1984), the rhizome need be cut at only one or two points along the runner, maintaining several internodes. Any foliage should be trimmed promptly to avoid water loss. The transplants should be kept moist. Whole culms are sometimes planted just under the surface of the ground (McClure, 1967).

Bamboo plantings should be set out just before the rainy season. In colder climates, however, spring is a better time of year to plant than fall. Winter transplants should be transplanted with a ball of soil around the roots; summer transplants should have wet straw around the roots and stem. Ideal conditions for planting are a cloudy day, followed by a rainy day (Farrelly, 1984).

In southeastern North America, *Arundinaria gigantea* is propagated through internodal sprouting, planting rhizomes with attached culms, and planting whole culms. A rooting hormone powder (Rootone, Green Light Co., San Antonio, Texas) can be applied at the base of each segment, planted in potting medium, placed outdoors in the shade, and watered daily (Platt and Brantley, 1993).

In prairie wetlands, sedge meadow species are almost always transplanted because these do not establish well from seed. Seeds can be planted in flats and then set out in the restoration site after two years. A natural wetland about to be destroyed is also an excellent source of transplant material. Plants should be dug with a sharp knife or spade, with as much root material as possible. Plant materials should be kept moist and transported rapidly to the restoration site. Depending on the rate of growth of the species, the plugs should be planted 0.5–1 m apart. Submerged aquatics should be planted underwater (approximately 0.5–1 m apart); tubers of species such as sago pondweed and wild celery can be planted (Galatowitsch and van der Valk, 1994). Two or three tubers of *Sagittaria latifolia* can be wrapped together with a rubber band and pushed into the soil (Marburger, 1993). Also, rhizomes, corms, and rootstocks of many species can be planted in cheesecloth bags and set in water about 5 to 13 cm deep (Warburton et al., 1985). When planting in water, it may be useful to carry plants on stretchers of 2 × 4 boards with wire mesh (Bowers, 1995).

Water levels need to be carefully monitored for the first year after planting. These can be raised slowly to keep shoots above the water level (Marburger, 1993). Species will sort themselves out to some extent at this point, depending on the water conditions in various portions of the restored wetland. However, deep flooding will eliminate almost all nonsubmerget species by preventing germination and the establishment of rhizomes and seedlings.

In tidal brackish and salt marsh wetlands, the site should be positioned at a high enough elevation so that the vegetated areas are well drained when the tide is out (Broome, 1990; Lewis, 1990; Garbisch, 1991). If the water level is too high, not only are these areas difficult to plant, but the plants may not survive because water salinity and temperatures will be too high (Lewis, 1990). Many tidal brackish and salt marsh species cannot tolerate flooded conditions during the dormant season, including *Pontederia cordata, Scirpus pungens, Scirpus tabernaemontani, S. alterniflora*, and *Typha latifolia* (McIninch and Garbish, 1991).

## REINCORPORATING FIRE

Fire is a natural part of the disturbance dynamics of many natural systems and has been prescribed in a wide range of restored wetland types around the world. Prairie potholes can be burned after one year and, following the guidelines for prairie restoration, revegetation may be encouraged by annual burning for five years. Early spring burns in March or April enhance growth by removing litter. For the following five years, half of the site can be burned, alternating with the other half on an annual basis to provide a refuge for insects and animals that are not fire tolerant. After the first ten years, burning should occur every three to four years (Galatowitsch and van der Valk, 1994).

For the first few years after restoration, it is sometimes best to protect young plants from fire. After that time, a suitable fire regime should be incorporated at the site. Heathland in Australia historically has burned as often as every 5 years or as infrequently as every 50 years, but neither of these extremes maintains species richness in the community. The specifics for maintaining particular sites are not well known, but fire frequency and intensity are certain to influence the site. Occasional clearing can be used instead of fire, but this method fails to stimulate seedling germination (Buchanan, 1989). Prescribed burns are also used in wetlands restoration sites near the Everglades (Joy Marburgher, personal communication).

## CONTROLLING EXOTICS

Natural sites that have been invaded by exotic species can sometimes be rejuvenated by encouraging the native species that are still on the site. The Bradley Method, used extensively in Australia, involves the gradual removal of weeds to allow native vegetation to revegetate. The first step is *primary weeding* of the site, which may be very time-consuming. To minimize disturbance, smaller tools such as clippers, trowels, and pliers (to pull roots) are used. *Consolidation* follows, with many visits to reweed the site. Good judgment should be exercised about the timing for removal of particular species; for example, annuals should be pulled before seed has matured. *Long-term maintenance* continues for a few months or even years, with infrequent visits to the site to pull scattered weeds. This method creates minimal disturbance and allows the native vegetation to recover the site through vegetative spreading and seed germination (Buchanan, 1989).

In some situations, it may be advisable to use herbicides, for example, areas of (1) dense weeds with no native plants, (2) rapidly proliferating weeds, or (3) erosion-prone sites with large trees that would be dangerous to cut or difficult to dig up by the root. It is possible to kill trees and shrubs by drilling a hole to the sapwood and filling it with herbicide. These holes should be angled down so that the herbicide stays in the tree. For the number and spacing of holes, refer to the manufacturer's recommendations (Buchanan, 1989).

## KEEPING OUT THE ANIMALS

Herbivores threaten the regeneration of species in newly growing forests by eating saplings and understory vegetation. In Britain, all of the five species of deer present browse in woodland habitats. They damage young trees by eating leaves and bark and by fraying the trunks with their antlers. Rabbits were likely introduced in Britain from the Mediterranean in the 11th century A.D. by the Normans. Young tree survivorship greatly decreases in response to increasing rabbit populations (Fig. 5-7; Gill et. al., 1995).

It is also important to keep the threat of herbivore damage to young

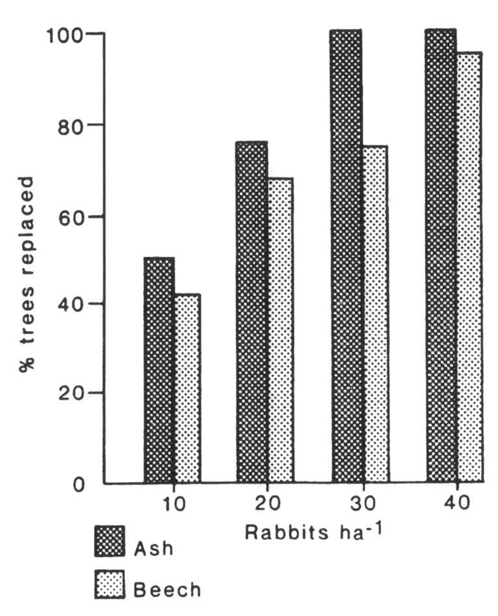


***Figure 5-7.*** *Increasing rabbit density and the effect on unprotected young ash and beech trees as a function of percent replaced. Rabbits were placed in densities of 10, 20, 30, and 40 individuals in hectare paddocks with 60 unprotected individuals of each species. The percentage of trees replaced are based on the mean of six replicate trials of trees damaged severely enough to be replaced after one year (from Gill et al., 1995; copyright © by Wiley, Chichester, by permission).*

plants in restoration areas in perspective. In a tidal restoration site on the Chesapeake Bay, muskrats and geese excavated the late season plantings of *S. alterniflora* during the winter following restoration. While damage appeared significant the following spring, by August of the next growing season, the plants had grown back. Geese prefer to graze the first-year growth of the restoration site and, thereafter, the leading edge of the marsh; therefore, if these areas can be protected, the geese are likely to make minimal use of the area (Garbisch et al., 1975). Once a sward is established, the geese are less likely to damage the area (Chung, 1982).

Disking after planting can reduce rodent activity. Acorns planted in the spring suffer fewer losses than those planted in the fall, when wildlife may eat them during the winter season while the seeds are dormant. Large fields typically suffer less seed loss to herbivores than small ones (Wilkins, 1994).

Low fences can be constructed to keep out muskrats (Fig. 5-8; Nancy Rorick, personal communication). A wire fence at least 2 feet below and 3 feet above ground is necessary (Garbisch, 1991). For submerged species, chicken wire fences can keep out fish and reduce wave action (Galatowitsch and van der Valk, 1994).

Beaver, as previously discussed, often scuttle water control structures by

***Figure 5-8.*** *Low fence to keep away muskrats until species establish at a newly planted restoration site of the Illinois Department of Transportation, Saline County, Illinois (photograph by Beth Middleton).*

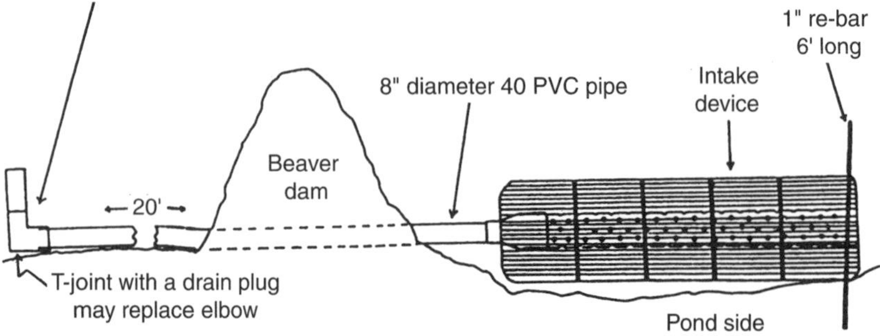


***Figure 5-9.*** *Clemson beaver pond leveler. The water is taken in at the deepest part of the pond through a perforated PVC pipe and drained to a point below the beaver dam or blocked culvert (from Dolbeer et al., 1994; copyright © by U.S. Department of Agriculture, Lincoln, Nebraska, by permission).*

using them for their own engineering projects. Culverts, drains, and other structures can be protected by building small fences around them. Beaver dams and the materials in the water control structure can be removed on a continuous basis (Dolbeer et al., 1994). I have personally attempted this approach many times and have always lost to the beaver. Shooting, trapping, or snaring the beaver, and dynamiting and/or bulldozing lodges, bank dens, and dams, sometimes discourages beaver (Dolbeer et al., 1994). Ro-Pel is used to repel beaver, as well as moose, elk, horses, rabbits, squirrels, bears, cattle, and rats (Garbisch, 1991).

Apparently, if the beaver pond can be drawn down, beaver can be controlled. A Clemson beaver pond leveler works by draining the beaver pond below the water level, so that the beaver cannot detect water flow. A perforated PVC pipe encased in hog wire is placed upstream of the beaver dam in the deepest part of the stream and run to a water control structure downstream (Fig. 5-9, Dolbeer et al., 1994).

Deer can be excluded from bottomland hardwood forests with complicated electric fences. One type has a wire about 2 feet off the ground with another traditional three-wire fence, 5 feet in height, within it (Marilyn Mathis and Warren Weaver, personal communication). Deer-Away repellent (IntAgra, Inc., Minneapolis, Minnesota) is also used (Garbisch, 1991).

Geese can be fenced from sites using low strings placed near the edge of the water about 0.3 m above the ground with additional strands hung 0.3 m apart. Apparently, geese view these as an obstruction to takeoff and landing (Bowers, 1995). Methiocarb repellent is also used to detract geese (Conover, 1985).

In sea grass planting, bioturbation, probably by rays, killed a large number of early season plantings in a study at Tampa Bay, Florida. Enclosure cages and later season planting (September through November) increased the survival of these sea grasses (Fonseca et al., 1996).

Until newly growing plants in restoration sites are established, it is important to keep out herbivores. In the early years of a restoration, heavy foraging by geese may necessitate replanting (Hammer, 1997). Because native herbivores are an important part of vegetation dynamics (see Chapter 2), there comes a time in the life of a restored site when native herbivores are a desirable and even a necessary part of the system.

# 6

# Case Histories

*One way to get a clear view of how wetland restoration is proceeding in the world today is to take a careful look at case histories. The solutions to restoration problems are usually complicated by biological, social, and political dilemmas and are never the same in two situations. This chapter presents the problems and solutions of researchers and practitioners engaged in wetland restoration around the globe. Four restoration projects are visited: the Kissimmee River, Florida, the United States; the Panshet Dam near Pune, India; the Murray-Darling River, Australia; and the Rhine River, Germany.*

## CASE STUDY 1—THE KISSIMMEE RIVER PROJECT, FLORIDA, THE UNITED STATES

Probably the most ambitious restoration project ever attempted, the Kissimmee River project is designed to reverse the environmental damage created by the channelization of the river between 1964 and 1971 (Glass, 1987; Koebel, 1995) and to recapture the original geomorphic processes (Toth, 1996). This project may reestablish flood pulsing and its associated hydrologic fluctuation to the system in a way that almost no other project has attempted. Early in the process of planning, it became clear that unless an attempt was made to restore some degree of the original hydrologic processes associated with riverine water movement and pulsing on the floodplain, little progress could be made in the restoration of the site. The high and low flow patterns of the river after alteration were out of phase with the original patterns, which contributed to limited floodplain inundation and fish reproduction (Toth et al., 1993).

While the channelization of the Kissimmee River did reduce flood damage in the region, the negative results of the project were immediate. Stable water levels replaced the naturally pulsing system (Loftin et al., 1990). Dry plains replaced marshes, and wildlife and water quality declined throughout the Kissimmee River floodplain, Lake Okeechobee, and the Everglades (Glass, 1987). The restoration project will return a huge amount of the original floodplain to wetland, notably by dechannelizing portions of the existing streambed and backfilling the canal (Koebel, 1995).

The Kissimmee River (Fig. 6-1) was once an extensive system with flood pulsing, similar to the large river-floodplain systems of the Amazon of South America and the Niger of Africa (Koebel, 1995). The original channel was somewhat braided and meandering, with point bars, backwater ponds, and sloughs. The floodplain consisted of 18,000 ha of wetlands (Toth, 1996), in a 1.5-3 km-wide corridor with a 166-km channel, south of the City of Orlando and draining into Lake Okeechobee in central Florida (Koebel, 1995). Lake Okeechobee originally overflowed to the south into the Everglades (Glass, 1987) but only for a few months of the year (Kushlan, 1991) in a shallow 65-to 97-km-wide sheet (National Research Council, 1992).

The hydrology of the Kissimmee River was characterized by continuously flowing water, with the highest discharges from September through November and seasonal water-level fluctuation typical of the subtropics (Toth, 1993). Prior to channelization, 94% of the floodplain was inundated 50% of the time, with water depths of 0.3–0.7 m and depths of over 1 m across 40% of the floodplain at least one-third of the time. In a 25-year record, a great deal of annual variation in flood duration occurred, so that 80% of the floodplain was wet for only 11 of these years. The entire floodplain was continuously inundated for two years during three intervals. Extensive drying was uncommon, though 84% of the floodplain was dry for five months during 1932, 1935, and 1955–1956 (Toth, 1990). River floodplains of the southeastern United States typically dry in late summer and autumn, with a high flood pulse inundating the floodplain in December–May (Smock and Gilinsky, 1992).

All of this changed after channelization was completed in 1971. Few people lived in the Kissimmee basin before 1940, but the population increased after World War II. The public pressed for flood control after a severe hurricane in 1947 and high water from 1947 to 1949, which prompted the channelization of the river in central Florida (Loftin, 1990;

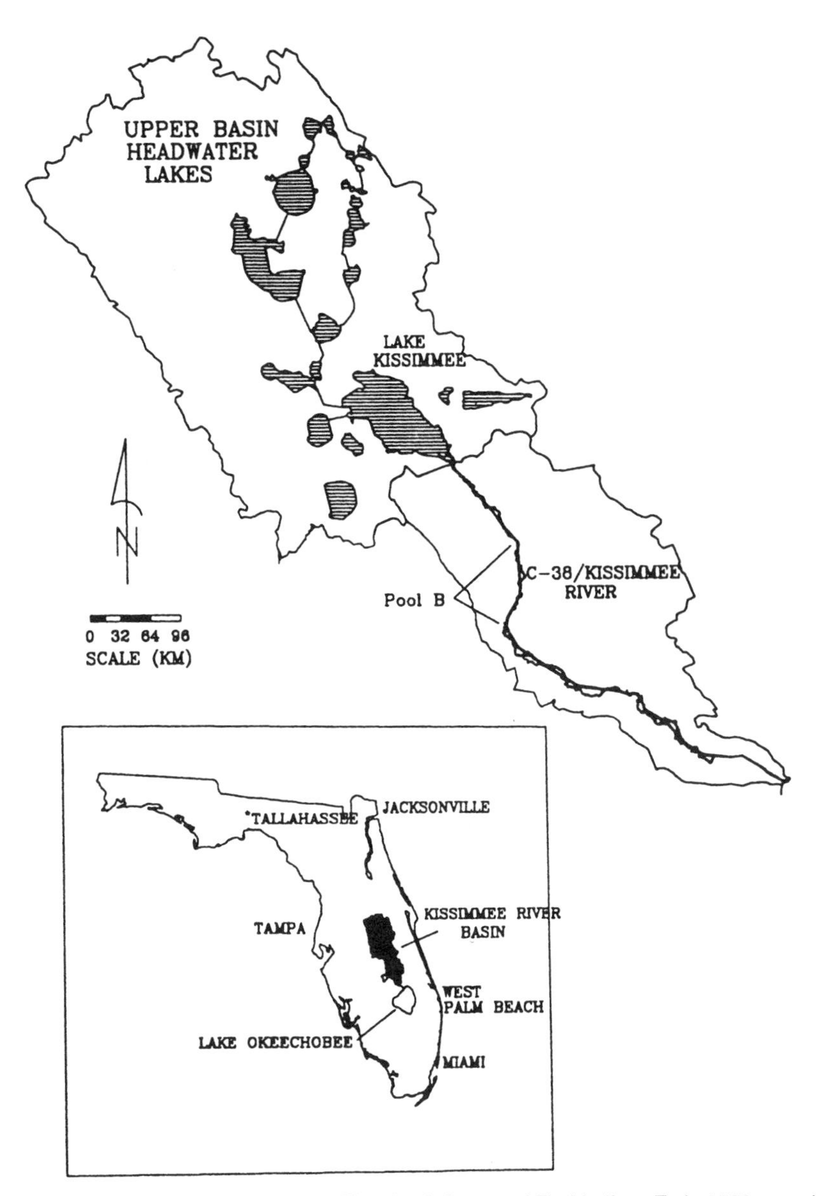


***Figure 6-1.*** *Location of the Kissimmee River basin in central Florida (from Toth, 1996; copyright © by John Wiley & Sons, Chichester, U.K., by permission).*

U.S. Army Corps of Engineers, 1992). The U.S. Congress authorized the construction of the Central and Southern Florida Project in 1948 and the Kissimmee River portion of this project in 1954.

Project planning occurred between 1954 and 1960 (Koebel, 1995) and construction between 1962 and 1971. The channel was straightened and deepened into a canal (C-38), and impounded reservoirs were created (Pools A–E; Fig. 6-2; Fig 6-3; Koebel, 1995; Toth, 1996). Six water control structures controlled the inflow of the water from the upper basin (S-65S; Koebel, 1995). Dams created stable water levels in stagnant reservoirs (Toth, 1996), and almost immediately, concerns were raised about the loss of fluctuating water levels (Table 6-1). Canals and other structures regulated water flow between the lakes of the upper basin (Koebel, 1995). In 1926, a levee was built to prevent Lake Okeechobee from draining into the Everglades (Glass, 1987).

Because the biotic characteristics of the Kissimmee system were visibly degraded by this project, a grass-roots movement emerged even while construction activities continued (Loftin, 1990; Koebel, 1995). In 1971, the U.S. Geological Survey published a report outlining environmental problems created by the flood control project (Koebel, 1995). In 1977, the Flor-

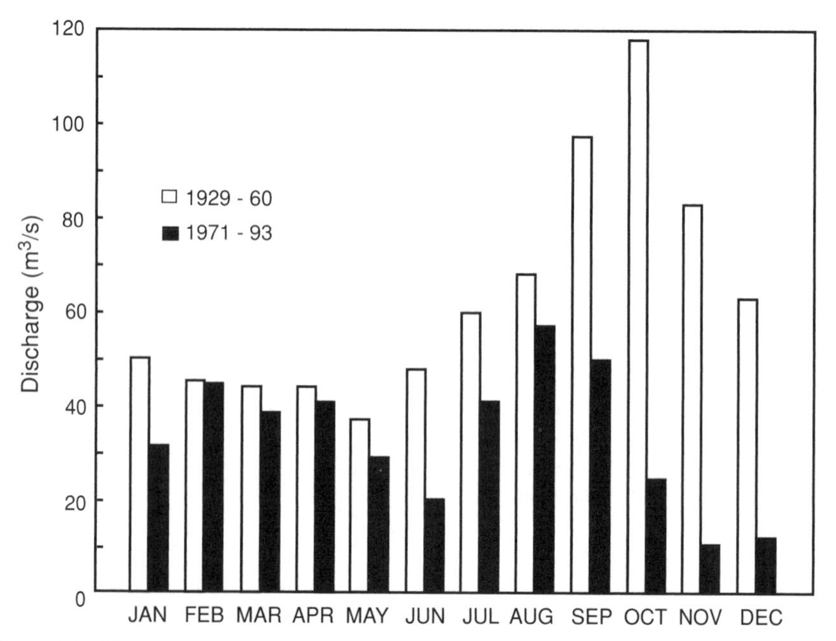


***Figure 6-2.*** *Mean monthly discharge for the Kissimmee River–C-38 system prior to 1929–1960 and following channelization in 1971–1993. Discharge was measured at a point near the current location of S-65E (see Fig 6-3) (from Koebel, 1995; copyright © by Restoration Ecology by permission).*

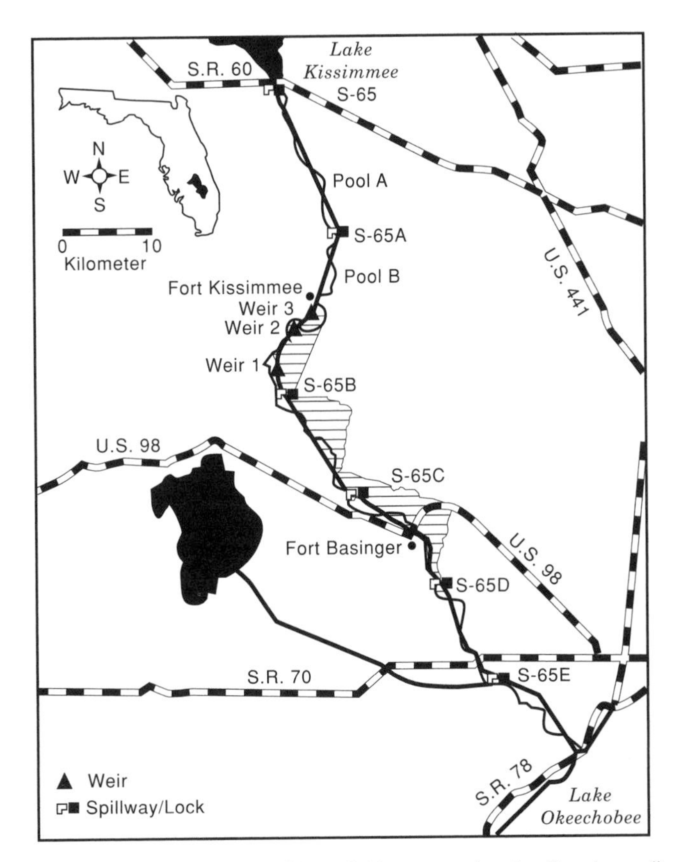


***Figure 6-3.*** *Location of the Kissimmee River–C-38 system showing the channelized route (C-38) superimposed upon the meandering river channel. The system is divided into a series of five pools (A–E) separated by water control structures (S–65s). Weirs 1–3, located in Pool B, were installed as part of the 1984 Kissimmee River Demonstration Project. The highlighted areas indicate the phased approach to restoration construction beginning in 1998 (from Koebel, 1995; copyright © by Restoration Ecology by permission).*

ida legislature created the Kissimmee River Coordinating Council by passing the Kissimmee River Restoration Act (Loftin, 1990; Koebel, 1995). The council focused on ways to restore the water level fluctuation characteristic of the natural hydrology of the upper basin lakes, reestablish the floodplain habitat, and create conditions favorable to native wetland biota. This effort led to restoration and planning projects by the Corps of Engineers and the South Florida Water Management District (Koebel, 1995).

The U.S. Army Corps of Engineers presented alternative restoration plans in their first federal feasibility study (1978–1985; Table 6-2). The public supported dechannelizing the C-38 by backfilling, and this was endorsed by the Coordinating Council in 1983. Other alternatives retained

**TABLE 6-1. Environmental Concerns and Objectives for Restoration of the Kissimmee River Set Forth in the First Federal Feasibility Study, 1978–1985**

| *Environmental Concerns* |
|---|
| Loss of naturally fluctuating water levels |
| Loss of large areas of wetlands |
| Deterioration of water quality in Lake Okeechobee and the Kissimmee Basin |
| Changes in land use resulting in increased drainage |
| Loss of natural river meanders |
| Low groundwater levels and reduced groundwater quality |
| Potential need for increased flood protection |
| Potential reduction in flood protection |
| Potential increase in mosquito populations |
| Reduced recreational navigational opportunities |
| *Feasibility Study Objectives for Restoration* |
| Restore wetland areas |
| Improve water quality |
| Restore river meanders and oxbows |
| Improve groundwater recharge |
| Maintain flood protection |
| Restore fluctuating water levels |
| Provide surface water supply |
| Maintain navigation |
| Meet recreational demands |

*Source:* Koebel (1995) from U.S. Army Corps of Engineers (1992).

through the planning stages include the creation of flow-through marshes, pool stage manipulations, and impounded wetlands, as well as the improvement of water quality through the best management practices. The plan also includes the restoration of the Paradise Run wetlands (Koebel, 1995).

## The Restoration Plan

Between 1984 and 1989, a demonstration project was set up in a 19.5-km section of the C-38 canal (Pool B). Flow increased in the original river channel with the use of three notched weirs to divert water from the canal (Toth, 1991; Toth et al., 1993). The project re-created prechannelization fluvial processes in a 9-km section of the remnant river channel, though the flows were not as high or regular as before channelization. Organic deposits were either covered with sand or flushed from the system (Toth et al., 1993). High-energy, erosive water movement emanates from the weirs, so these have some shortcomings in restoration (Toth, 1996).

Weirs by themselves could not reintegrate the floodplain with the channel; modeling suggested backfilling of the canal and dechannelization as

**TABLE 6-2. Alternative Restoration Plans for Restoring the Ecological Integrity of the Kissimmee River**

| Alternative Restoration Plan | Action |
|---|---|
| No action | Operate and maintain existing flood control and navigation systems. |
| Modify lake regulation schedule | Increase flood storage in the upper basin by modifying lake regulation schedules. |
| Additional lake control structure | Install structure above Lake Kissimmee to regulate Lakes Cypress, Hatchineh, and Kissimmee at different levels. |
| Complete backfilling | Fill entire length of C-38 and remove structures. |
| Partial backfilling[a] | Fill middle section of C-38, remove appropriate structures and install flow-through elements in Pools A and B. |
| Plugging | Place earthen plugs at points along C-38 to divert flow to portions of remnant river channel. |
| Flow-through marshes[a] | Construct wetlands adjacent to C-38 and below S-65A, S-65B, S-65C, and S-65D. |
| Pool stage manipulations[a] | Modify S-65A, S-65B, S-65C, S65-D, and S-65E to accommodate higher stages to increase wetlands. |
| Impounded wetlands[a] | Create wetlands through a series of separate elements, including flow-through marshes, tributary impoundments, and pool stage manipulations. |
| Enhance existing system | Remove or reshape excavated material along C-38. |
| Paradise Run[a] | Restore Paradise Run wetlands. |
| Best management practices[a] | Use best management practices to improve water quality and restore wetlands. |
| Minimum maintenance | Restore prechannelization conditions through lack of maintenance. |
| Dual watercourse | Create a riverine system adjacent to C-38. |

[a] Plans that were advanced for further consideration.

*Source:* Koebel (1995) fron U.S. Army Corps of Engineers (1992).

solutions to the problem (Loftin et al., 1990; Toth et al., 1995). A major goal of the plan is to reestablish fluctuating water levels (U.S. Department of the Interior, 1996). The initial restoration attempts did not restore hydroperiods on the floodplain comparable to those that existed before the drainage activity because the canal quickly drained the floodplain. A spoil pile from the digging of the canal impeded the movement of water across the floodplain. When water levels were low in the canal (discharge less than 28 $m^3$ per second), the weirs were not adequate to divert water into the remnant channel (National Research Council, 1992).

Thus, in 1994, the Pilot Test-Fill Project began by filling a 300-m section of Pool B. The backfilled channel diverts river water to the old stream course. Because the spoil bank separating the canal and the floodplain was removed, a 5.2-ha portion of the former floodplain became reintegrated with the river channel (Toth, 1996).

Vegetation and wildlife quickly recolonized the flooded areas in this demonstration project. Wetland species quickly regenerated naturally in

**TABLE 6-3. Pre- and Postchannelization Wetland Plant Community Coverage on the Kissimmee River Floodplain**

| Vegetation Community Type | Prechannelization (ha) | Postchannelization (ha) | Percentage Change |
|---|---|---|---|
| Broadleaf marsh | 8892 | 1238 | −86 |
| Wet prairie | 4126 | 2128 | −48 |
| Wetland shrub | | | |
| Willow | 733 | 693 | −6 |
| Buttonbush | 1335 | 310 | −77 |
| Wetland forest | 150 | 243 | +62 |
| Switchgrass | 252 | 193 | −23 |
| Other wetlands | 281 | 726 | +158 |
| Total | 15769 | 5532 | −65 |

*Source:* Toth et al. (1995); prechannelization from Pierce et al. (1982); postchannelization from Mileson et al. (1980).

sites with inundation periods of one to two years (Toth, 1991). Flow created a 20-m-wide, 3-m-deep channel in the borrow area, which provided habitat diversity for fish and wildlife (Toth, 1996).

## Pre- and Postchannelization Biotic Structure

After the channelization of the Kissimmee River, the biotic characteristics of the canal and remnant river channels resembled those of stagnant reservoirs, unlike the former riverine system. In these waterways, biological degradation is directly related to low dissolved oxygen levels due to low flow and high temperatures, which occur frequently in the summer (June and September; Toth et al., 1993). The channelization of the river damaged the biota in a way that was both immediate and striking.

The reengineering of the system cut the connection from the channel to the floodplain. This created conditions on the floodplain of low energy transfer to fish, invertebrates, and plants, as well as poor nutrient filtration and absorption. The cessation of flood pulsing also damaged the accessibility of channel fish to the floodplain for breeding, feeding, and nursery usage (Toth et al., 1993).

Channelization massively altered the vegetation of the area, reducing the overall wetland from 15,769 to 5,532 ha (Table 6-3; Fig. 6-4). The narrow edges of the canal were inhabited by *Hydrilla verticillata, Hydrocotyle umbellata, Nuphar luteum, Polygonum densiflorum, Sacciolepis striata*, and *Typha domingensis*. The deeper parts of the C-38 canal supported floating tropical exotics such as *Eichhornia crassipes* and *Pistia stratiotes* (Toth et al., 1995). About 44% of the floodplain was converted to pasture. Unimproved pasture was dominated by *Axonopus affinis* or *Ax-*

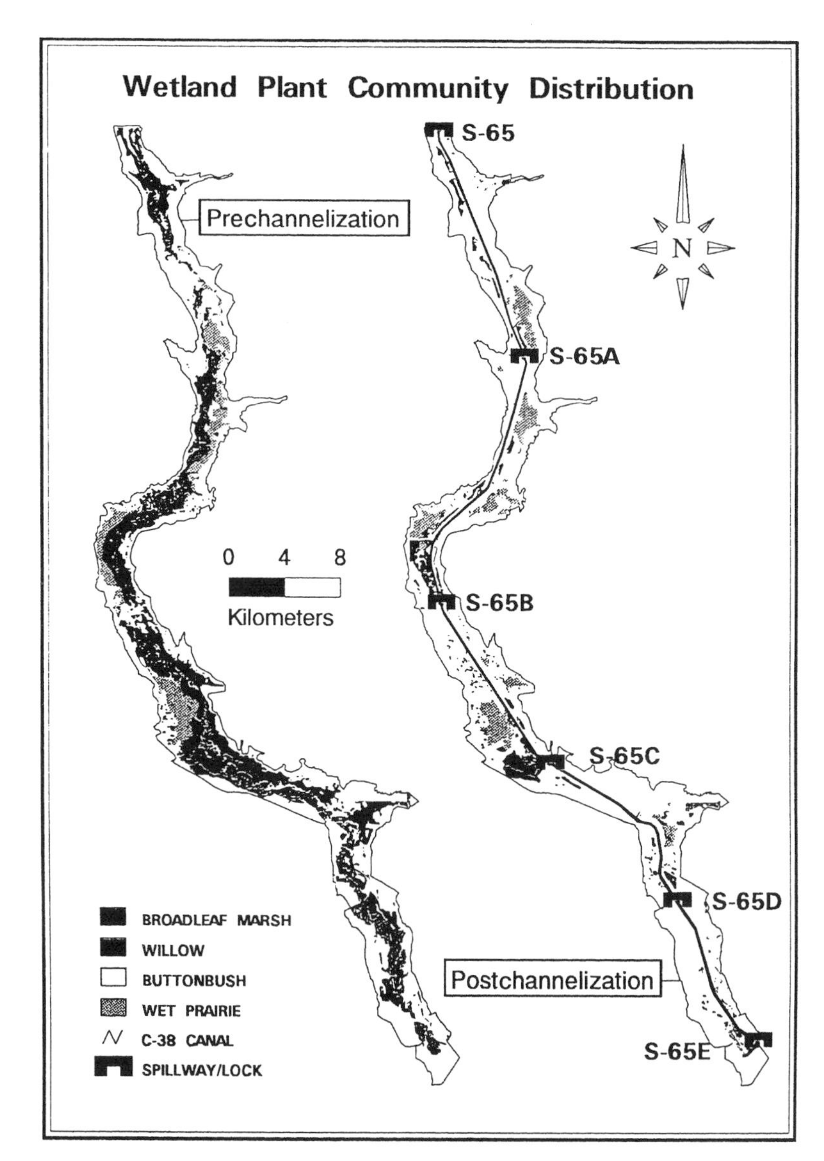


***Figure 6-4.*** *Pre- and postchannelization distributions of dominant wetland plant communities on the Kissimmee River floodplains (from Koebel, 1995; copyright © by Restoration Ecology by permission).*

*onopus compressus* and improved pasture by *Paspalum notatum* (Milleson et al., 1980).

The recovery of the habitat is a measure of the success of the Kissimmee River restoration project (Toth et al., 1995). After the hydrologic changes to the floodplain, mesophytic and xeric species including *Centella asiatica,*

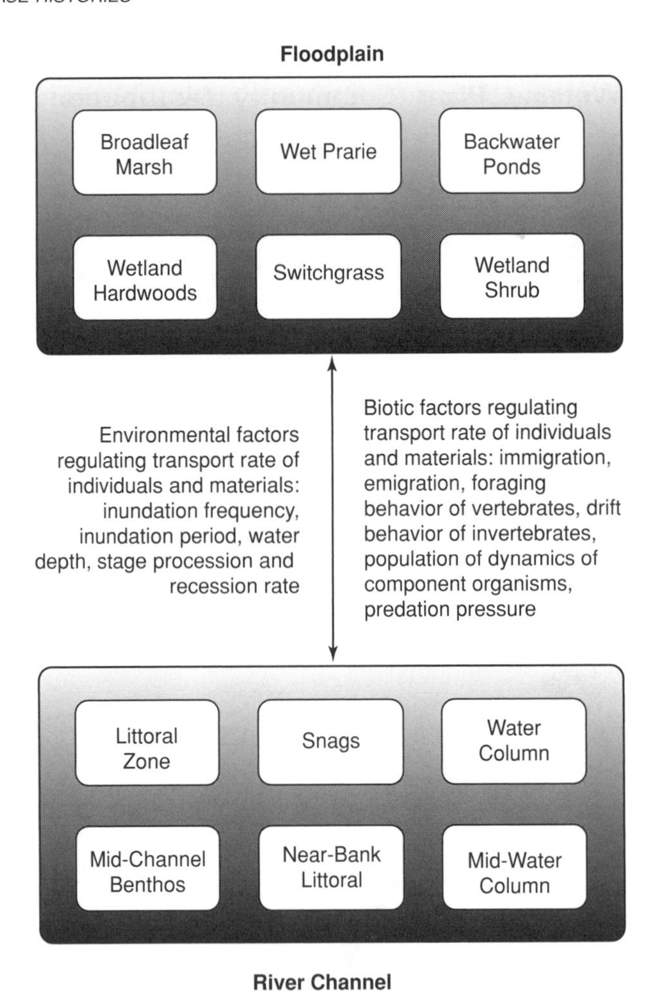


***Figure 6-5.*** *Factors influencing the transfer of aquatic invertebrates, as well as organic materials (both fine & coarse particulate organic matter), between the river channel and floodplain (from Harris et al., 1995; copyright © by Restoration Ecology by permission).*

*Paspalum conjugatum*, and *Sambucus canadensis* declined (Fig. 6-4) and aquatic species including *Panicum hemitomon, Polygonum punctatum*, and *Alternanthera philoxeroides* increased (Toth, 1990).

Flood pulsing on floodplains was critical to the recovery of trophic systems because it reinvigorated invertebrate populations. Invertebrates are linked to fish, wildlife, and waterfowl populations (Toth, 1993; Harris et al., 1995). The restoration of flow in remnant water channels allowed the colonization of insect taxa characteristic of rivers rather than lakes (Fig. 6-5). Thus, lentic species such as *Chaoborus* sp. (Diptera: Chaoboridae) were

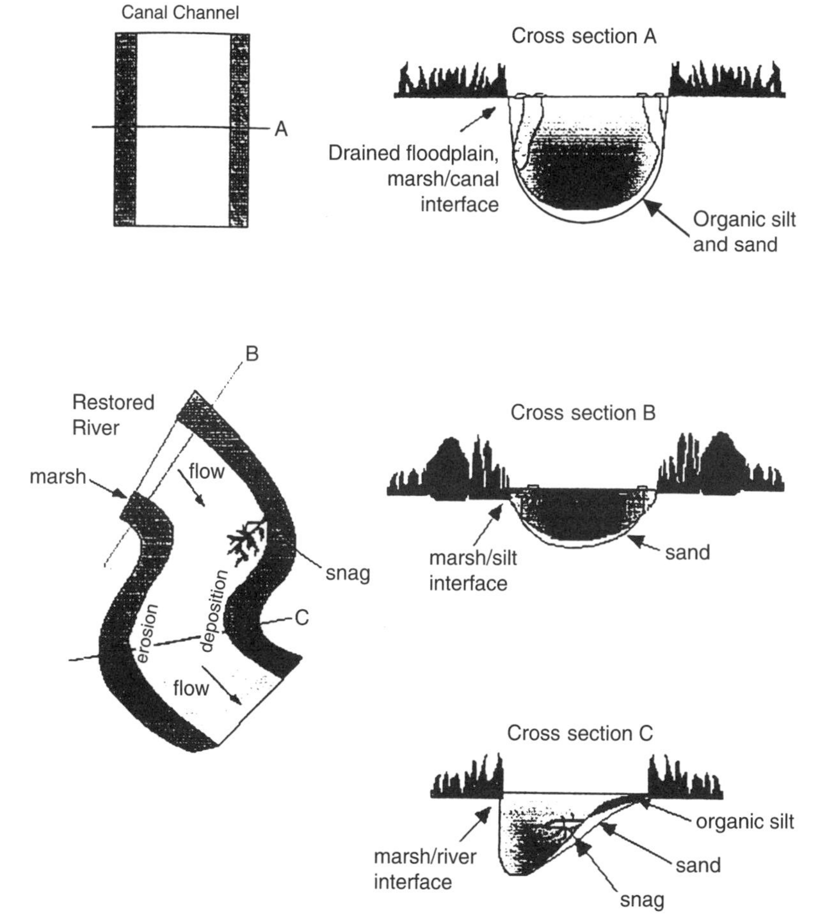


***Figure 6-6.*** *Restoration and anticipated invertebrate habitat changes. Much of the floodplain will be transformed from pasture to aquatic marsh habitat (from Harris et al., 1995; copyright © by Restoration Ecology by permission).*

replaced by *Stenacron* sp. (Ephemeroptera: Heptageniidae), *Cheumatopsyche* (Trichoptera: Hydropsychidae), and *Rheotanytarsus* sp. (Diptera: Chironomidae; Toth, 1993).

Increased flow may flush organic matter from the system so that the substrate will become more sandy in the center of the river channel. The outer margin of meanders may become a more heterogeneous invertebrate habitat (Fig. 6-6; Harris et al., 1995). In the demonstration project area, these projected changes in substrate were evident (Toth, 1991).

Woody debris is a common and important part of invertebrate habitat diversity in unchannelized rivers (Maser and Sedell, 1994). In the restored

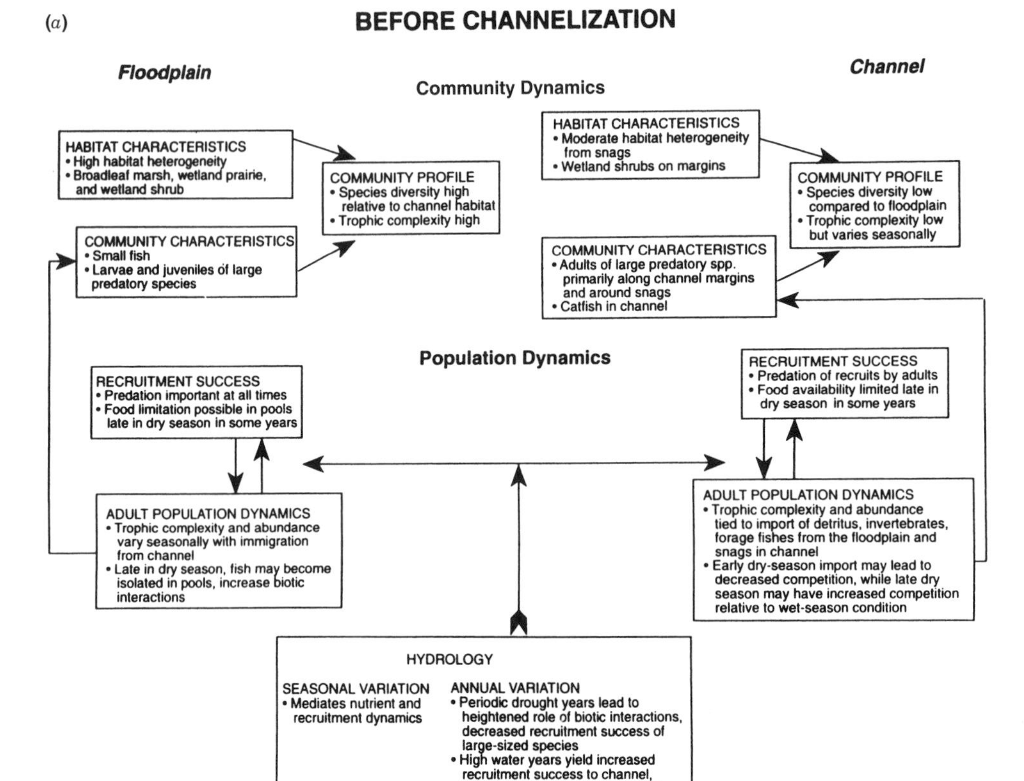


***Figure 6-7.*** *Conceptual model for the Kissimmee River fish community (a) prior to channelization and (b) in its channelized state (from Trexler, 1995; copyright © by Restoration Ecology by permission).*

Kissimmee River system, snags will likely develop from *Salix* sp., *Acer* sp., and *Quercus* sp. Snag habitats support filtering and gathering collectors, as well as scrapers (Harris et al., 1995).

The main problem for fish in the channelized river is the small seasonal fluctuation in water levels and the lack of upstream floodplain inundation (Fig. 6-7a, b; Trexler, 1995). The portions of the Kissimmee River with reintroduced flow hosted a larger number of sunfish (*Lepomis punctatus* and *Lepomis auritus*). Game fish increased by 20.0–43.7% (Wullschleger et al., 1990; Bull et al., 1991).

Because wading and waterfowl populations are indicators of habitat quality, their presence can help measure the success of the restoration of the Kissimmee River (Weller, 1995). After the floodplain was inundated, wading bird and waterfowl populations increased. While the restored Pool B area constituted less than 40% of the floodplain, 70% of the waterfowl and

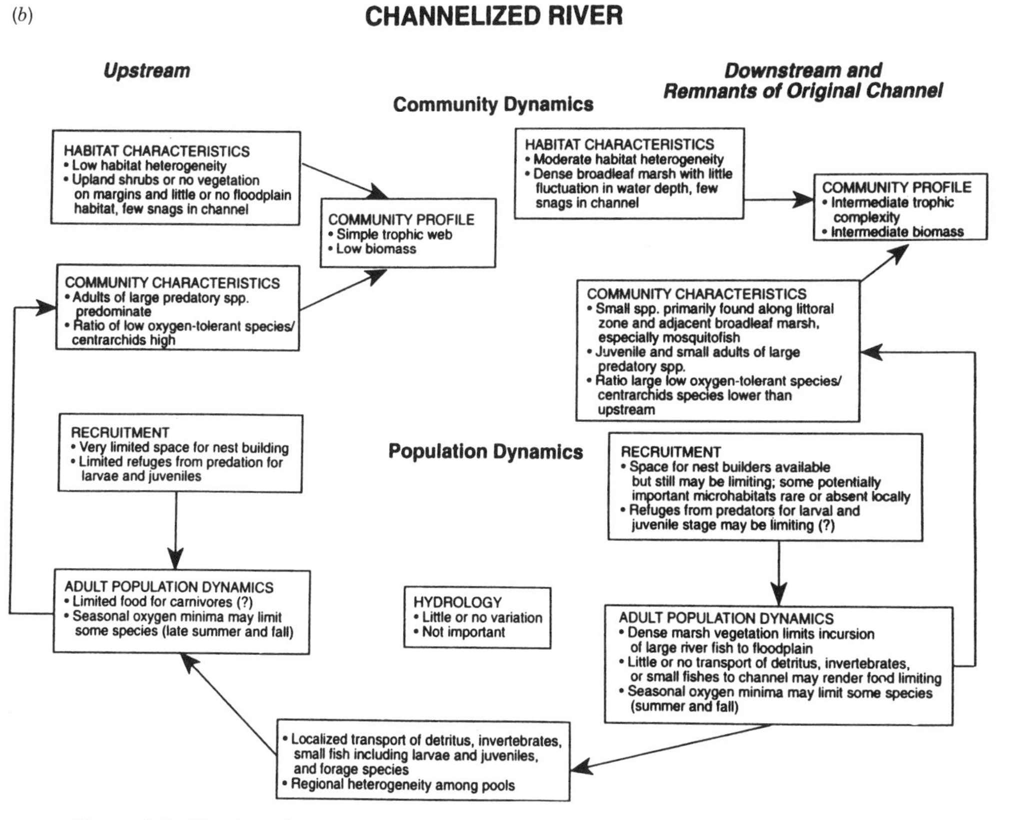


***Figure 6-7.*** *(Continued)*

66% of the wading birds occupied this portion of the Kissimmee River system (Toth et al., 1993).

## CASE STUDY 2—THE ECOLOGICAL SOCIETY OF INDIA RESTORATION PROJECTS, PUNE, INDIA

Until recently, most wetland restoration projects have been in developed countries of temperate regions. Very large wetlands still exist in some developing countries of the tropics, which, as a whole, still retain about half of their original wetlands (Maltby, 1986). Wetland restoration and rehabilitation, however, is a brand new idea in the tropics (Gopal, 1992), and there is a growing need for wetland restoration and rehabilitation to fit the needs and situations of developing countries.

Rural people are often dependent on wetlands for all or part of their

(a)

***Figure 6-8.*** *Rural village adjacent to a wetland in northern India (photograph by Beth Middleton).*

livelihood (Fig. 6-8; Box 6-1; Gopal, 1992), and they recognize the economic value of natural wetland products. Unfortunately, development projects increasingly focus on wetlands for drainage, dam, and embankment construction; polder creation; expansion of agriculture; urbanization and mining (Roggeri, 1995).

Restoration projects in developing countries are few but notable and mark the beginning of an important new trend. The Ecological Society of India initiated a restoration project for both tropical wetlands and forests in response to the environmental deterioration created by the construction of the Panshet Dam near Pune in 1985 (Fig. 6-9 Prakash Gole, personal communication).

Immediately downstream from the dam, water flow and streamside habitat were so reduced after construction that the soil stood barren for most of the year. Recently, the Ecological Society of India converted this site into a wildlife habitat by utilizing downstream barrow pits and mounds related to the construction of the dam. To impound water, water leaking from a pipe below the dam (Fig. 6-10a) was directed into ditches (Fig. 6-10b) feeding depressions with low "bunds" (embankments) positioned to impound water (Fig. 6-10c Prakash Gole, personal communication). The vegetation and wildlife responded immediately. The impoundments now are vegetated with aquatic vegetation such as *Typha* sp., *Polygonum* sp., *Jus-*

*siaea* sp., *Cyperus* sp., and *Scirpus* sp. Chestnut Bitterns, Blackheaded Munias and a large variety of fish and frogs have colonized the area (Prakash Gole, personal communication).

The Ecological Society is creating a mosaic of habitats at this site including dry and wet meadows, marshlands, reed swamps, and, on drier sites, upland shrub and forest. Selected trees have been planted in strategic positions to encourage tree-nesting birds (Prakash Gole, personal communication). The idea of this project is not to restore the original site conditions but rather to re-create a lost habitat for wildlife.

Upstream from the dam, about 7000 farmers were displaced when the floodplains along the river were flooded (Prakash Gole, personal communication). Many of these displaced people moved to the uplands directly adjacent to the river corridor. Quickly, overgrazing, tree lopping, and farming stripped the hillsides of vegetation (Prakash Gole, personal communication).

***BOX 6-1***

## PARTIAL RESTORATION OF TROPICAL FLOODPLAIN FUNCTION

In developing countries, the destruction of wetlands, particularly through the development of dams, creates hardship for local people by destroying fishing, agricultural, and/or grazing areas. In 1982, in Gounougou, Cameroon, the Lagdo reservoir was created on the Benue River for hydroelectric power. The fishermen downstream suffered severe ecological and socioeconomic problems following its construction.

To restore partial floodplain function, the Projet Pisciculture Lagdo has managed the remaining water on the floodplain to increase fish and vegetable production. Community input is essential in this project. Gounougou has many immigrants and more than 20 ethnic groups. Local participation is solicited through meetings with the 11 local chiefs and other interested people. The beneficiaries of the project determine what will be done, so the plan of action is adjusted often (Slootweg and van Schooten, 1995). In addition, extension workers live and work in the community and so are not estranged from the activities of the community. For such wetland management projects to be successful, it is necessary for the local people to be involved early. Furthermore, they must feel they are partners in the effort (Roggeri, 1995).

*continued*

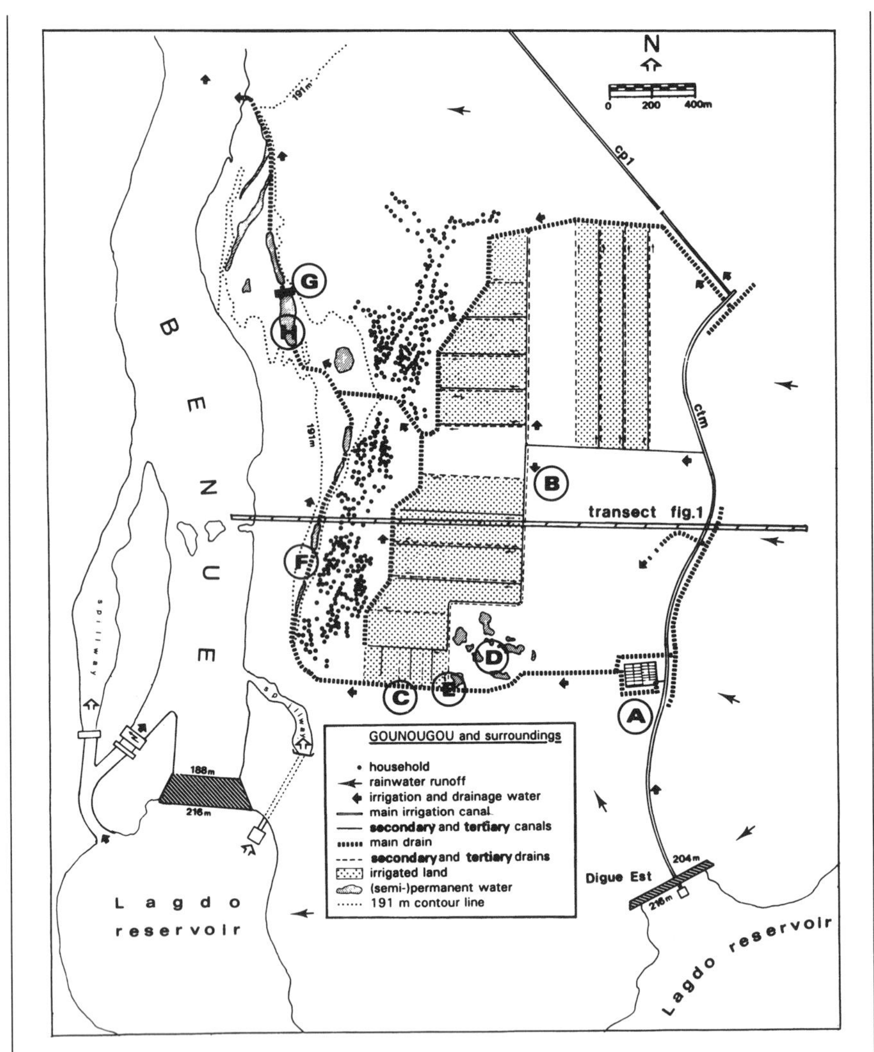


*Project area and interventions in the drainage system around the pilot village of Gounougou. Circled areas A–G (but not B) follow a drainage system designed to maximize water usage for fisheries and vegetable farming in the dried floodplain below the Lagdo Reservoir along the Benue River, Cameroon, Africa (from Slootweg and van Schooten, 1995; copyright © by Kluwer, Dordrecht, the Netherlands, by permission).*

In the degraded upland area adjacent to the dam, the Ecological Society of India began a pilot restoration project in 1986. While this is not the story of a wetland restoration, it does illuminate the relationship between river alteration and the surrounding landscape. Initially, in some of these upland restoration plots, a number of trees and shrubs were planted. However,

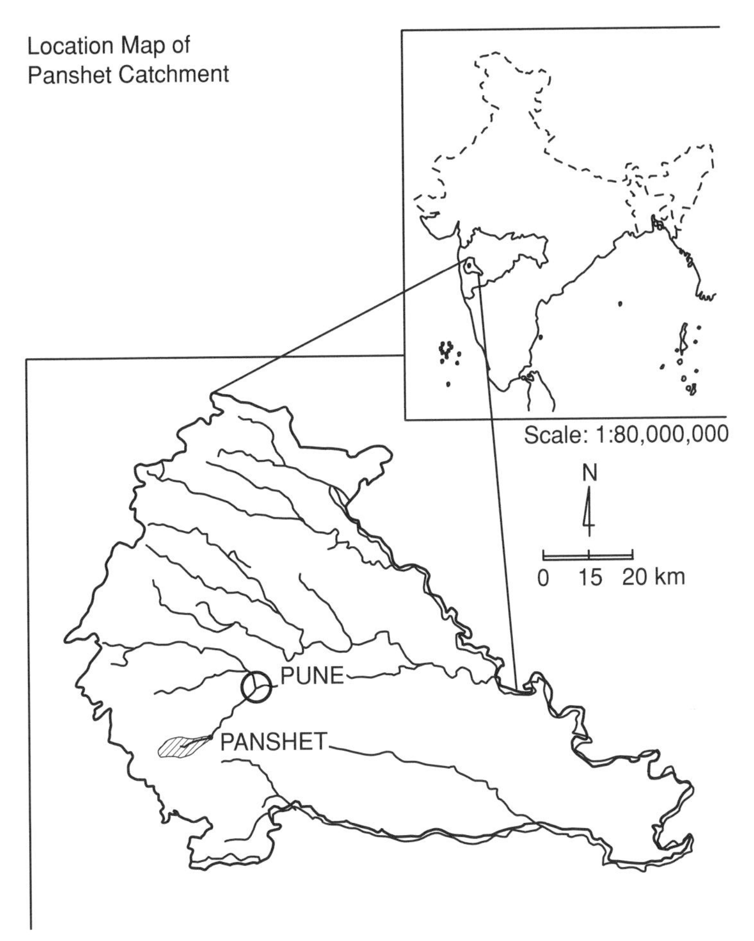


*Figure 6-9.* *Location of the Panshet catchment area near Pune, India (map courtesy of the Ecological Society of India).*

watering the plantings throughout the dry season proved difficult (Gole, 1990). In the end, natural regeneration of the upland sites was cheaper and more successful. The key problem was to exclude cattle. Mounds were built to fence them out. After two to three years behind these exclosed areas, trees, shrubs, grasses, and herbs regenerated naturally on the gravelly hillsides. In certain patches, villagers are sometimes allowed to cut grasses for their cattle (Gole, 1990). Wildlife species have also returned to the site, including barking deer (*Muntiacus muntjak*), black-naped hares (*Lepus nigricollis*), Indian fox (*Vulpes bengalensis*), common mongoose (*Herpestes edwardsi*), small Indian civet (*Viverricula indica*), wild boar (*Sus scrofa*), and leopard (Prakash Gole, personal communication).

(a)

(b)

**Figure 6-10.** *Wetland restoration project below the Panshet Dam near Pune, India. (a) Water leaking from a pipe below the dam leading into a series of ditches directing water to impounded barrow areas, (b) Water from a pipe directed to a series of "bunds" (small dams), and (c) restored wetland with Polygonum, Cyperus,* and *Scirpus* (photographs by Prakash Gole).

(c)

*Figure 6-10.* (Continued)

## CASE STUDY 3—MURRAY-DARLING RIVER, AUSTRALIA

The Murray-Darling River in Australia is a relatively well-studied example of a hydrologically and biotically altered ecosystem (Fig. 6-11). While this river is not yet the site of extensive restoration, because of the clear understanding of the problems it is poised for future activities. Any restoration activity there will be greatly complicated by the huge demand for water for development activities along the river, water that is already overcommitted. Along the semiarid lower river, water is abstracted for irrigation and drinking water (Shiel, 1996) and has been a critical factor in the development of the country as a whole (Mackay, 1990). Presently, 1.8 million people, or 10% of the population of Australia, live in the basin (Shiel, 1996).

### Alteration along the Murray-Darling River

Even before the arrival of Europeans in the basin in the 1820s (Shiel, 1996), humans had occupied the region for at least 40,000 years (Mackay, 1990). Aboriginal practices likely changed the site little, but their fire management practices may have increased *Eucalyptus camaldulensis* (red gum; Shiel, 1996).

The subsequent history of the Murray-Darling River is one of massive

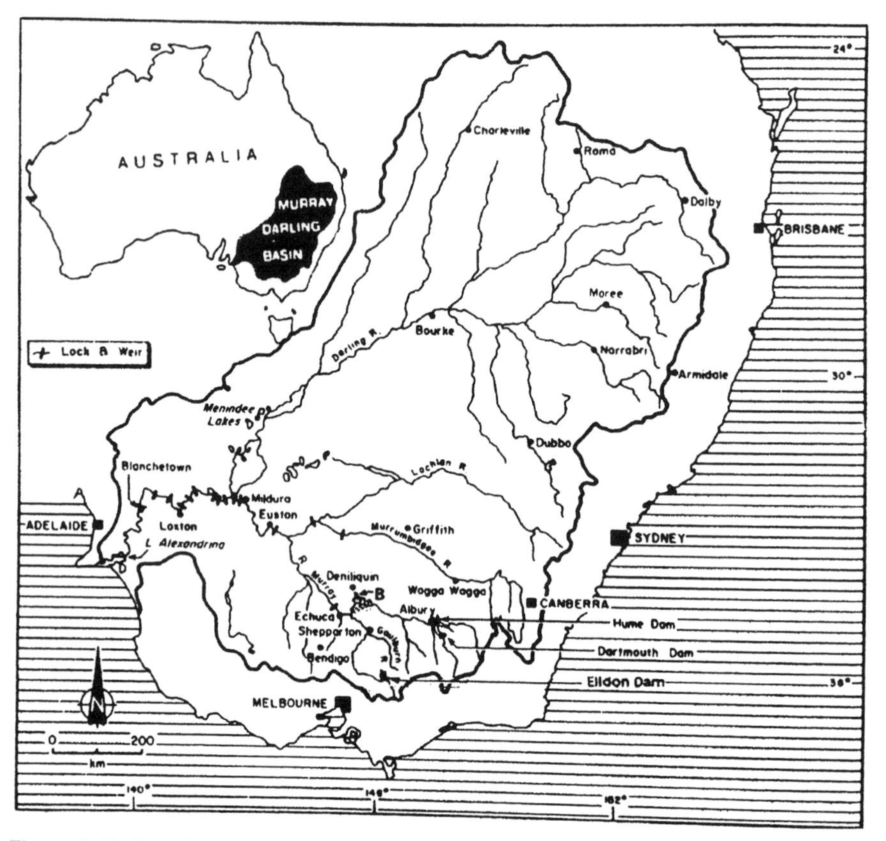


*Figure 6-11. Location of the Murray-Darling River Basin, Australia (from Shiel, 1996, as modified from Walker, 1985; copyright © by GeoJournal by permission).*

alteration. Cattle operations began in 1838, and in 1865, 350,000 sheep were brought to the region. Cattle damaged the system because they waded into the river, ate the plants, and disturbed the stream banks (Shiel, 1996).

Activities along the river have damaged red gum forests. Presently, red gum regeneration is patchy, especially along the lower stretches of the river, and grazing is partially to blame (Smith and Smith, 1990). In the Barmah Forest, Victoria, cattle have limited the regeneration of red gum (Chesterfield, 1986). Rabbits, introduced in the 1870s, killed seedlings and ringed trees. Myxomytosis began to control rabbit populations to some extent in the 1950s (Smith and Smith, 1990). Logging intensified between the 1850s and the 1890s. Riverboats became an important route of transportation in the last part of the 1800s, and these boats used red gum for fuel (Shiel, 1996).

Deforestation led to rising water tables. Salination of the soils is common in the southern part of the region because the rising water table moved through marine sediments and carried salt to the surface (Walker and Thoms, 1993; Shiel, 1996). As tree removal increased, groundwater and salt moved closer to the soil surface (Fig. 6-12, Shiel, 1996).

Hydrologic alteration grew with the human population. In the 1870s, farmers began to use water from the Murray-Darling River for irrigation (Walker, 1986), and this water use conflicted with navigation until railroads linked the region in the 1880s (Jacobs, 1990). By the 1900s, irrigation along the river was common (Table 6-4), and this activity began to affect the frequency and duration of inundation on the floodplain. Through reengineering, the water storage capacity of the river grew slowly after 1922 and then rapidly in the mid-1950s. By 1990, 10,000 to 11,000 gigaliters of water had been diverted from the river (Close, 1990). Dryland cotton has been grown along the upper Darling since the 1830s, but irrigated cotton did not become important until the 1960s (Arthington, 1996).

The flow regime in the basin was affected by hydrologic alterations in three major ways:

1. Regulated flow peaked in summer and autumn and unregulated flow in winter and spring.
2. Flood flows decreased and drought flows increased.
3. Increased water storage led to increased water diversion by allowing more water to be supplied during peak demand and drought months (Close, 1990).

Before regulation, flows were highly variable. After regulation, low flows were decreased by 500% and moderate flows increased by 200% (Walker and Thoms, 1993).

Ten low weirs were constructed in the Murray River in 1922–1935 to facilitate riverboat navigation (Fig. 6-13; Walker et al., 1992; Thoms and Walker, 1993). Not surprisingly, sediment is building up in the pools behind the weirs and erosion is occurring below them (Walker et al., 1992). These weirs are also referred to as **locks** because they have collapsible panels to allow boat passage (Thoms and Walker, 1993). Later, these weirs became important as irrigation impoundments for water storage (Fig. 6-13; Walker and Thoms, 1993). Similarly, the Keepit Dam in the Namoi Valley facilitates irrigation along the Darling River (Arthington, 1996). Presently, the weirs along the Lower Murray maintain a stable pool depth ($\pm$ 50 mm). Water travels downriver only when the storage capacity is exceeded (Thoms and Walker, 1993). While a single weir may have little negative impact on the hydrology of a river, a succession of them may have as much influence

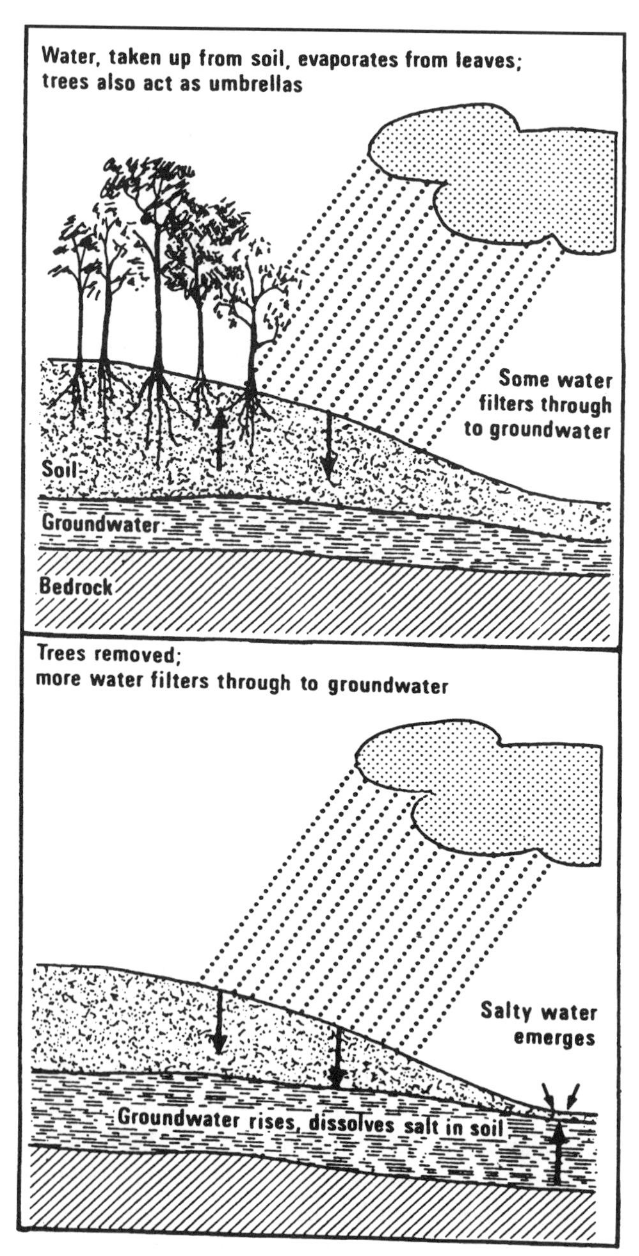


***Figure 6-12.*** *Devegetation leading to dryland salinization (from Shiel, 1996, as modified from Anonymous, 1975 © by ECOS (CSIRO by permission).*

**TABLE 6-4. History of Water Development Along the River Murray, Australia, from 1860 to the Present**

| Stage | Setting | Typical Year | Development Activity |
|---|---|---|---|
| 1 | Natural | 1860 | Natural flow regime, no storage or diversion |
| 2 | Pre–Lake Victoria | 1926 | Irrigation on Murray and tributaries, weirs on Murray; Goulburn Weir and Eildon Dam on Goulburn River |
| 3 | Pre–Hume Dam | 1929 | Lake Victoria regulator |
| 4 | Post–Hume Dam | 1939 | Hume Dam, weirs on Murray, increased irrigation |
| 5 | Post–Yarrawonga | 1948 | Yarrawonga Weir, river mouth barrages, increased irrigation |
| 6 | Post–Hume Weir | 1962 | Hume Weir, increased irrigation, changed operating rules |
| 7 | Post–Menindee Lakes | 1977 | Menindee Lake, diversion from Snowy River to Murray, increased irrigation |
| 8 | Post–Dartmouth Dam | 1990 | Dartmouth Dam (Mitta Mitta River) |

*Source:* Maheshwari et al. (1995).

as or more influence than a dam (Thoms and Walker, 1993: Walker et al., 1995).

The weirs altered **billabongs**, especially in the Valley and Gorge sections of the Lower Murray, with 75% of these now permanently flooded (Walker and Thoms, 1993). The hydrology of these wetlands is altered and is presently of four types, including

1. permanent wetland connected even during low, regulated flow,
2. seasonal wetland connected to the river only during high flow (irrigation season),
3. wetland filled above regulated flow with surplus flow, and
4. wetland filled above regulated flow with irrigation return water and surplus flow (Pressey, 1990).

Many billabongs were lost as they were filled, drained, or used for garbage dumps, waste stabilization ponds, or sheep dips (Smith and Smith, 1990).

An erratic flood pulse characterized the natural flow regime of the lower Murray. Because of hydrologic alterations, an increasingly stable water regime prevailed (Walker and Thoms, 1993). Unpredictable summer monsoon rains fed the Darling River, and the original water source for the Murray River was catchment precipitation in the winter and spring (Walker, 1986).

The water in the Murray River was not enough to fill the demands. So, beginning in 1949 and continuing until 1974 (Jacobs, 1990), water was diverted from the Headwaters Tract of the Murray River from the Snowy River via the Snowy Mountains Hydro-electric Scheme (Walker, 1986).

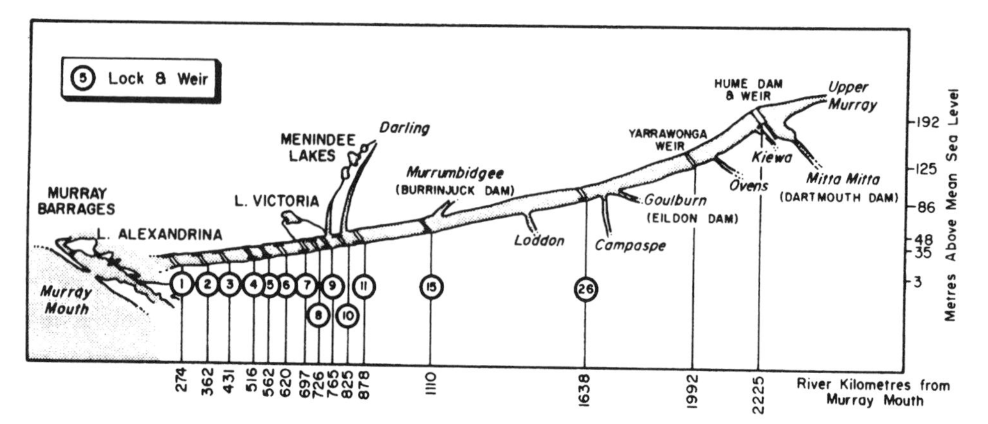


*Figure 6-13.* Longitudinal profile of the Murray River, Australia, showing the main tributaries, weirs (Locks 1–10), and impoundments (from Davies and Walker, 1986; copyright © by Junk Publishers by permission).

Water has been extracted from the Darling River to Lake Victoria since 1968 (Walker and Thoms, 1993).

The Billabong Tract, located at the confluence of the Murray and Mitta Mitta rivers, is flooded by the Hume Dam (Fig. 6-11, Walker, 1986), which was constructed in 1936 (Baker and Wright, 1990). Through the Mallee Plains Tract of semidesert from Swan Hill to Mildura, the Murray River joins the Murrumbidgee and the Darling. The weirs extend from approximately Mildura to Blanchetown. The South Australia Tract extends from Mildura to Lake Alexandrina, which is adjacent to the ocean (Walker, 1986).

The Murray-Darling outlet to the sea via Lake Alexandrina was severed by a barrage in 1939–1940 (Walker, 1986; Baker and Wright, 1990). Tidal obstructions in the river mouths alter physical, biological, and chemical processes that are vital to the functioning of estuaries (Pressey and Middleton, 1982). Barrages (artificial obstructions in a waterway used to raise water depths for irrigation) can impede the movement of fish to nursery areas, as well as the downstream flow of detritus and nutrients (Coiaccetto, 1996).

## Biotic Alteration

For the most part, the biota cannot tolerate the altered water regimes of the Lower Murray River, and probably the majority of species are in decline (Walker and Thoms, 1993). Whether a species can survive depends on its evolutionary experience with the spatiotemporal environment created by a disturbance. The resilience of the species of the basin is low, probably because of the disparity between the natural and regulated water regimes (Poff and Ward, 1990). Currently, below the Hume Dam, 2 plant species are believed to be extinct and 16 others are listed as rare or threatened by the Australian National Parks and Wildlife Service. In addition, 33% of all of the plants on the floodplain are exotic compared to 10% of the overall flora of Australia (Smith and Smith, 1990).

Stands of Moira grass (*Pseudoraphis spinescens*) once lined the Murray River south of the Hume Dam. Along portions of the river with stable water (Fig. 6-14), Moira grass has disappeared from much of its range, replaced by *Juncus ingens* (Chesterfield, 1986; Smith and Smith, 1990; Bren, 1992) or *Myriophyllum propinquum* (Bren, 1992). In some areas, the dominance of *J. ingens* was reduced (Chesterfield, 1986). While a small portion of the floodplain was more flooded after the Hume Dam was constructed, much of the floodplain was no longer flooded at all (Bren, 1992). Thus, since 1945, red gum has advanced from the drier side of the floodplain into stands of Moira grass (Bren, 1992).

The red gum forests are suffering from changes in hydrology. The natural hydrologic condition consists of winter/spring flooding for two or three months in most years. After extensive regulation along the Murray River

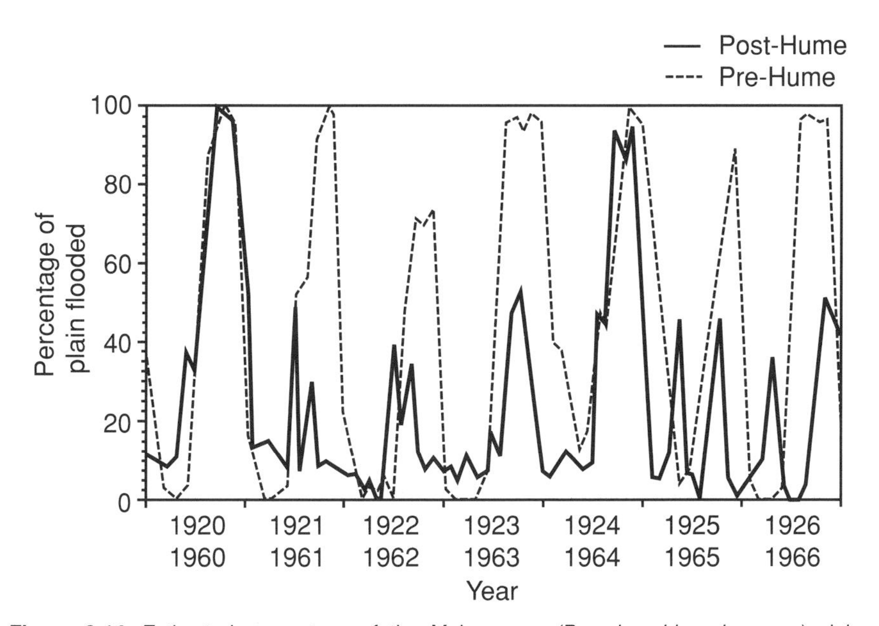


***Figure 6-14.*** *Estimated percentage of the Moira grass (Pseudoraphis spinescens)* plains flooded either during the pre–Hume Dam (1920–1926) or post–Hume Dam (1960–1966) period (from Bren, 1992; copyright © by Australian Journal of Ecology by permission).

for irrigation, localized summer flooding became common along with increased streamflow (Chesterfield, 1986; Dexter et al., 1986; Bren, 1992). After river alteration at Tocumwal, flow increased by 20% between December and May after 1895–1933 (Dexter et al., 1986).

## The Future of the Murray-Darling River

Because of the dependence of the region on the regulated water regime, particularly along the lower reaches of the Murray River, it is unlikely that natural flows can be restored (Walker and Thoms, 1993). However, the restoration of seasonal flow patterns could result in reduced algal blooms and improved water and biotic quality (Arthington, 1996). Water from the river is critical for human settlements and agriculture (Bren, 1992), especially for Adelaide and the Iron Triangle towns of South Australia (Baker and Wright, 1990), but restrained development may be required to improve river quality (Arthington, 1996). In fact, current levels of water usage along the Murray may be unsustainable (Walker et al., 1995). The restoration of natural flow is probably the only option for reintroducing natural function on the floodplain, particularly along the lower reaches (Walker and Thoms, 1993).

Pressey (1990) has suggests several alternatives for wetland management along the Murray. One is to do nothing, which is untenable given that the river is likely to undergo further highly undesirable changes. Another is to completely restore the natural environment, which, while serving as a useful model, is hardly likely to be adopted as a management strategy given the importance of the water to the economy of the region. A third option is special-purpose management, which consists of partial rehabilitation directed at specific problems along the river (Pressey, 1990).

Much thought has been given to reducing salinization in the Lower Murray. Groundwater intercepted by tube wells could reduce salinization, but this method is too expensive. Weir pool levels could be lowered, but this could lead to hydrologic readjustment and higher rates of salinization for up to 30 years (Walker and Thoms, 1993). Although extensive surveys and reviews exist, strategic research directed to resolving environmental problems along the river is mostly lacking (Walker et al., 1992). Many gaps exist in the knowledge required to restore the river (Shiel, 1996).

Operational changes in weir management could reduce turbidity and unnatural water levels in the irrigation system along Locks 1–10. Because water from the Darling is more turbid than from the Murray (Walker et al., 1992), turbidity could be reduced by a change in the amount of Darling versus Murray water used for irrigation (Walker et al., 1992). Also, the narrow range of allowable water fluctuation in the pools behind weirs (50 mm) needs to be reconsidered. Desnagging operations have been reduced recently, which has improved the environment for fish, animals, and plants (Walker et al., 1992).

One advantage in the future management of the lower river is that it is controlled by one political and economic unit as part of South Australia. The river as a whole is regulated by state government departments, as well as by the federal Murray-Darling Commission (Walker and Thoms, 1993), first established in 1917 as the River Murray Commission (Baker and Wright, 1990; Jacobs, 1990). River regulation can be coordinated and is not as diffuse as that of many other rivers in the world. Nevertheless, integrated management suffers from territorial jealousies involving state water authorities, universities, and research facilities (Shiel, 1996).

## CASE STUDY 4—RHINE RIVER, GERMANY

By the 1960s and 1970s, the Rhine River was by most accounts dead. The lower reaches were channelized (Fig. 6-15) and diked to improve the channel as a shipping route, and the upper portions downstream from Lake Constance were highly altered by hydroelectric dams (Friedrich and Müller, 1984). The breaking point came in November 1986 when 30 tonnes of toxic chemicals (insecticides, herbicides, and fungicides) were accidentally re-

***Figure 6-15.*** *Channelized portion of a river near Münden, Germany (photograph by Beth Middleton).*

leased during an industrial fire in Basel, Switzerland (Schulte-Wülwer-Leidig, 1995). The water used to put out the fire became contaminated with pesticides from the chemical plant and then flowed into the Rhine (Lelek, 1989). Eels, fish, and animals died, and a drinking water alert was posted for 60 million people from Basel to Amsterdam (Drozdiak, 1996). After this episode, a very serious effort began to revitalize the Rhine (Schulte-Wülwer-Leidig, 1995).

## History of the Degradation of the Rhine River

In Roman times, the Rhine connected an extensive network of Roman and Teutonic towns, some of which still exist, including Basel, Cologne, and Strasburg (Fig. 6-16, Table 6-5; Friedrich and Müller, 1984). To improve the Rhine as a transportation route, the Romans built the Drusus Canal in the lower Rhine. By the Middle Ages, water mills, timber floating, and irrigation of small meadows for cultivation had altered smaller tributaries of the Upper Rhine Valley (Kern, 1992a).

In 1707, canals were constructed in the Rhine Delta (Fig. 6-17) to ensure a more equal distribution of water in the three branches of the Rhine—the Waal, Neder Rijn and Ijssel of the Netherlands (Friedrich and Müller, 1984; van Urk, 1984; van Urk and Smit, 1989). Sluice gates were built at Volkerak and eventually at Haringvliet to protect against storm tides and prevent salt water intrusion into the estuary (Hofius, 1991). Now, only one river mouth to the sea is not cut off by a sluice gate (Lelek, 1989).

Dike construction began in the lower Rhine in about 1450 (van Urk and Smit, 1989), but many of these dikes were low, so high flooding still inundated the floodplain (Dister et al., 1990). These embankments created pasture in the peat swamps in the western part of the Netherlands between the branches of the rivers Vecht, Oude Rijn, Lek, Merwede, and Meuse (van Urk, 1984).

Changes along the Rhine were minor until the Tulla-Rhine Rectification in 1817–1876. The main objective of this major channelization and embankment project was to create more agricultural land along the Lower Rhine and to facilitate navigation and flood control (Friedrich and Müller, 1984; Zinke and Gutzweiler, 1990). The channel became a single waterway in both the braided Upper Rhine and the meandering Middle Rhine (Fig. 6-18; Lelek, 1989; Dister et al., 1990; Zinke and Gutzweiler, 1990), and it was deepened in the Middle and Lower Rhine (Friedrich and Müller, 1984). After the rectification, the Rhine no longer changed course (Hofius, 1991).

The erosion of the channel bed was much deeper than Tulla predicted, so that large areas of the floodplain were drained along the southern Upper Rhine (Dister et al., 1990; Hofius, 1991). Downstream of Basel, downcutting was so pronounced that, after the rectification, the floodplain almost

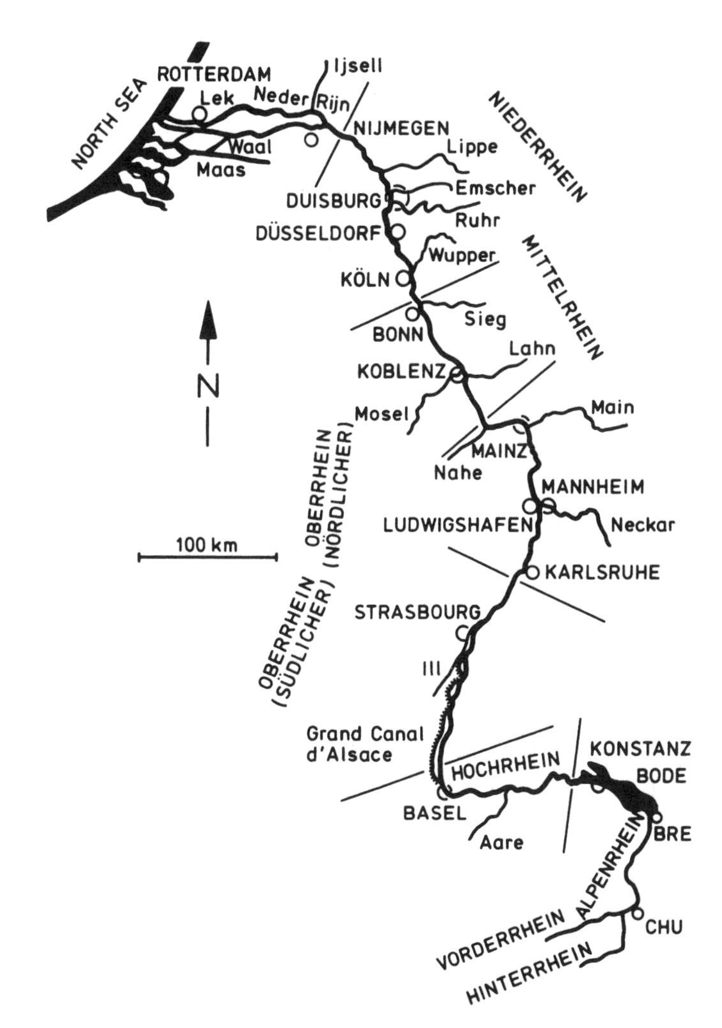


*Figure 6-16.* Map of the Rhine River (from Friedrich and Müller, 1984; copyright © by Blackwell Scientific Publications by permission).

never flooded. Because inundated floodplains regenerate groundwater, the water table in this region has dropped (Dister et al., 1990; Larson, 1995) by up to 10 m (Dister et al., 1990).

Starting after the rectification, many of the backwaters of the Rhine no longer had any water, so the vegetation changed along the watercourse and fisheries declined (Lelek, 1989). The connection between the channel and the floodplain was severed (Larson, 1995), and the natural flood pulse ceased to function. Because the channel is not integrated with its floodplain, cities along the Rhine are no longer well protected from 200-year floods (Ward and Stanford, 1995a).

Following the Treaty of Versailles in 1919, France, with the sole right

**TABLE 6-5. Hydrologic Divisions of the Rhine**

| Main Section | Section | Zone | Boundary Points | Km | Altitude (m) | Slope (%) |
|---|---|---|---|---|---|---|
| | | | Sources of Vorder Rhein and Hinter Rhein | | | |
| | Alpine Rhine | | Entry to Bodensee (Lake Constance) | | 395 | |
| | Bodensee | | | | | 0.0 |
| Alpine Rhine | | | Konstanz | 0 | 395 | |
| | Lake Rhine | | | | | 0.04 |
| | | | Stein | 25 | 395 | |
| | Hochrhein | | | | | 1.0 |
| | | | Basel | 170 | 244 | |
| | Upper Rhine | Furcation zone | | | | 0.7 |
| | | | Karlsruhe | 360 | 100 | |
| Upper Rhine | | Meander zone | | | | 0.1 |
| | | | Mainz | 500 | 80 | |
| | Rhein | | | | | 0.1 |
| | | | Bingen | 530 | 77 | |
| | Middle Rhein | Transverse valley | | | | 0.2 |
| | | | Koblenz | 590 | 59 | |
| | | Neuwied basin | | | | 0.3 |
| Middle Rhein | | | | | | |
| | | | Andernach | 615 | 52 | |
| | | Broadening valley | | | | 0.1 |
| | | | Bonn | 655 | 45 | |
| Lower Rhein | | | | | | 0.1 |
| | | | Pannerden | 870 | 10 | 0.1 |
| | | Estuary, inclusive of branches in the Nether-lands | | | | |
| | | | North Sea | 1000 | | |

*Source:* Freidrich and Müller (1984), as modified from Solmsdorf et al. (1975).

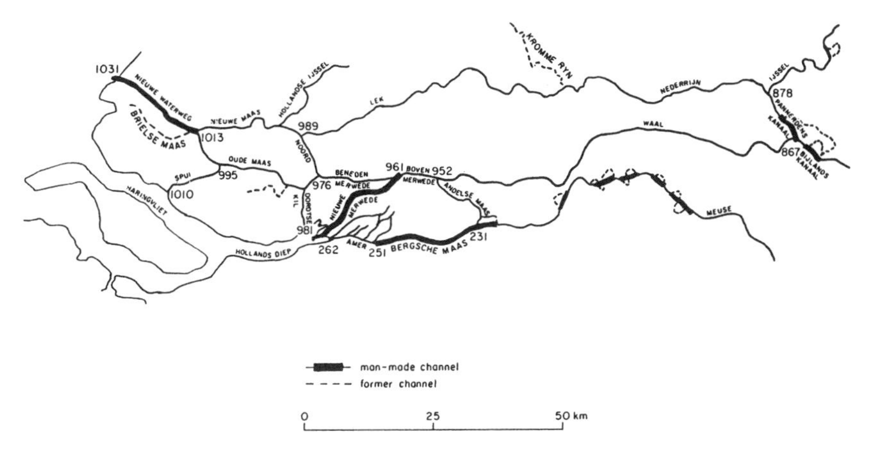


*Figure 6-17.* Map showing man-made canals (Pannerdens and Bijlands Kanaals) in the main system of the Rhine-Meuse, together with former river channels (from van Urk, 1984; copyright © by Blackwell Scientific Publications by permission).

to develop hydroelectric power on the Upper Rhine, dug the Grand Canal d'Alsace, a lateral canal used for shipping and hydroelectric power generation that withdraws most of the water from the Rhine (Dister et al., 1990). As a result, 87% of the 183-km stretch of the Rhine in France have been canalized (Klein et al., 1996). The Aare River and a tributary of the Aare have been diverted through three lakes in the alpine foothills to increase water retention for hydroelectric power generation in this portion of the Rhine. Between Schaffhausen and Basel, the Aare and the Hochrein have been altered to produced electricity, which has resulted in considerable changes in water depth and flow characteristics (Friedrich and Müller, 1984). This hydrodevelopment has resulted in an increased danger of flooding along the southern Upper Rhine (Zinke and Gutzweiler, 1990).

## The Toxic River

The Rhine is the core of a megalopolis and the heart of transportation in the region (Jongman, 1992). The character of this riverine landscape changed over time as settlements, agriculture, industry, and roads expanded across Central Europe (Hofius, 1991). By 1989, 40 million people lived along the Rhine, one of the most heavily populated regions of the world (van Urk, 1984; Lelek, 1989).

After the turn of the century, municipal and industrial sewage increased along with the population. Pollution reached a maximum in the 1960s (Lelek, 1989). Fish consumption became hazardous, and the water smelled of phenol in some areas by the 1970s (Lelek, 1989). Many fish species that

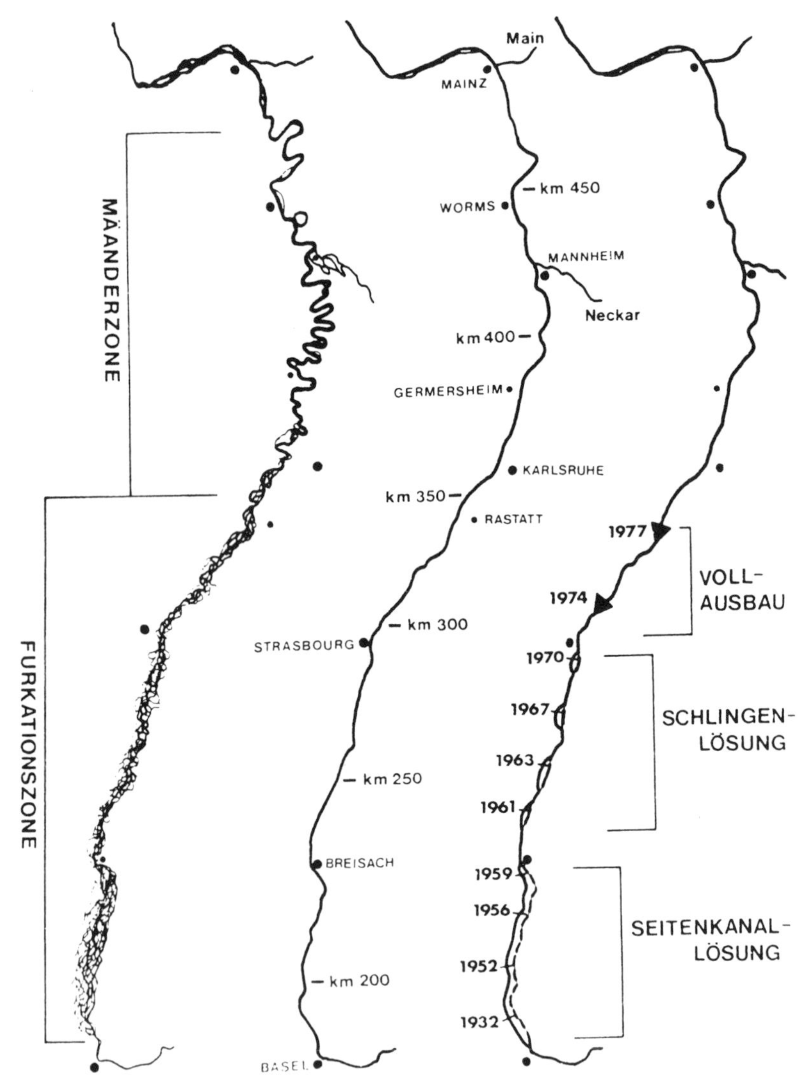


***Figure 6-18.*** *The Upper Rhine before (left) and after Tulla's Rectification (middle) and following modern development (right) (from Dister et al., 1990; copyright © by [Regulated Rivers: Research and Management] by permission).*

had once been common disappeared, along with the entire salmon fishery around 1935 (Fig. 6-19; Schulte-Wülwer-Leidig, 1995). Water from the Rhine was toxic to salmonids, according to a study in 1980 (Poels et al., 1980).

Tracking recent water quality along the Rhine, one finds fairly high quality in the upper portions of the river, but it rapidly declines downstream

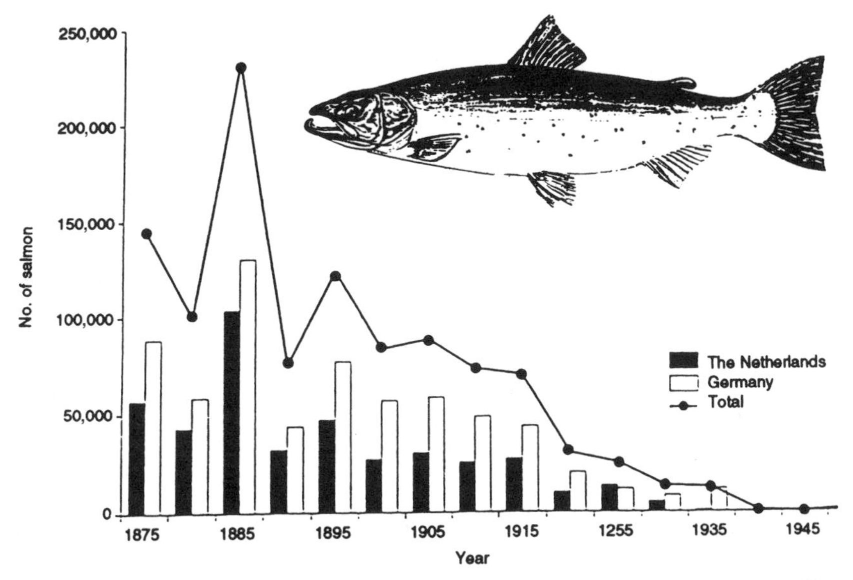


***Figure 6-19.*** *Number of salmon caught in Germany and the Netherlands between 1875 and 1950 (from Schulte-Wülwer-Ledig, 1995; copyright © by Wiley, Chichester, by permission).*

(Fig. 6-20). At Basel the Rhine flows north (Fig. 6-16), where the Neckar and Main add a load of pollution to it. In the Lower Rhine, the Emscher, a river with wastes from the industrial Ruhr District, joins the river (Friedrich and Müller, 1984). As power capacity has increased, thermal pollution has become more pronounced near the power plants (Friedrich and Müller, 1984).

Between 1976 and 1980 (Fig. 6-20), water quality improved somewhat along the Rhine (Friedrich and Müller, 1984). By 1992, 92% of the residences were connected to sewage treatment facilities in the state of Burden-Württenberg (Kern, 1992). A truly concerted effort began after 1986. The industrial accident at Basel, Switzerland, so stunned Europe that in 1987 the Rhine Action Committee, with representatives from the countries bordering the Rhine, was formed to address the extreme water quality problems facing the region.

Less than 1% of the original natural area still remains along the Rhine (Hügin, 1981). Natural floodplain remnants remain along abandoned river courses in the Upper Rhine (Dister et al., 1990; Zinke and Gutzweiler, 1990) and in the Lower Rhine between Basel and the Dutch border. The canalization along the French Rhine dried out the floodplain there, so that oak forest became common in the remaining natural reserves (Klein et al., 1996). The periodically drying oxbows and backwaters of the Lower Rhine are still very diverse (Friedrich and Müller, 1984), although agricultural

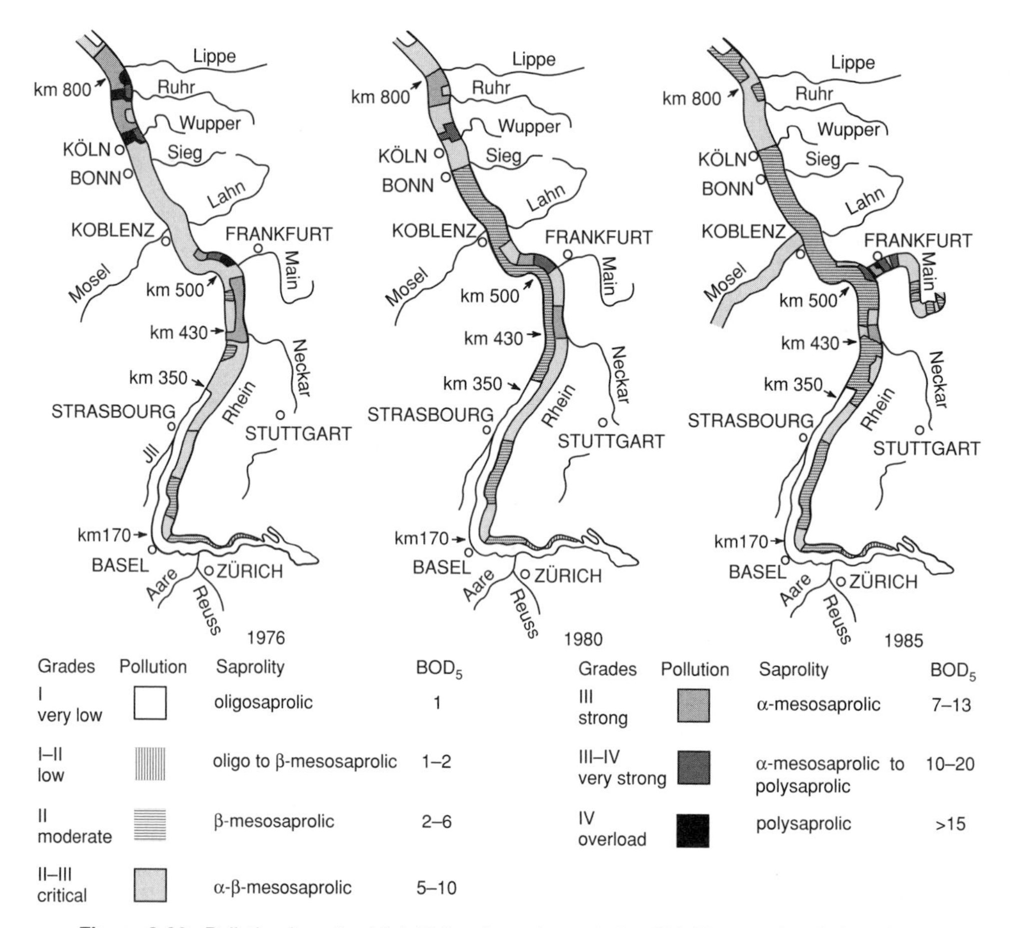


***Figure 6-20.*** *Pollution from the High Rhine downstream to km 800 (German frontier), as based on three parameters. From 1976 to 1985 some improvement was recorded particularly on the Upper Rhine (from Lelek, 1989, as modified from Deutsches Gewässerkundliches Jahrbuch (1974) Ländergemeinschaft Wasser (1985); copyright © by Ontario Ministry of Natural Resources by permission).*

expansion in this part of the Rhine in the nineteenth century removed much of the original vegetation (van Urk and Smit, 1989). The vegetation of the Middle Rhine was never so diverse because it was composed of a confined river bed (Friedrich and Müller, 1984).

Both willow and elm were once common along the Rhine (Ellenberg, 1988), but Dutch elm disease has almost totally destroyed the elm forests. Agriculture and recreational activities have reduced natural grasslands. Because the Rhine is an international shipping route, a great many introduced species can be found near harbors and in sludge (Friedrich and Müller, 1984). Nevertheless, the river still functions as an ecological corridor for native species to some extent (Jongman, 1992).

## The Future of the Rhine

The rejuvenation of the Rhine has very little to do with reconstructing pristine conditions, as is often the goal in North American projects. It also has little to do with striving for the German idea of the *Leitbild*, or the ideal solution to the problem. Instead, the current work on the Rhine is concerned with the amelioration of severe pollution and environmental problems (Larson, 1995) or the optimal solution (Kern, 1992a) within the restrictions of the Rhine River setting. The American ideal of restoration of a pristine state is untenable in this and many other situations. Nevertheless, considerable benefit can come from the rehabilitation of very altered systems (Larson, 1995).

In 1987, in response to the disaster at Basel, the Rhine Action Committee developed a set of goals for the Rhine in the year 2000. These included the following:

1. Create suitable habitat for species such as salmon along the Rhine.
2. Make the water of the Rhine usable again as drinking water by reducing noxious chemicals.
3. Decrease toxins in the sediments.
4. Protect the North Sea from pollution (Schulte-Wülwer-Leidig, 1995).

Industrial and municipal pollution, as well as nonpoint atmospheric and agricultural pollution, has been reduced (Schulte-Wülwer-Leidig, 1995). Industrial accidents are fairly rare now, with international patrols on the watch and industries that handle dangerous materials moved away from the river. Lead, mercury, and dioxin levels have been cut by 70%, chrome, nickel and some other heavy metals by 50% (Drozdiak, 1996). The reduction in the load of heavy metals such as cadmium and lead in the Rhine has been precipitous from the mid-1970s to the early 1990's (Figs. 6-21 and 6-22; Schulte-Wülwer-Ledig, 1995).

Following the Ecological Master Plan for the Rhine (Salmon 2000), the main stream will be restored as a migratory route for salmon and other migratory fish. Important fish habitats (spawning grounds, migratory resting areas) in the Rhine and its valley will be protected and preserved. The passability of barrages can be improved by creating better-quality fish ladders. In particular, the sluice gates at Haringvliet and Ijsselmeer need to be adapted and the fish ladders at the Iffezheim and Gambsheim power stations improved so that the fish can pass. Farming on the floodplain should be reduced so that the interaction of the river and floodplain can be increased (Schulte-Wülwer-Leidig, 1995).

Since stream dynamics rely on flow, rehabilitation should allow flow and lateral movement of the channel, with frequent inundation of the floodplain

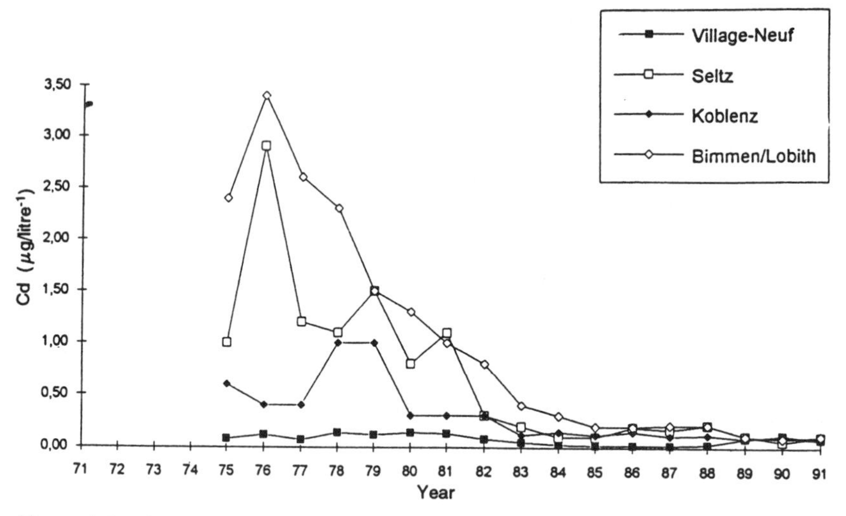


***Figure 6-21.*** *Average annual concentration of total cadmium (µg litre$^{-1}$) (from Schulte-Wülwer-Ledig, 1995; copyright © by John Wiley & Sons, Chichester, U.K., by permission).*

(Kern, 1992). Because the natural dynamics of the river have been altered, no new succession is occurring on newly formed meanders (Dister et al., 1990).

Along the Rhine, river banks and alluvial floodplains need restoration since 90% of the alluvial area between Basel and Kalruhe have been cut off by dikes (Schulte-Wülwer-Leidig, 1995). In 1983, dikes collapsed near Darmstadt on the northern Upper Rhine, providing an opportunity to observe the effects of dike removal. The floodplain meadows redeveloped

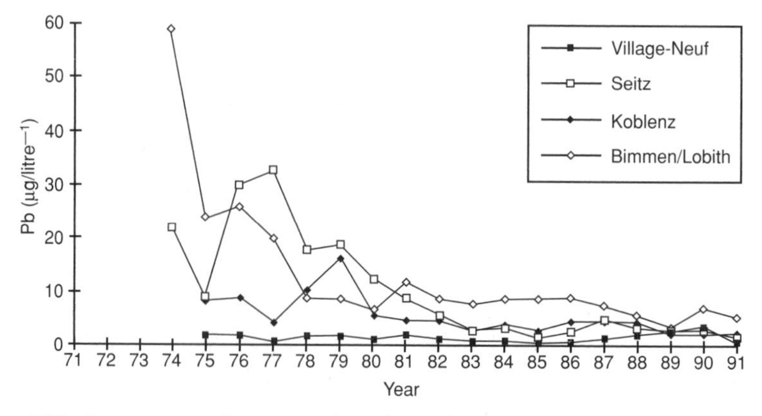


***Figure 6-22.*** *Average annual concentration of total lead (µg litre$^{-1}$) (From Schulte-Wülwer-Ledig, 1995; copyright © by John Wiley & Sons, Chichester, U.K., by permission).*

naturally and rapidly. Dikes were purposely opened near Breisach on the Upper Rhine. Now the floodplain water can reenter the channel after the flood peak has passed, contributing to flood prevention downstream. Opening dikes offers a promising solution to the problem of regenerating floodplain forests along the Rhine (Zinke and Gutzweiler, 1990).

The Rhine was once a free-flowing salmon river, but it seemed hopelessly altered until very recently (Lelek, 1989; Schulte-Wülwer-Leidig, 1995). Thus far, the concerted efforts of the basin countries along the Rhine have improved water quality and fish habitat. Forty of 47 indigenous fish and **cyclostome** species have returned to the Rhine but not sturgeon. A few Alice shad, sea trout, and sea lamprey remain in the river, but in very low numbers (Lelek and Buhse, 1992).

The good news? A few trout and salmon were found in the Upper Rhine near the Iffizhen Dam in 1995 for the first time in 50 years. These individuals had been marked by Dutch scientists in the Lower Rhine, so it is certain that they migrated through the Rhine channel from the Netherlands to the Upper Rhine (Drozdiak, 1996). It is a cause for celebration that salmon can once again live in the Rhine.

# *Appendix 1*

# *Dispersal (Dispersion) of Wetland Species*

| Species | Propagule Type | Dispersal Mechanism | | | | | | References |
|---|---|---|---|---|---|---|---|---|
| | | | | Animals | | | | |
| | | Wind | Water | Bird | Fish | Animal | Other | |
| *Acacia nilotica* | Seeds | | | | | + | | Middleton and Mason (1992) |
| *Acer barbatum* | Seeds | + | | + | | + | | McKnight (1980) |
| *Acer negundo* | Seeds | + | + | + | | + | | McKnight (1980); Middleton (1995a) |
| *Acer platanoides* | Seeds | + | | | | | | Rydin and Borgegård (1991) |
| *Acer rubrum* | Seeds | + | + | + | | + | | Hutnik and Yauney (1965) |
| *Acer saccharum* | Seeds | + | + | + | | + | | Weitzman and Hutnik (1965) |
| *Aconitum septentrionale* | Seeds | | + | | | | | Danvind and Nilsson (1997) |
| *Acorus* sp. | Seeds | | ? | | | + | | Cook (1990a) |
| *Aeschynomeme* sp. | Seeds, whole plants | | + | | | | | Cook (1990a) |
| *Aglaodorum* sp. | Seeds | | + | | | | | Cook (1990a) |
| *Aldrovanda* sp. | Whole plants | | + | | | | | Cook (1990a) |
| *Alisma* sp. | Seeds | | + | | | + | | Cook (1990a) |
| *Alisma plantago-aquatica* | Seeds, fruits | o | + | + | | + | | Arber (1920); Sculthorpe (1967); Cook (1990a) |
| *Alisma subcordatum* | Achenes | | + | | | | | Kaul (1978) |
| *Alnus tenuifolia* | Seeds | + | | | | | | Walker et al. (1986) |
| *Alternanthera* sp. | Seeds, stem fragments | | + | | | | | Cook (1990a) |
| *Alternanthera hassleriana* | Whole plant | | + | | | | | Junk (1986) |
| *Amaranthus spinosus* | Seeds | | | | | + | | Middleton and Mason (1992) |
| *Ammannia* sp. | Seeds | | | | | ? | | Cook (1990a) |
| *Amphibolus* sp. | Seedlings | | + | | | | | van der Pijl (1982); Cook (1990a) |
| *Anaphyllopsis* sp. | Seeds | | + | | | ? | | Cook (1990a) |
| *Angelica archangelica* | Fruits | | + | | | | | Danvind and Nilsson (1997) |
| *Anubias* sp. | Seeds | | | | | ? | | Cook (1990a) |
| *Apalanthe* sp. | Seeds | | ? | | | | | Cook (1990a) |
| *Apium* sp. | Seeds | | ? | | | ? | | Cook (1990a) |
| *Aponogeton* spp. | Seeds, adults | | + | | | | | Cook (1987, 1990a) |
| *Appertiella* sp. | Seeds | | + | | | | | Cook (1990a) |
| *Arundinaria gigantea* | Internodes | | + | | | | | Middleton (1996) |
| *Arundo* spp. | Seeds | + | | | | | | Cook (1987, 1990a) |

| | | | | | | | |
|---|---|---|---|---|---|---|---|
| *Asimina triloba* | Seeds | | | + | | + | McKnight (1980) |
| *Aster tenuifolius* | Seeds | | | + | | | Vivian-Smith and Stiles (1994) |
| *Aster tripolium* | Seedlings | | + | | | | van der Pijl (1982) |
| *Asterochaete* sp. | Seeds | | | | | ? | van der Pijl (1982) |
| *Astragalus alpinus* | Seeds, pods | | + | | | | Danvind and Nilsson (1997) |
| *Avicennia germinans* | Propagules | | + | | | | McKee (1995a) |
| *Avicennia marina* | Propagules | | + | | | | Clarke (1993) |
| *Azolla* sp. | Whole plants, spores | | + | | + | | Cook (1990a) |
| *Azolla filiculoides* | Whole plant | | + | | | | Junk (1986) |
| *Bacopa* sp. | Seeds | | + | | | | Cook (1990a) |
| *Baldellia* sp. | Seeds | | + | | | ? | Cook (1990a) |
| *Baldellia ranunculoides* | Seedling | | + | | | | Sculthorpe (1967) |
| *Barclaya* sp. | Seeds | | | | | + | Cook (1990a) |
| *Bartsia alpina* | Seeds | | + | | | | Danvind and Nilsson (1997) |
| *Beckmannia eruciformis* | Seeds | | ? | ? | | | Cook (1990a) |
| *Beckmannia syzigachne* | Seeds | | ? | ? | | | Cook (1990a) |
| *Berchemia scandens* | Seeds | + | | | | | Sharitz et al. (1990) |
| *Berula* sp. | Seeds | | ? | | ? | | Cook (1990a) |
| *Berula erecta* | Seeds | o | + | ? | ? | ? | Cook (1990a) |
| *Betula nigra* | Seeds | + | + | | | | Putnam et al. (1960) |
| *Betula pubescens* | Seeds | + | | | | | Wiegers (1985) |
| *Bidens* sp. | Seeds | | + | | | | Cook (1990a) |
| *Bidens discoidea* | Seeds | | + | | | | Middleton (1995a,b) |
| *Bidens frondosa* | Seeds | | + | | | | Middleton (1995a,b) |
| *Bolbitis* sp. | Spores | + | | | | | Cook (1990a) |
| *Bothriocline* sp. | Seeds | | ? | | | | Cook (1990a) |
| *Brasenia* sp. | Seeds | | + | | | ? | Cook (1990a) |
| *Brunnichia ovata* | Seeds | | + | | | | Middleton (1995a,b) |
| *Burnatia* sp. | Seeds, rhizome fragments | | + | | | + | Cook (1990a) |
| *Butomus* sp. | Rhizome fragments | | ? | | | | Cook (1990a) |
| *Cabomba* sp. | Seeds | | + | ? | | | Cook (1990a) |
| *Cadiscus* spp. | Seeds | o | | | | + | Cook (1987, 1990a) |
| *Caladium* sp. | Seeds | | | | | ? | Cook (1990a) |
| *Caldesia* sp. | Seeds, bulbils | | + | | | ? | Cook (1990a) |

| Species | Propagule Type | Dispersal Mechanism | | | | | | References |
|---|---|---|---|---|---|---|---|---|
| | | | | Animals | | | | |
| | | Wind | Water | Bird | Fish | Animal | Other | |
| *Calla* sp. | Seeds | | | | | + | | Cook (1990a) |
| *Callitriche* spp. | Detached shoot, fruits, seeds | o | + | + | + | + | | Arber (1920); Cook (1990a) |
| *Calluna vulgaris* | Seeds | + | | | | | | Fenner (1985) |
| *Caltha* sp. | Seeds | | + | | | | | Cook (1990a) |
| *Cardamine* sp. | Vegetative fragments | | ? | | | | | Cook (1990a) |
| *Carex* sp. | Achenes | | | + | | | | Vivian-Smith and Stiles (1994) |
| *Carex* sp. | Inflated utricle (if present) | | ? | | | | | Cook (1990a) |
| *Carex bigelowii* | Fruit | | + | | | | | Danvind and Nilsson (1997) |
| *Carex normalis* | Achenes | | + | | | | | Middleton (1995a,b) |
| *Carpinus carolinianal* | Seeds | | + | + | | | | McKnight (1980) |
| *Carum* sp. | Seeds | | + | | | ? | | Cook (1990a) |
| *Carya aquatica* | Seeds | + | | + | | + | | Johnson and Beaufait (1965b) |
| *Carya illinoinensis* | Seed | | | + | | + | | McKnight (1981) |
| *Carya lacinosa* | Seed | | + | + | | + | | Merz (1965); Middleton (1995a,b) |
| *Cecropia* sp. | Seeds | | | | + | | | Goulding (1993) |
| *Celtis laevigata* | Seeds | + | | + | | + | | McKnight (1980) |
| *Celtis occidentalis* | Seeds | + | | + | | + | | McKnight (1980) |
| *Centella* sp. | Seeds | | ? | | | ? | | Cook (1990a) |
| *Cephalanthus occidentalis* | Seed, stem fragments | + | + | | | | | McKnight (1980); Middleton (1995a,b, 1996) |
| *Ceratophyllum* sp. | Seeds, adults | | ? | | | ? | | Cook (1990a) |
| *Ceratophyllum demersum* | Detached shoots, seeds turions, whole plants | o | + | + | | | | Arber (1920); Sculthorpe (1967); Waisel (1971); Junk (1986) |
| *Ceratopsis* sp. | Seedlings | | + | | | | | Cook (1987) |
| *Ceratopteris pteridoides* | Whole plant | | + | | | | | Junk (1986) |
| *Chamaecyparis thyoides* | Seeds | | + | | | | | Little (1965) |
| *Chara* sp. | Oospores | o | + | + | | | | Proctor (1962); Kamat (1967); Wade (1990) |
| *Chara aspera* | Tubers, shoots | | + | | | | | Kautsky (1990) |
| *Chara baltica* | Shoot chains, shoots | | + | | | | | Kautsky (1990) |

| | | | | | | | |
|---|---|---|---|---|---|---|---|
| *Chara fragilis* | Shoots | | + | | | | Kautsky (1990) |
| *Chara tomentosa* | Tubers, shoots | | + | | | | Kautsky (1990) |
| *Cicuta* sp. | Seeds | | ? | | ? | | Cook (1990a) |
| *Cladium mariscus* | Achene | | + | | | | Alexander (1971) |
| *Colocasia* sp. | Seeds | | | | + | | Cook (1990a) |
| *Commelina forskalii* | Seeds | | | | + | | Middleton and Mason (1992) |
| *Coix* sp. | Seeds | + | + | | | | van der Pijl (1982) |
| *Cornus drummondii* | Seeds | | | + | + | | McKnight (1980) |
| *Cornus florida* | Seeds | | | + | + | | Putnam et al. (1960) |
| *Cotula* sp. | Seeds | | | | ? | | Cook (1990a) |
| *Cymodocea* sp. | Seeds | | ? | | | | Cook (1990a) |
| *Cynodon dactylon* | Seeds | | | | + | + | Middleton and Mason (1992); Lonsdale and Lane (1994) |
| *Cynosciadium* sp. | Seeds | | ? | | ? | | Cook (1990a) |
| *Cyrptocoryne* sp. | Seeds | | + | | ? | | Cook (1990a) |
| *Cyrtosperma* sp. | Seeds | | + | | ? | | Cook (1990a) |
| *Cyperus erythrorhyzos* | Seeds | | + | | | | Middleton (1995a,b,c) |
| *Cyperus rotundus* | Seeds | | | | + | | Middleton and Mason (1992) |
| *Dactoloctenium aegyptium* | Seeds | | | | + | | Middleton and Mason (1992) |
| *Damosonium* sp. | Seeds | | ? | | ? | | Cook (1990a) |
| *Diospyros virginiana* | Seeds | | | + | + | | Morris (1965) |
| *Discolobium* sp. | Seeds | ? | | | ? | | Cook (1990a) |
| *Distichilis spicata* | Seeds | | | + | | | Vivian-Smith and Stiles (1994) |
| *Donax* sp. | Seeds | | + | | | | Cook (1990a) |
| *Dopatrium* sp. | Seeds | ? | ? | | | | Cook (1990a) |
| *Dracontioides* sp. | Seeds | | + | | ? | | Cook (1990a) |
| *Echinochloa* sp. | Seeds | | ? | | ? | | Cook (1990a) |
| *Echinochloa crus-galli* | Seeds | | + | | + | | Middleton (1995c); Middleton and Mason (1992) |
| *Echinodorus* sp. | Seeds, bulbils | | + | | + | | Cook (1990a) |
| *Echinodorus rostratus* | Seeds | | + | | | | Kaul (1978) |
| *Eclipta* spp. | Seeds | | + | | + | | Cook (1987, 1990) |
| *Egeria* sp. | Vegetative fragments | | + | | | | Cook (1990a) |
| *Egleria* sp. | Seeds | | + | | | | Cook (1990a) |

| Species | Propagule Type | Dispersal Mechanism | | | | | | References |
|---|---|---|---|---|---|---|---|---|
| | | | | Animals | | | | |
| | | Wind | Water | Bird | Fish | Animal | Other | |
| *Eichhornia* sp. | Seeds, whole plant, vegetative fragments | | + | | | | | Cook (1990a) |
| *Eichhornia crassipes* | Whole plant | | + | | | | | Junk (1986) |
| *Elatine* sp. | Seeds | | ? | | | | | Cook (1990a) |
| *Eleocharis acicularis* | Seeds | o | + | + | | | | Sculthorpe (1967); Cook (1990a) |
| *Elodea* sp. | Vegetative fragments | | + | | | | | Cook (1990a) |
| *Elodea canadensis* | Vegetative fragments, turions | o | + | + | | + | | Arber (1920); Sculthorpe (1967); van der Pijl (1982); Nichols and Shaw (1986) |
| *Eleocharis* sp. | Seeds | | | | | + | | Cook (1990a) |
| *Eleocharis variegata* | Whole plant | | + | | | | | Junk (1986) |
| *Elymus virginicus* | Seeds | | | + | | | | Vivian-Smith and Stiles (1994) |
| *Enhalus acoroides* | Seeds, seedlings | | + | | | | | van der Pijl (1982); Cook (1990a) |
| *Enhydra* spp. | Seeds | | | | | + | | Cook (1987) |
| *Ephemerum crassinervi* | Spores | | + | | | | | Conrad (1997) |
| *Epilobium anagallidifolium* | Seeds | | + | | | | | Danvind and Nilsson (1997) |
| *Equisetum* sp. | Spores | + | | | | | | Cook (1990a) |
| *Equisetum fluviatile* | Vegetative diaspores, internode sprouting | | + | | | | | Johansson and Nilsson (1993); Pearce and Cordes (1988) |
| *Erigeron* sp. | Seeds | | ? | | | | | Cook (1990a) |
| *Eriocaulon compressum* | Seeds | + | + | | | | | Conti and Gunther (1984) |
| *Eriophorum* spp. | Seeds | + | | | | | | Cook (1987, 1990a) |
| *Euphorbia prostrata* | Seeds | | | | | + | | Middleton and Mason (1992) |
| *Ephrasia frigida* | Seeds | | + | | | | | Danvind and Nilsson (1997) |
| *Euryale ferox* | Seeds | | + | ? | ? | | | van der Pijl (1982); Cook (1990a) |
| *Fagus grandiflora* | Seeds | | | | | + | | Rushmore (1965) |
| *Ficus* sp. | Seeds | | | | + | | | Goulding (1983) |
| *Forestiera acuminata* | Seeds | | + | + | | | | Maisenhelder (1958) |
| *Fraxinus caroliniana* | Seeds | + | + | | | | | Putnam et al. (1960) |

| | | | | | | | |
|---|---|---|---|---|---|---|---|
| *Fraxinus excelsior* | Seeds | + | | | | | Rydin and Borgegård (1991) |
| *Fraxinus pennsylvanica* | Seeds | + | + | | | | Middleton (1995a,b) |
| *Fraxinus profunda* | Seeds | + | + | | | | Hosner and Boycel (1962); Middleton (1995a) |
| *Galium aparine* | Seeds | | + | | + | + | Rydin and Borgegård (1991); Hodkinson and Thompson (1997) |
| *Gentiana nivalis* | Seeds | | + | | | | Danvind and Nilsson (1997) |
| *Geum rivale* | Fruits | | + | | | | Danvind and Nilsson (1997) |
| *Gleditsia aquatica* | Seeds | | + | + | + | | Putnam et al. (1960); Middleton (1995a,b) |
| *Gleditsia tricanthos* | Seeds | + | + | + | + | | Funk (1965) |
| *Glyceria* sp. | Seeds | ? | + | + | | | Arber (1920); Sculthorpe (1967) |
| *Groenlandica densa* | Seeds, vegetative fragments | | + | ? | | | Cook (1990a) |
| *Gynerium sagittatum* | Seeds | + | | | | | Cook (1987, 1990a) |
| *Gymnocornonis* sp. | Seeds | o | | | | | Cook (1990a) |
| *Halodule* sp. | Seeds | | + | | | | Cook (1990a) |
| *Halophila* sp. | Seeds | | + | | | | Cook (1990a) |
| *Hanguana* sp. | Seeds | | | | + | | Cook (1990a) |
| *Heteranthera* sp. | Seeds | | + | | | | Cook (1990a) |
| *Heterozostera tasmanica* | Seeds | | ? | | | | Cook (1990a) |
| *Hionanthera* sp. | Seeds | | | | ? | | Cook (1990a) |
| *Hippuris vulgaris* | Fruits, fragments of rhizomes | + | + | | | | Arber (1920); Bartley and Spence (1987) |
| *Hottonia palustris* | Seeds via water?, seedlings, detached shoots | + | + | + | | | Arber (1920); Sculthorpe (1967); Brock et al. (1989); van der Pijl (1972) |
| *Hydrilla verticillata* | Seeds, vegetative fragments, turions | | + | | | | Langeland and Sutton (1980); Cook (1990a) |
| *Hydrocera* sp. | Berries, seeds | | + | | + | | Cook (1990a) |
| *Hydrocharis* sp. | Seeds, turions | | + | | + | | Cook (1990a) |
| *Hydrocharis morsus-ranae* | Seeds, turions, seedlings | | + | + | ? | | Sculthorpe (1967) Bartley and Spence (1987); Cook (1990a) |
| *Hydrocotyle* sp. | Seeds, adults | | ? | | ? | | Cook (1990a) |
| *Hydropectus* spp. | Seeds | | | | ? | | Cook (1987, 1990a) |
| *Hydrostachys* sp. | Seeds | | ? | | | | Cook (1990a) |
| *Hydrothrix* sp. | Seeds | | + | | | | Cook (1990a) |
| *Hygrochloa* sp. | Seeds | | ? | | ? | | Cook (1990a) |
| *Hygrophila* sp. | Seeds | | | | | + | Cook (1990a) |

| Species | Propagule Type | Dispersal Mechanism | | | | | | References |
|---|---|---|---|---|---|---|---|---|
| | | | | Animals | | | | |
| | | Wind | Water | Bird | Fish | Animal | Other | |
| *Hygroryza* sp. | Seeds | | | | | + | | Cook (1990a) |
| *Hypericum elodes* | Adults | | + | | | | | Cook (1990a) |
| *Hypsela* sp. | Seeds | | | | | + | | Cook (1990a) |
| *Ilex cassine* | Seeds | | + | + | | | | Conti and Gunther (1984) |
| *Ipomoea aquatica* | Seeds, whole plants | | | | | + | | Middleton and Mason (1992) |
| *Iris pseudacorus* | Seeds | | + | | | | | Rydin and Borgegård (1991) |
| *Isoetes* sp. | Spores | | + | | + | | | Cook (1990a) |
| *Jaegeria* spp. | Seeds | | ? | | | ? | | Cook (1987); Cook (1990a) |
| *Juglans cinerea* | Seeds | | + | | | | | Middleton (1995a) |
| *Juglans nigra* | Seeds | | | | | + | | Brinkman (1965) |
| *Juncus* sp. | Seeds, vegetative fragments, seedlings, bulbils | | + | + | + | + | | Arber (1920); van der Pijl (1972); Cook (1990a) |
| *Juncus biglumis* | Seeds | | + | | | | | Danvind and Nilsson (1997) |
| *Juncus inflexus* | Seeds | + | | | | | | Fenner (1985) |
| *Jussisaea* sp. | Seeds | | + | | | | | Middleton (1995c) |
| *Justicia* sp. | Seeds | | | | | | + | Cook (1990a) |
| *Juncus trifidus* | Seeds | | + | | | | | Danvind and Nilsson (1997) |
| *Lagarosiphon* sp. | Seeds | | ? | | | | | Cook (1990a) |
| *Lagenandra* sp. | Seeds | | + | | | ? | | Cook (1990a) |
| *Languncularia racemosa* | Propagules | | + | | | | | McKee (1995a) |
| *Lasiomorpha* sp. | Seeds | | ? | | | ? | | Cook (1990a) |
| *Leersia hexandra* | Whole plant | | + | | | | | Junk (1986) |
| *Leersia oryzoides* | Seeds | | + | | | | | Middleton (1995a) |
| *Lemna* sp. | Whole plant, turions, seeds | | + | + | | ? | | Sculthorpe (1967); Bartley and Spence (1987); Cook (1990a) |
| *Lemna aequinoctialis* | Whole plant | | + | | | | | Junk (1986) |
| *Lemna valdiviana* | Whole plant | | + | | | | | Junk (1986) |
| *Licania* sp. | Seeds | | | | + | | | Goulding (1983) |
| *Lilaea* sp. | Seeds | | ? | | | | | Cook (1990a) |

| | | | | | | | |
|---|---|---|---|---|---|---|---|
| *Lilaeopsis* sp. | Seeds, vegetative fragments | | + | | | | Cook (1990a) |
| *Limnobium* spp. | Seeds, seedlings | | + | ? | | ? | Cook (1987) |
| *Limnobium spongium* | Whole plant, plant bases | | + | | | | Middleton (1996) |
| *Limnobium stolonifera* | Whole plant | | + | | | | Junk (1986) |
| *Limnocharis* sp. | Seeds | | + | | | + | Cook (1990a) |
| *Limnocharis flava* | Follicles (seeds sink) | | + | | | | Kaul (1978) |
| *Limnophylla* sp. | Seeds | ? | + | | | | Cook (1990a) |
| *Limnophyton* sp. | Seeds | | + | | | | Cook (1990a) |
| *Limnopoa meeboldii* | Seeds | | ? | | | | Cook (1990a) |
| *Limnosciadium* sp. | Seeds | | ? | | | | Cook (1990a) |
| *Limonium carolinianum* | Seeds | | | + | | | Vivian-Smith and Stiles (1994) |
| *Liparophyllum* sp. | Seeds | | | | | ? | Cook (1990a) |
| *Liriodendron tulipifera* | Seeds | + | | | | | Renshaw and Doolittle (1965) |
| *Lobelia* sp. | Seeds | | | | | + | Cook (1990a) |
| *Ludwigia* sp. | Seeds | | + | | | ? | Cook (1990a) |
| *Ludwigia natans* | Whole plant | | + | | | | Junk (1986) |
| *Luronium* sp. | Seeds, adults | | + | | | | Cook (1990a) |
| *Luronium natans* | Seeds, vegetative fragments | o | + | | | | Cook (1990a) |
| *Lycopus* sp. | Seeds | | + | | | ? | Cook (1990a) |
| *Lycopus europeas* | Seeds | | + | | | | Cook (1987) |
| *Lysimachia thyrisiflora* | Vegetative diaspores | | + | | | | Johansson and Nilsson (1993) |
| *Lythrum salicaria* | Seedlings | | + | | | | van der Pijl (1982) |
| *Lyonia lucida* | Seeds | + | + | | | | Conti and Gunther (1984) |
| *Maidenia* sp. | Seeds | | + | | | | Cook (1990a) |
| *Marsilea* sp. | Spores, adults | + | + | | ? | | Cook (1990a) |
| *Maundia* sp. | Seeds, root tubers | | ? | | | | Cook (1990a) |
| *Megalodonta* spp. | Seeds, turions(?) | | + | | | + | Cook (1987, 1990a) |
| *Melochia corchorifolia* | Seeds | | | | | + | Middleton and Mason (1992) |
| *Mensanthemum* sp. | Seeds | ? | | | | | Cook (1990a) |
| *Mentha* sp. | Seeds | | + | | | | Cook (1990a) |
| *Menyanthes trifoliata* | Vegetative diaspores | | + | | | | Johansson and Nilsson (1993) |
| *Mimulus* sp. | Seeds | | ? | | | | Cook (1990a) |
| *Mimulus guttatus* | Seeds | ? | + | | | | Waser et al. (1982) |
| *Mimulus luteus* | Seedlings | | + | | | | van der Pijl (1982) |

| Species | Propagule Type | Dispersal Mechanism | | | | | | References |
|---|---|---|---|---|---|---|---|---|
| | | | | Animals | | | | |
| | | Wind | Water | Bird | Fish | Animal | Other | |
| *Monochoria* sp. | Seeds | | + | | | | | Cook (1990a) |
| *Morus rubra* | Seeds | | | + | | + | | Putnam et al. (1960) |
| *Mourirai* sp. | Seeds | | | | + | | | Goulding (1983) |
| *Myrica cerifera* | Seeds | | + | + | | | | Conti and Gunther (1984) |
| *Myriophyllum* sp. | Seeds, vegetative fragments, turions, rhizomes, fruits, rhizome and root and/or tuber | o | + | + | | + | | Sculthorpe (1967); Aiken et al. (1979) Nichols and Shaw (1986); Coble and Vance (1987); Cook (1990a); Kautsky (1990) |
| *Myriophyllum exalbescens* | Turions | | + | | | | | Aiken and Walz (1979) |
| *Najas* sp. | Seeds | | + | | | | | Cook (1990a) |
| *Navarretia* sp. | Spiny inflorescence | | | | | ? | | Cook (1990a) |
| *Nelumbo* sp. | Seeds | | ? | | | ? | | Cook (1990a) |
| *Nelumbo lutea* | Fruits, seedlings | | + | | | | | van der Pijl (1982) |
| *Neostapfia* sp. | Seeds | | | | | ? | | Cook (1990a) |
| *Neptunia* sp. | Seeds | | | | | + | | Cook (1990a) |
| *Neptunia oleraceae* | Whole plant | | + | | | | | Junk (1986) |
| *Nesaea* sp. | Seeds | | | | | ? | | Cook (1990a) |
| *Nuphar* sp. | Seeds | | + | | | + | | Cook (1990a) |
| *Nuphar lutea* | Seeds in fruit, fragments of rhizomes | o | + | + | + | | | Heslop-Harrison (1955); van der Pijl (1982); Barbe (1984); Conti and Gunther (1984); Brock et al. (1987); Smits et al. (1989) |
| *Nymphaea* sp. | seeds; birds and bulbils | + | | | + | | | Cook (1990a) |
| *Nymphaea alba* | Seeds, fragments of rhizomes | o | + | + | + | | | Sculthorpe (1967); van der Pijl (1982); Smits et al. (1989) |
| *Nymphaea odorata* | Seeds | | + | | | | | Conti and Gunther (1984) |
| *Nymphoides* sp. | Seeds | | + | | | + | | Cook (1990a) |
| *Nymphoides orbiculata* | Seeds | | + | | | | | van der Pijl (1982) |
| *Nymphoides peltata* | Seeds, fragments of rhizomes, seedlings | o/+ | + | + | + | + | | van der Velde and van der Heijden (1981); Cook (1987, 1990b); Smits et al. (1989) |

| | | | | | | | | |
|---|---|---|---|---|---|---|---|---|
| *Nyssa aquatica* | Seeds | | + | | | | | Middleton (1995a,b) |
| *Nyssa sylvatica* var. *biflora* | Seeds | | + | + | | + | | DeBell and Hook (1969); Conti and Gunther (1984) |
| *Odinea* sp. | Seeds | | + | | | ? | | Cook (1990a) |
| *Odontelytrum* sp. | Seeds | | | ? | | ? | | Cook (1990a) |
| *Oenanthe* sp. | Seeds | | + | | | ? | | Cook (1990a) |
| *Orcuttia* sp. | Seeds | | | | | ? | | Cook (1990a) |
| *Orontium* spp. | Seeds, root and rhizome fragments | | + | | | ? | | Cook (1987, 1990a) |
| *Orontium aquaticum* | Seeds | | + | | | | | Conti and Gunther (1984) |
| *Oryza* sp. | Seeds | + | + | | | | | van der Pijl (1982) |
| *Oryza sativa* | Seeds | | | | | | + | Lonsdale and Lane (1994) |
| *Ottelia* sp. | Seeds | | + | | + | ? | | Cook (1990a) |
| *Oxypolis* sp. | Seeds | | ? | | | ? | | Cook (1990a) |
| *Oxyria digyna* | Fruits | | + | | | | | Danvind and Nilsson (1997) |
| *Pacourina* sp. | Seeds | | | | | ? | | Cook (1990a) |
| *Panicum* sp. | Seeds | | + | | | | | Middleton (1995c) |
| *Panicum virgatum* | Seeds | | | + | | | | Vivian-Smith and Stiles (1994) |
| *Paratheria* sp. | Seeds | | | | | ? | | Cook (1990a) |
| *Paspalum distichum* | Seeds, whole plants | | | | + | + | | Middleton (1989); Middleton and Mason (1992) |
| *Paspalidium flavidum* | Whole plants | | + | | | | | Personal observation (India) |
| *Paspalum repens* | Whole plant | | + | | | | | Junk (1986) |
| *Pedicularis sceptrum—carolinum* | Seeds | | + | | | | | Danvind and Nilsson (1997) |
| *Peltandra* sp. | Seeds | | + | | | ? | | Cook (1987, 1990a) |
| *Peplis portula* | Seeds | | + | | | | | van der Pijl (1982) |
| *Phalaris arundinacea* | Achenes, fragments of rhizomes | | + | + | o | | | Sculthorpe (1967); De Vlaming and Proctor (1968); Cook (1990a); Vivian-Smith and Stiles (1994) |
| *Phragmites* sp. | Seeds | + | | | | | | Cook (1990a) |
| *Phragmites australis* | Achenes, fragments of rhizomes | + | + | o | | | | De Vlaming and Proctor (1968); Barbe (1984); Bartley and Spence (1987); Cook (1987, 1990b) |
| *Phyllanthus fluitans* | Whole plant | | + | | | | | Junk (1986) |

| Species | Propagule Type | Dispersal Mechanism | | | | | | References |
|---|---|---|---|---|---|---|---|---|
| | | | | Animals | | | | |
| | | Wind | Water | Bird | Fish | Animal | Other | |
| *Picea glauca* | Seeds | + | | | | | | Walker et al. (1986) |
| *Pinus elliottii* var. *elliottii* | Seeds | + | | | | | | McKnight (1980) |
| *Pinus glabra* | Seeds | + | + | | | | | McKnight (1980) |
| *Pinus serotina* | Seeds | + | | | | | | Wenger (1965) |
| *Pinus taeda* | Seeds | + | | | | | | Wahlenberg (1960) |
| *Pistia* sp. | Seeds, seedlings, adults | | + | | | | | Cook (1987, 1990a) |
| *Pistia stratiotes* | Whole plant | | + | | | | | Junk (1986) |
| *Planera aquatica* | Seeds | | + | | | | | Middleton (1995a,b) |
| *Platanus occidentalis* | Seeds | + | + | + | | | | Merz (1965); Middleton (1995a) |
| *Polygonum* sp. | Achenes | | + | + | | + | | Staniforth and Cavers (1976); Cook (1990a) |
| *Polygonum plebeium* | Achenes | | | | | + | | Middleton and Mason (1992) |
| *Polygonum spectabile* | Whole plant | | + | | | | | Junk (1986) |
| *Polypogon monspeliensis* | Seeds | | | | | + | | Middleton and Mason (1992) |
| *Pontederia* sp. | Seeds | | + | | | ? | | Cook (1990a) |
| *Pontederia rotundifolia* | Whole plant | | + | | | | | Junk (1986) |
| *Populus balsamifera* | Seeds | + | | | | | | Walker et al. (1986) |
| *Populus deltoides* var. *deltoides* | Seeds | + | + | | | | | Maisenhelder (1958) |
| *Posidonia* sp. | Seeds | | ? | | | | | Cook (1990a) |
| *Potamogeton* sp. | Seeds | | + | + | | | | Cook (1990a) |
| *Potamogeton crispus* | Turions, seeds | o | + | | | | | Nichols and Shaw (1986); Kunii and Maeda (1989) |
| *Potamogeton indicus* | Seeds | | | | + | | | Middleton (1989) |
| *Potamogeton natans* | Seeds, turions, vegetative fragments | | + | + | + | | | Arber (1920); Smits et al. (1989); Wiegleb and Brux (1991) |
| *Potamogeton pectinatus* | Seeds, turions, rhizomes, tuber, vegetative fragments | o | + | + | + | | | Guppy (1897); De Vlaming and Proctor (1968); van Wijk (1983, 1986, 1989); Smits et al. (1989); Kautsky (1990) |

| | | | | | | | |
|---|---|---|---|---|---|---|---|
| *Potamogeton perfoliatus* | Turions, achenes, shoots | o | + | | | | Arber (1920); Kautsky (1990) |
| *Potentilla crantzii* | Seeds | | + | | | | Danvind and Nilsson (1997) |
| *Potentilla supina* | Seeds | | | | | + | Middleton and Mason (1992) |
| *Primula japonica* | Seedlings | | + | | | | van der Pijl (1982) |
| *Prosopus juliflora* | Seeds | | | | | + | Middleton and Mason (1992) |
| *Prunus serotina* | Seeds | | | + | | + | Hough (1965) |
| *Ptilmnium* sp. | Seeds | | ? | | | | Cook (1990a) |
| *Quercus* sp. | Seeds | + | | | | | Sharitz et al. (1990) |
| *Quercus alba* | Seeds | | | | | + | Minckler (1965) |
| *Quercus bicolor* | Seeds | | + | | | | Middleton (1995a,b) |
| *Quercus falcata* var. *pagodifolia* | Seeds | | | + | | + | Lotti (1965a) |
| *Quercus laurifolia* | Seeds | | + | + | | + | Halls (1977) |
| *Quercus lyrata* | Seeds | | + | + | | + | Morris (1965); Middleton (1995a,b) |
| *Quercus michauxii* | Seeds | | | | | + | Lotti (1965b) |
| *Quercus nigra* | Seeds | | + | + | | + | Toole (1965a) |
| *Quercus palustris* | Seeds | | + | + | | + | Minckler (1965); Middleton (1995a) |
| *Quercus phellos* | Seeds | | + | + | | + | Toole (1965b) |
| *Quercus shumardii* | Seeds | | + (rare) | | + | | Lotti (1965) |
| *Quercus stellata* var. *paludosa* | Seeds | | + | + | | + | McKnight (1980) |
| *Quercus virginiana* | Seeds | | | + | | + | Woods (1965) |
| *Ranalisma* sp. | Seeds, adults | | + | | | ? | Cook (1990a) |
| *Ranunculus* sp. | Seeds, vegetative fragments fragments of rhizomes | | + | + | | ? | Arber (1920); Cook (1990a) |
| *Ranunculus baudotti* | Vegetative fragments, rhizomes | | + | | | | Kautsky (1990) |
| *Ranunculus circinatus* | Vegetative fragments | | + | | | | Kautsky (1990) |
| *Ranunculus lingua* | Rhizomes | | + | | | | Johansson and Nilsson (1993) |
| *Reussia* sp. | Seeds | | + | | | ? | Cook (1990a) |
| *Rhizophora mangle* | Propagules | | + | | | | McKee (1995a) |
| *Rhus copallina* | Seeds | + | | | | | Sharitz et al. (1990) |
| *Rhyncoryza* sp. | Seeds | | ? | | | | Cook (1990a) |
| *Rhyncospora corniculata* | Seeds | | + | | | | Middleton (1995a) |
| *Riccia hirta* | Seeds | | + | | | | Conrad (1997) |

| Species | Propagule Type | Dispersal Mechanism | | | | | | References |
|---|---|---|---|---|---|---|---|---|
| | | | | Animals | | | | |
| | | Wind | Water | Bird | Fish | Animal | Other | |
| *Riccia hubenariana* subsp. *sullivantii* | Spores | | + | | | | | Conrad (1997) |
| *Ricciocarpus natans* | Spores, vegetative fragments, whole plants | | + | | | | | Junk (1986); Conrad (1997) |
| *Rosa palustris* | Seeds | | + | | | | | Middleton (1995a) |
| *Rotala* sp. | Seeds | | | + | | + | | Cook (1990a) |
| *Rumex* sp. | Seeds | | ? | | | ? | | Cook (1990a) |
| *Rumex dentatus* | Seeds | | | | | + | | Middleton and Mason (1992) |
| *Rumex orbiculatus* | Seeds | | + | | | | | Middleton (1995a) |
| *Ruppia* sp. | Seeds | | ? | | | | | Cook (1990a) |
| *Ruppia spiralis* | Plant fragment, rhizome | | + | + | | | | Kautsky (1990) |
| *Saccharum* spp. | Seeds | + | | | | | | Cook (1987, 1990a) |
| *Saccharum benghalensis* | Seeds | + | + | | | | | Dinerstein (1989) |
| *Saccharum spontaneum* | Seeds | + | + | | | | | Dinerstein (1989) |
| *Sacciolepis* sp. | Seeds | | ? | | | ? | | Cook (1990a) |
| *Sagittaria* sp. | Seeds, seedlings | | + | | | + | | van der Pijl (1982); Cook (1990a) |
| *Sagittaria sagittifolia* | Seeds, fruits, tubers | o | + | + | | | | Arber (1920); Sculthorpe (1967); De Vlaming and Proctor (1968); Cook (1990a) |
| *Salicornia herbacea* | Seedlings | | + | | | | | van der Pijl (1982) |
| *Salix* sp. | Seeds | | + | | | | | Middleton (1995a) |
| *Salix alaxensis* | Seeds | + | + | | | | | Putnam et al. (1960) |
| *Salix fragilis* | Seeds, floating twigs | + | + | | | | | Rydin and Borgegård (1991) |
| *Salix nigra* | Seeds | + | + | | | | | McKnight (1965) |
| *Salvinia* sp. | Seedlings, vegetative fragments | | + | | | | | Cook (1987, 1990a) |
| *Salvinia auriculata* | Whole plant | | + | | | | | Junk (1986) |
| *Salvinia minima* | Whole plant | | + | | | | | Junk (1986) |
| *Salvinia sprucei* | Whole plant | | + | | | | | Junk (1986) |
| *Sassafras albidium* | Seeds | | | + | | | | Halls (1977) |

| | | | | | | | |
|---|---|---|---|---|---|---|---|
| *Saururus* sp. | Seeds | | + | | | | Cook (1990a) |
| *Saussurea alpina* | Fruits | | + | | | | Danvind and Nilsson (1997) |
| *Saxifraga aizoides* | Seeds | | + | | | | Danvind and Nilsson (1997) |
| *Scirpus* sp. | Seeds | | | | | ? | Cook (1990a) |
| *Scirpus cubensis* | Whole plant | | + | | | | Junk (1986) |
| *Scirpus tuberosus* | Seeds | | | | + | + | Middleton (1989); Middleton and Mason (1992) |
| *Sclerolepis* spp. | Seeds | | ? | | | ? | Cook (1987, 1990a) |
| *Scrophularia aquatica* | Seedlings | | + | | | | van der Pijl (1982) |
| *Scutellaria lateriflora* | Seeds | | + | | | | Middleton (1995) |
| *Shinnersia* spp. | Seeds | | | | | + | Cook (1987) |
| *Sium* sp. | Seeds | | + | | | ? | Cook (1990a) |
| *Solanum nigrum* | Seeds | | | | | + | Middleton and Mason (1992) |
| *Sonchus asper* | Seeds | | | + | | | Vivian-Smith and Stiles (1994) |
| *Sonchus palustris* | Seeds | + | + | | | | van der Pijl (1982) |
| *Sparganium* sp. | Seeds, fragments of rhizomes | | + | | | + | Cook (1990a) |
| *Spartina alterniflora* | Seeds | | | + | | | Vivian-Smith and Stoles (1994) |
| *Spirodela* sp. | Seeds, whole plant | | + | | | + | Cook (1990a) |
| *Spirodela intermedia* | Whole plant | | + | | | | Junk (1986) |
| *Spirodela polyrhiza* | Whole plant, turions | | + | + | o | + | Hillman (1961); Bartley and Spence (1987) |
| *Stachys palustris* | Seedlings | | + | | | | van der Pijl (1982) |
| *Stratiotes* sp. | Adults | | + | | | | Cook (1990a) |
| *Subularia* sp. | Seeds | | + | | | | Cook (1990a) |
| *Sueda fructicosa* | Seeds | | | | | + | Middleton and Mason (1992) |
| *Syringodium* sp. | Seeds | | ? | | | | Cook (1990a) |
| *Taxodium ascendens* | Seeds | | + | | | | Conti and Gunther (1984) |
| *Taxodium distichum* | Seeds | | + | | | | Middleton (1995a,b) |
| *Tetragastris* sp. | Seeds | | | | + | | Goulding (1983) |
| *Thalassia* sp. | Seeds | | ? | | | | Cook (1990a) |
| *Thalassodendron* sp. | Seeds, seedlings | | + | | | | Cook (1990a) |
| *Thallia* sp. | Seeds | | | | | ? | Cook (1990a) |
| *Tilia cordata* | Seeds | + | | | | | Rydin and Borgegård (1991) |
| *Tococa* sp. | Seeds | | | | + | | Goulding (1983) |
| *Trapa* sp. | Seeds | | + | | | + | Cook (1990a) |

| Species | Propagule Type | Dispersal Mechanism | | | | | | References |
|---|---|---|---|---|---|---|---|---|
| | | | | Animals | | | | |
| | | Wind | Water | Bird | Fish | Animal | Other | |
| *Trapella* sp. | Seeds | | ? | | | + | | Cook (1990a) |
| *Trianthema portulacastrum* | Seeds | | | | | + | | Middleton and Mason (1992) |
| *Triglochin* sp. | Seeds | | + | | | | | Cook (1990a) |
| *Trollius europaeus* | Seeds | | + | | | | | Danvind and Nilsson (1997) |
| *Typha* sp. | Seeds, fragments of stolons, seedlings | + | + | | + | + | | Barbe (1984); Cook (1987, 1990a); Shipley et al. (1989) |
| *Typha latifolia* | Seeds | + | | | | | | Matlack (1987) |
| *Typhonodorum* sp. | Seeds | | + | | | | | Cook (1990a) |
| *Ulmus alata* | Seeds | + | + | | | | | McKnight (1980); Middleton (1995a) |
| *Ulmus americana* | Seeds | + | + | | | | | Guilkey (1965) |
| *Ulmus crassifolia* | Seeds | + | + | | | | | McKnight (1980) |
| *Ulmus glabra* | Seeds | + | | | | | | Rydin and Borgegård (1991) |
| *Ulmus rubra* | Seeds | + | + | | | | | Scholz (1965) |
| *Urosphatha* sp. | Seeds | | + | | | ? | | Cook (1990a) |
| *Utricularia* sp. | Turions, seeds | ? | + | | | + | | Arber (1920); Sculthorpe (1967); Cook (1990a) |
| *Utricularia foliosa* | Whole plant | | + | | | | | Junk (1986) |
| *Vallisneria* spp. | Seeds | | + | | | | | Cook (1987) |
| *Veronica* sp. | Seeds | | + | | | + | | Cook (1990a) |
| *Vetiveria nigritana* | Seeds | | | | | ? | | Cook (1990a) |
| *Vetiveria zizanioides* | Seeds | | | | | ? | | Cook (1990a) |
| *Victoria* sp. | Seeds | | + | | | ? | | Cook (1987, 1990a) |
| *Vitus aestivalis* | Seeds | + | | | | | | Sharitz et al. (1990) |
| *Vitus rotundifolia* | Seeds | + | | | | | | Sharitz et al. (1990) |
| *Vossia cuspidata* | Seeds | | | | | ? | | Cook (1990a) |
| *Websteria* sp. | Seeds | | | | | + | | Cook (1990a) |
| *Wiesneria* sp. | Seeds | | ? | | | | | Cook (1990a) |
| *Wolffia* sp. | Seeds, whole plant | | + | | | + | | Cook (1990a) |

| | | | | | | |
|---|---|---|---|---|---|---|
| *Wolffiella* sp. (including *Pseudowolfiella* and *Wolffiopsis*) | Seeds, whole plant | | + | | ? | Cook (1990a) |
| *Wolffiella lingulata* | Whole plant | | + | | | Junk (1986) |
| *Wolffiella tropica* | Whole plant | | + | | | Junk (1986) |
| *Wolffiella oblonga* | Whole plant | | + | | | Junk (1986) |
| *Xyris smalliana* | Seeds | + | + | | | Conti and Gunther (1984) |
| *Zannichellia* sp. | Seeds | | + | | ? | Cook (1990a) |
| *Zannichellia major* | Vegetative fragments | | + | | | Kautsky (1990) |
| *Zizania* spp. | Seeds | | + | ? | ? | Cook (1987, 1990a) |
| *Zizania aquatica* | Seeds | | | + | | Vivian-Smith and Stiles (1994) |
| *Zizaniopsis* sp. | Seeds | | | | ? | Cook (1990a) |
| *Ziziphus mauritiana* | Seeds | | | | + | Middleton and Mason (1992) |
| *Zostera* sp. | Seeds | | ? | | | Cook (1990a) |
| *Zostera marina* | Vegetative fragments | | + | | | Kautsky (1990) |
| *Zosterella* sp. | Seeds, vegetative fragments | | ? | | | Cook (1990a) |

o = absent, + = present, blank = not documented, ? = uncertain.

*Source:* Based on Barrat-Segretain and Amoros (1996) and others.

# *Appendix 2*

# *Seed Germination Requirements*

| Species | Germinates Under Water | Germinates in Moist, Unflooded | Salinity | Reference | Comment |
|---|---|---|---|---|---|
| *Acer barbatum* | — | Yes | — | McKnight (1980) | Germinates well on moist, mineral soil. |
| *Acer negundo* | — | Yes | — | McKnight (1980) | Germinates well on moist, mineral soil. |
| *Acer rubrum* | No | Yes | — | Hosner (1957); Hutnik and Yauney (1965); Schneider and Sharitz (1986) | Germinates well on moist, mineral soil. |
| *Acer rubrum* var. *drummondii* | No | Yes | — | Hosner (1957) | Germination rapid after soaking. |
| *Acer saccharinum* | No | Yes | — | Hosner (1957) | Germination rapid after soaking. |
| *Acer saccharum* | — | Yes | — | Weitzman and Hutnik (1965) | Germinates well on moist, mineral soil. |
| *Acnida cannabina* | Yes | Yes | — | Leck and Graveline (1979) | Depth of flooding not given. |
| *Acorus calamus* | — | Yes | — | Shipley and Parent (1991); Shipley et al. (1989) | |
| *Acrostichum danaeifolium* | No | Yes | — | van der Valk and Rosburg (1997) | |
| *Aeschynomene indica* | No | Yes | — | Mason (1996); Schneider and Sharitz (1986) | |
| *Agrostis perennans* | — | Yes | — | Wisheu and Keddy (1991) | |
| *Agrostis stolonifera* | — | Yes | — | Meredith (1985); Shipley et al. (1989); Shipley and Parent (1991); von der Valk and Verhoeven 1998 | |
| *Alisma plantago-aquatica* | Yes | Yes | — | Shipley et al. (1989); Stockey and Hunt (1992); Willis and Mitsch (1995) | Germination limited unless subjected to a flood/wet/flood treatment. |
| *Alternanthera sessilis* | No | Yes | — | Mason (1996) | |
| *Amaranthus australis* | No | Yes | — | van der Valk and Rosburg (1997) | |
| *Ammannia auriculata* | Yes | Yes | — | Haukos and Smith (1994); Mason (1996) | |
| *Ammannia baccifera* | Yes | Yes | — | Mason (1996) | |
| *Ammania coccinea* | — | Yes | — | Schneider and Sharitz (1986) | |
| *Ammannia multiflora* | Yes | Yes | — | Mason (1996) | |
| *Amischophacelus axillaris* | No | Yes | — | Mason (1996) | |
| *Ampelopsis arborea* | — | Yes | — | Schneider and Sharitz (1986) | |

| | | | | | |
|---|---|---|---|---|---|
| *Agrostis scabra* | — | Yes | — | Keddy and Reznicek (1982) | |
| *Artemisia campestris* | — | Yes | Yes | Galinato and van der Valk (1986) | Germinate in salinities from 0 to 5000 mg $l^{-1}$. |
| *Asclepias incarnata* | — | Yes | — | Shipley and Parent (1991); Willis and Mitsch (1995) | |
| *Aster brachyactis* | No | Yes | — | Pederson (1981) | Very limited germination under 2–3 cm of water. |
| *Aster tradescantii* | — | Yes | — | Wisheu and Keddy (1991) | |
| *Aster valhii* | — | Yes | — | Raffaele (1996) | |
| *Atriplex patula* | No | Yes | Yes | Hopkins and Parker (1984); Galinato and van der Valk (1986); van der Valk and Pederson (1989) | Seed bank samples from a brackish water site in Manitoba, Canada. Germinates in salinities from 0–5000 mg $l^{-1}$ |
| *Avicennia germinans* | — | — | — | McKee (1995a) | Propagules root only at higher elevations in the interidal zone |
| *Avicennia marina* | — | — | — | Clarke (1993) | Propagules must disperse within one week; remain in tides or submerged for up to five to seven months. Note: seeds germinate on trees before dispersal. |
| *Azolla* sp. | — | — | — | Cook (1990a) | Megapores germinate on the water surface. |
| *Baccharis halimifolia* | — | Yes | — | Schneider and Sharitz (1986) | |
| *Baccharis pilularis* var. consanguineus | — | Yes | — | Hopkins and Parker (1984) | Salt marsh seed bank watered with fresh water. |
| *Berchemia scandens* | — | Yes | — | Schneider and Sharitz (1986) | |
| *Bergia ammanniodes* | Yes | Yes | — | Mason (1996) | |
| *Berula erecta* | — | Yes | — | van der Valk and Verhoeven (1988) | |
| *Betula pubescens* | — | Yes | — | Wiegers (1985); van der Valk and Verhoeven (1988) | Seed germination is highest at ~10% soil moisture. |
| *Bidens cernua* | No | Yes | — | van der Valk and Davis (1978); Keddy and Reznicek (1982) | |
| *Bidens frondosa* | — | Yes | — | Schneider and Sharitz (1986); Shipley and Parent (1991) | |

| Species | Germinates Under Water | Germinates in Moist, Unflooded | Salinity | Reference | Comment |
|---|---|---|---|---|---|
| *Bidens laevis* | Yes | Yes | — | Leck and Graveline (1979) | Depth of flooding not given. |
| *Bidens pilosa* | Yes | No | Yes | Reddy and Singh (1992) | Germinates at 100m*M* but not at 200m*M*; burial at 10 cm eliminates germination. |
| *Blechnum penna-marina* | — | Yes | — | Raffaele (1996) | |
| *Blumea obliqua* | No | Yes | — | Mason (1996) | |
| *Boehmeria cylindrica* | — | Yes | — | Schneider and Sharitz (1986) | |
| *Brachyactis angusta* | No | Yes | — | Smith and Kadlec (1983) | |
| *Caesulia axillaris* | No | Yes | — | Mason (1996) | |
| *Calamagrostis canadensis* | — | Yes | — | Keddy and Reznicek (1982) | |
| *Calamagrostis canescens* | — | Yes | — | Meredith (1985); van der Valk and Verhoeven (1988) | |
| *Calamagrostis epigejos* | — | Yes | — | Meredith (1985) | |
| *Callitriche heterophylla* | Yes | Yes | — | Leck and Graveline (1979) | Depth of flooding not given. |
| *Caltha palustris* | — | Yes | — | van der Valk and Verhoeven (1988) | |
| *Calystegia sepium* | — | Yes | — | Meredith (1985) | |
| *Cardamine palustris* | — | Yes | — | Meredith (1985) | |
| *Cardamine pratensis* | — | Yes | — | van der Valk and Verhoeven (1988) | |
| *Cardamine valdiviana* | — | Yes | — | Raffaele (1996) | |
| *Carex* sp. | No | Yes | — | van der Valk and Davis (1978) | |
| *Carex atherodes* | No | Yes | — | Pederson (1981) | |
| *Carex bigelowii* | — | Yes | — | McGraw et al. (1991) | Seeds deeply buried below solifluction lobe. |
| *Carex buxbaumii* | — | Yes | — | Keddy and Reznicek (1982) | |
| *Carex crinita* | — | Yes | — | Shipley et al. (1989); Shipley and Parent (1991) | |
| *Carex curta* | — | Yes | — | van der Valk and Verhoeven (1988) | |
| *Carex diandra* | — | Yes | — | van der Valk and Verhoeven (1988) | |
| *Carex disticha* | — | Yes | — | van der Valk and Verhoeven (1988) | |

| | | | | | |
|---|---|---|---|---|---|
| *Carex echinata* | — | Yes | — | Keddy and Reznicek (1982); Wisheu and Keddy (1991) | |
| *Carex folliculata* | — | Yes | — | Shipley and Parent (1991) | |
| *Carex lasiocarpa* | — | Yes | — | Keddy and Reznicek (1982) | |
| *Carex lenticularis* | — | Yes | — | Wisheu and Keddy (1991) | |
| *Carex lupulina* | — | Yes | — | Shipley and Parent (1992) | |
| *Carex projecta* | — | Yes | — | Shipley et al. (1989); Shipley and Parent (1991) | |
| *Carex pseudocyperus* | — | Yes | — | van der Valk and Verhoeven (1988) | |
| *Carex retrorsa* | — | Yes | — | Shipley and Parent (1991) | |
| *Carex rostrata* | — | Yes | — | van der Valk and Verhoeven (1988) | |
| *Carex scoparia* | — | Yes | — | Keddy and Reznicek (1982) | |
| *Carex subantartica* | — | Yes | — | Raffaele (1996) | |
| *Carex tuckermani* | — | Yes | — | Shipley and Parent (1991) | |
| *Carex vulpinoidea* | — | Yes | — | Shipley et al. (1989) | |
| *Cephalanthus occidentalis* | — | Yes | — | Schneider and Sharitz (1986) | |
| *Ceratophyllum demersum* | Yes | No | — | van der Valk and Davis (1978); Kimber et al. (1995); Mason (1996) | |
| *Ceratopteris thalictroides* | Yes | Yes | — | Finlayson et al. (1990) | |
| *Chamaedaphne calyculata* | — | Yes | — | Keddy and Reznicek (1982) | |
| *Chara* spp. | Yes | — | Yes | Kautsky (1990); Kimber et al. (1995); Mason (1996) | Germinates in brackish water. |
| *Chara fibrosa* | Yes | Yes | — | van der Valk and Rosburg (1997) | |
| *Chenopodium album* | No | Yes | — | Smith and Kadlec (1983) | |
| *Chenopodium glaucum* | No | Yes | — | Pederson (1981) | |
| *Chenopodium polyspermum* | — | Yes | — | Shipley and Parent (1991) | |
| *Chenopodium rubrum* | No | Yes | Yes | Pederson (1981); Smith and Kadlec (1983); Galinato and van der Valk (1986) | Germination reduced at salinities above 1000 mg $l^{-1}$. Limited germination under 2–3 cm of water. |
| *Cicuta maculata* | No | Yes | — | Pederson (1981) | Limited germination under 2–3 cm of water. |

| Species | Germinates Under Water | Germinates in Moist, Unflooded | Salinity | Reference | Comment |
|---|---|---|---|---|---|
| *Cirsium arvense* | No | Yes | — | Pederson (1981) | |
| *Cirsium dissectum* | — | Yes | — | Meredith (1985) | |
| *Cirsium palustre* | — | Yes | — | Meredith (1985); van der Valk and Verhoeven (1988) | |
| *Cladium jamaicensis* | No | Yes | — | van der Valk and Rosburg (1997) | |
| *Cladium mariscoides* | — | Yes | — | Keddy and Reznicek (1982); Wisheu and Keddy (1991) | |
| *Cladium mariscus* | — | Yes | — | Alexander (1971) | |
| *Cochlearia cochleariodes* | No | Yes | — | Mason (1996) | |
| *Commelina benghalensis* | No | Yes | — | Mason (1996) | |
| *Commelina forskalii* | No | Yes | — | Mason (1996) | |
| *Commelina virginica* | — | Yes | — | Schneider and Sharitz (1986) | |
| *Coreopsis rosea* | — | Yes | — | Wisheu and Keddy (1991); Shipley and Parent (1991) | |
| *Coreopsis tinctoria* | — | Yes | — | Haukos and Smith (1994) | |
| *Crypsis schoenoides* | No | Yes | — | Mason (1996) | |
| *Cuscuta* sp. | Yes | Yes | — | Leck and Graveline (1979) | Depth of flooding not given. |
| *Cynodon dactylon* | No | Yes | — | Mason (1996) | |
| *Cyperus aristatus* | — | Yes | — | Shipley and Parent (1991) | |
| *Cyperus compressus* | No | Yes | — | Mason (1996) | |
| *Cyperus dentatus* | — | Yes | — | Wisheu and Keddy (1991) | |
| *Cyperus diandrus* | — | Yes | — | Shipley and Parent (1991) | |
| *Cyperus difformis* | No | Yes | — | Mason (1996) | |
| *Cyperus digitatus* | Yes | Yes | — | Finlayson et al. (1990) | |
| *Cyperus eragrostis* | — | Yes | — | Hopkins and Parker (1984) | Salt marsh seed bank watered with fresh water. |
| *Cyperus erythrorhizos* | No | Yes | — | Baskin et al. (1993) | Annual: flooded seeds will germinate in winter if brought out of flooding; flooding prevents germination. |
| *Cyperus esculentus* | — | Yes | — | Shipley and Parent (1991) | |

| | | | | | |
|---|---|---|---|---|---|
| *Cyperus ferruginescens* | — | Yes | — | Willis and Mitsch (1995) | |
| *Cyperus flavicomus* | No | Yes | — | Baskin et al. (1993) | Annual: flooded seeds will germinate in winter if brought out of flooding; flooding prevents germination. |
| *Cyperus iria* | No | Yes | — | Mason (1996) | |
| *Cyperus odoratus* | No | Yes | — | van der Valk and Davis (1978); van der Valk and Rosburg (1997) | |
| *Cyperus rivularis* | — | Yes | — | Shipley and Parent (1991) | |
| *Cyperus rotundus* | No | Yes | — | Mason (1996) | |
| *Cystopteris fragilis* (spore) | — | Yes | — | Raffaele (1996) | |
| *Dactyloctenium aegyptium* | No | Yes | — | Mason (1996) | |
| *Digitaria ischaemum* | — | Yes | — | Schneider and Sharitz (1986); Shipley and Parent (1991) | |
| *Distichilis spicata* | No | Yes | — | Smith and Kadlec (1983); Hopkins and Parker (1984) | |
| *Drosera* sp. | — | Yes | — | Wisheu and Keddy (1991) | |
| *Drosera intermedia* | — | Yes | — | Keddy and Reznicek (1982) | |
| *Drosera rotundifolia* | — | Yes | — | Keddy and Reznicek (1982); van der Valk and Verhoeven (1988) | |
| *Dulichium arundinaceum* | — | Yes | — | Shipley et al. (1989); Shipley and Parent (1991) | |
| *Echinochloa crus-galli* | Yes | Yes | — | Kennedy et al. (1987); Haukos and Smith (1994); Mason (1966) | |
| *Echinochloa crus-pavoni* | No | yes | — | Kennedy et al. (1987) | |
| *Echinochloa walteri* | — | Yes | — | Willis and Mitsch (1995) | |
| *Eclipta alba* | No | Yes | — | Schneider and Sharitz (1986); Mason (1996) | |
| *Elatine triandra* | No | Yes | — | Mason (1996) | |
| *Eleocharis* sp. | No | Yes | — | van der Valk and Davis (1978); Finlayson et al. (1990) | |
| *Eleocharis acicularis* | — | Yes | — | Wisheu and Keddy (1991) | |

| Species | Germinates Under Water | Germinates in Moist, Unflooded | Salinity | Reference | Comment |
|---|---|---|---|---|---|
| *Eleocharis atropurpurea* | No | Yes | — | Mason (1996) | |
| *Eleocharis elongata* | No | Yes | — | van der Valk and Rosburg (1997) | |
| *Eleocharis erythropoda* | — | Yes | — | Shipley et al. (1989); Shipley and Parent (1991) | |
| *Eleocharis obtusa* | — | Yes | — | Shipley and Parent (1991) | |
| *Eleocharis palustris* | No | Yes | — | Pederson (1981); Smith and Kadlec (1983); van der Valk and Verhoeven (1988) | Limited germination under water. |
| *Eleocharis parishii* | Yes | No | — | Smith and Kadlec (1983) | |
| *Eleocharis smallii* | — | Yes | — | Shipley et al. (1989); Shipley and Parent (1991) | |
| *Eleocharis tenuis* | — | Yes | — | Wisheu and Keddy (1991) | |
| *Elodea canadensis* | Yes | — | — | Kimber et al. (1995) | |
| *Elymus virginicus* | — | Yes | — | Shipley et al. (1989) | |
| *Epilobium adenocaulon* | — | Yes | — | Smith and Kadlec (1983); Wisheu and Keddy (1991) | |
| *Epilobium ciliatum* | — | Yes | — | Shipley and Parent (1991) | |
| *Epilobium glandulosum* | No | Yes | — | Pederson (1981) | |
| *Epilobium hirsutum* | — | Yes | — | van der Valk and Verhoeven (1988) | |
| *Epilobium nivale* | — | Yes | — | Raffaele (1996) | |
| *Epilobium palustre* | — | Yes | — | van der Valk and Verhoeven (1988) | |
| *Epilobium parviflorum* | — | Yes | — | van der Valk and Verhoeven (1988) | |
| *Eragrostis hypnoides* | — | Yes | — | Shipley and Parent (1991) | |
| *Eragrostis tenella* | No | Yes | — | Mason (1996) | |
| *Erigeron canadensis* | — | Yes | — | Schneider and Sharitz (1986) | |
| *Eriocaulon compressum* | Yes | Yes | — | Conti and Gunther (1984) | |
| *Eriocaulon septangulare* | — | Yes | — | Keddy and Reznicek (1982); Shipley and Parent (1991) | |
| *Eriocaulon setaceum* | Yes | Yes | — | Finlayson et al. (1990) | |
| *Eriochloa procera* | No | Yes | — | Mason (1996) | |

| | | | | | |
|---|---|---|---|---|---|
| *Eupatorium cannabinum* | — | Yes | — | Meredith (1985); van der Valk and Verhoeven (1988) | |
| *Eupatorium capillifolium* | — | Yes | — | Schneider and Sharitz (1986) | |
| *Eupatorium maculatum* | — | Yes | — | Shipley and Parent (1991) | |
| *Eupatorium perfoliatum* | — | Yes | — | Shipley and Parent (1991) | |
| *Euthemia galetorum* | — | Yes | — | Shipley and Parent (1991) | |
| *Filipendula ulmaria* | — | Yes | — | Meredith (1985) | |
| *Fimbristylis aestivalis* | Yes | Yes | — | Finlayson et al. (1990) | |
| *Fimbristylis autumnalis* | No | Yes | — | Baskin et al. (1993) | Annual: flooded seeds will germinate in winter if brought out of flooding; flooding prevents germination. |
| *Fimbristylis vahlii* | No | Yes | — | Baskin et al. (1993) | Annual: flooded seeds will germinate in winter if brought out of flooding; flooding prevents germination. |
| *Frangula alnus* | — | Yes | — | Meredith (1985) | |
| *Frankenia grandifolia* | — | Yes | — | Hopkins and Parker (1984) | Salt marsh seed bank watered with fresh water. |
| *Fraxinus caroliniana* | — | — | — | Titus (1991) | Common in seed bank in spring, does not persist until fall |
| | — | Yes | — | Schneider and Sharitz (1986) | |
| *Fraxinus pennsylvanica* | Yes | Reduced | — | DuBarry (1963) | 46% germination under water, 5% in saturated conditions. |
| *Galium palustre* | — | Yes | — | Meredith (1985); van der Valk and Verhoeven (1988) | |
| *Galium trifidum* | No | Yes | — | Pederson (1981); Smith and Kadlec (1983) | |
| *Gleditsia aquatica* | — | Yes | — | Putnam et al. (1960) | Germinates well on moist mineral soil. |
| *Glinus lotoides* | No | Yes | — | Mason (1966) | |
| *Glinus oppositifolius* | Yes | Yes | — | Finlayson et al. (1960); Mason (1996) | |
| *Glossostigma spathulatum* | Yes | Yes | — | Mason (1996) | |
| *Gnaphalium luteo-album* | — | Yes | — | Hopkins and Parker (1984) | Salt marsh seed bank watered with fresh water. |
| *Gnaphalium polycaulon* | No | Yes | — | Mason (1996) | |
| *Gnaphalium purpureum* | — | Yes | — | Schneider and Sharitz (1986) | |

| Species | Germinates Under Water | Germinates in Moist, Unflooded | Salinity | Reference | Comment |
|---|---|---|---|---|---|
| *Gnaphalium uliginosum* | — | Yes | — | Keddy and Reznicek (1982); Wisheu and Keddy (1991) | |
| *Grangea maderaspatana* | No | Yes | — | Mason (1996) | |
| *Gratiola aurea* | — | Yes | — | Wisheu and Keddy (1991); Shipley and Parent (1991) | |
| *Gratiola virginica* | — | Yes | — | Schneider and Sharitz (1986) | |
| *Grindelia humilis* | — | Yes | — | Hopkins and Parker (1984) | Salt marsh seed bank watered with fresh water. |
| *Gunnera magellanica* | — | Yes | — | Raffaele (1996) | |
| *Heliotropium indicum* | Yes | Yes | — | Finlayson et al. (1990) | |
| *Hemiadelphus polysper-mum* | No | Yes | — | Mason (1996) | |
| *Heteranthera dubia* | Yes | No | — | Haukos and Smith (1994) | |
| *Hibiscus lobatus* | No | Yes | — | Mason (1996) | |
| *Holcus lanatus* | — | Yes | — | van der Valk and Verhoeven (1988) | |
| *Hordeum jubatum* | No | Yes | Yes | Pederson (1981); van der Valk and Pederson (1989); Galinato and van der Valk, 1986. | Seed bank samples from a brackish water site in Manitoba, Canada. Germinate in salinities from 0 to 5000 mg $l^{-1}$. |
| *Hottonia palustris* | Yes | Yes | — | Brock et al. (1989) | Germination lower when submersed than on moist substrate. |
| *Hydrilla verticillata* | Yes | No | — | Mason (1996) | |
| *Hydrocotyle verticillata* | — | Yes | — | Schneider and Sharitz (1986) | |
| *Hydrocotyle vulgaris* | — | Yes | — | Meredith (1985); van der Valk and Verhoeven (1988) | |
| *Hydrolea zeylandica* | No | Yes | — | Mason (1996) | |
| *Hygrochloa aquatica* | Yes | Yes | — | Finlayson et al. (1990) | |
| *Hymenachne acutigluma* | No | Yes | — | Finlayson et al. (1990) | |
| *Hypericum boreale* | — | Yes | — | Keddy and Reznicek (1982); Wisheu and Keddy (1991) | |

| | | | | | |
|---|---|---|---|---|---|
| *Hypericum canadense* | — | Yes | — | Keddy and Reznicek(1982) | |
| *Hypericum ellipticum* | — | Yes | — | Shipley and Parent (1991) | |
| *Hypericum hypericoides* | — | Yes | — | Schneider and Sharitz (1986) | |
| *Hypericum majus* | — | Yes | — | Keddy and Reznicek (1982) | |
| *Hypericum walteri* | — | Yes | — | Schneider and Sharitz (1986) | |
| *Ilex cassine* | Yes | Yes | — | Conti and Gunther (1984) | Germinates in light or dark. |
| *Impatiens capensis* | No | Yes | — | Pederson (1981) | |
| *Iris pseudoacorus* | — | Yes | — | Meredith (1985); van der Valk and Verhoeven (1988) | |
| *Iris versicolor* | — | Yes | — | Shipley et al. (1989); Shipley and Parent (1991) | |
| *Isoetes coromandelina* (spores) | Yes | Yes | — | Finlayson et al. (1990) | |
| *Itea virginica* | — | Yes | — | Schneider and Sharitz (1986) | |
| *Jaumea carnosa* | — | Yes | — | Hopkins and Parker (1984) | Salt marsh seed bank watered with fresh water. |
| *Juncus articulatus* | — | Yes | — | Meredith (1985); van der Valk and Verhoeven (1988) | |
| *Juncus brevicaudatus* | — | Yes | — | Wisheu and Keddy (1991) | |
| *Juncus bufonius* | — | Yes | — | van der Valk and Verhoeven (1988); Shipley and Parent (1991) | |
| *Juncus canadensis* | — | Yes | — | Keddy and Reznicek (1982); Wisheu and Keddy (1991); Raffaele (1996) | |
| *Juncus chilensis* | — | Yes | — | Raffaele (1996) | |
| *Juncus effusus* | — | Yes | — | van der Valk and Verhoeven (1988); Wisheu and Keddy (1991); Shipley and Parent (1991) | |
| *Juncus filiformis* | — | yes | — | Shipley et al. (1989); Wisheu and Keddy (1991); Shipley and Parent (1991) | |
| *Juncus gerardii* | Yes | — | Yes | Kautsky (1990) | Germinates in brackish water. |
| *Juncus ingens* | — | Yes | — | Sorrell and Armstrong (1994) | Seeds germinated in waterlogged conditions. |

| Species | Germinates Under Water | Germinates in Moist, Unflooded | Salinity | Reference | Comment |
|---|---|---|---|---|---|
| *Juncus pelocarpus* | — | Yes | — | Keddy and Reznicek (1982); Wisheu and Keddy (1991) | |
| *Juncus stipulatus* | — | Yes | — | Raffaele (1996) | |
| *Juncus subnodulosus* | — | Yes | — | Meredith (1985) | |
| *Lactuca* sp. | — | Yes | — | Schneider and Sharitz (1986) | |
| *Laggera aurita* | No | Yes | — | Mason (1996) | |
| *Laguncularia racemosa* | — | — | — | McKee (1995a) | Propagules root only at higher elevations in the intertidal zone. |
| *Leersia lenticularis* | — | Yes | — | Schneider and Sharitz (1986) | |
| *Leersia oryzoides* | Yes | Yes | — | van der Valk and Davis (1978); Keddy and Reznicek (1982); Smith and Kadlec (1983) | Seeds submerged to 1 cm germinate. |
| *Lemna* sp. | Yes | Yes | — | van der Valk and Rosburg (1997) | |
| *Leucothoe racemosa* | — | Yes | — | Conti and Gunther (1984) | |
| *Limnophila brownii* | Yes | Yes | — | Finlayson et al. (1990) | |
| *Limnophila indica* | No | Yes | — | Mason (1996) | |
| *Limnophyton obtusifolium* | No | Yes | — | Mason (1996) | |
| *Linaria canadensis* | — | Yes | — | Schneider and Sharitz (1986) | |
| *Lindernia dubia* | — | Yes | — | Willis and Mitsch (1995) | |
| *Lindernia parviflora* | Yes | Yes | — | Mason (1996) | |
| *Linum striatum* | — | Yes | — | Keddy and Reznicek (1982) | |
| *Liquidambar styraciflua* | — | Yes | — | Putnam et al. (1960); Schneider and Sharitz (1986) | Germinates well on mineral soil in full sunlight. |
| *Lobelia dortmanna* | — | Yes | | Wisheu and Keddy (1991) | |
| *Ludwigia decurrens* | — | Yes | — | Schneider and Sharitz (1986) | |
| *Ludwigia leptocarpa* | — | Yes | — | Schneider and Sharitz (1986) | |
| *Ludwigia palustris* | — | Yes | — | Leck and Graveline (1979); Keddy and Reznicek (1982); Schneider and Sharitz (1986); Willis and Mitsch (1995) | |

| | | | | | |
|---|---|---|---|---|---|
| *Ludwigia perennis* | Yes | Yes | — | Mason (1996) | |
| *Luzula parviflora* | — | Yes | — | Bennington et al. (1991) | Long seed longevity (up to 197 years). |
| *Lychnis flos-cuculi* | — | Yes | — | van der Valk and Verhoeven (1988) | |
| *Lycopus americanus* | — | Yes | — | Shipley and Parent (1991) | |
| *Lycopus asper* | No | Yes | — | Pederson (1981) | |
| *Lycopus europeas* | — | Yes | — | Thompson (1974); van der Valk and Verhoeven (1988) | |
| *Lycopus lucidus* | No | Yes | — | Smith and Kadlec (1983) | |
| *Lycopus uniflorus* | — | Yes | — | Keddy and Reznicek (1982); Shipley et al. (1989); Wisheu and Keddy (1991) | |
| *Lyonia lucida* | Yes | Yes | — | Conti and Gunther (1984) | |
| *Lysimachia terrestris* | — | Yes | — | Keddy and Reznicek (1982) | |
| *Lysimachia thrysiflora* | — | Yes | — | van der Valk and Verhoeven (1988) | |
| *Lythrum salicaria* | — | Yes | — | Meredith (1985); van der Valk and Verhoeven (1988); Shipley et al. (1989); Shipley and Parent (1991) | |
| *Maidenia rubra* | Yes | Yes | — | Finlayson et al. (1990) | |
| *Marsilea minuta* (spore) | Yes | Yes | — | Mason (1996) | |
| *Melochia corchorifolia* | No | Yes | — | van der Valk and Davis (1978) | |
| *Mentha aquatica* | — | Yes | — | Meredith (1985); van der Valk and Verhoeven (1988) | |
| *Mentha arvensis* | No | Yes | — | Pederson (1981) | |
| *Menyanthes trifoliata* | — | Yes | — | van der Valk and Verhoeven (1988) | |
| *Mercardonia acuminata* | — | Yes | — | Schneider and Sharitz (1986) | |
| *Mikania scandens* | — | Yes | — | Leck and Graveline (1979); Schneider and Sharitz (1986); van der Valk and Rosburg (1997) | |
| *Mimulus parviflorus* | — | Yes | — | Raffaele (1996) | |
| *Mimulus ringens* | — | Yes | — | Shipley and Parent (1991) | |
| *Molinia caerulea* | — | Yes | — | Meredith (1985) | |
| *Monochoria vaginalis* | No | Yes | — | Mason (1996) | |
| *Muhlenbergia uniflora* | — | Yes | — | Keddy and Reznicek (1982); Wisheu and Keddy (1991) | |

| Species | Germinates Under Water | Germinates in Moist, Unflooded | Salinity | Reference | Comment |
|---|---|---|---|---|---|
| *Myrica cerifera* | Yes | Yes | — | Conti and Gunther (1984) | No germination in dark. |
| *Myriophyllum spicatum* | Yes | Yes | — | Coble and Vance (1987); Kimber et al. (1995) | Germinates in water depths of less than 50 cm, but only under bright light. |
| *Najas graminea* | Yes | No | — | Mason (1996) | |
| *Najas flexilis* | Yes | No | — | van der Valk and Davis (1978); Kimber et al. (1995) | |
| *Najas marina* | Yes | — | Yes | Kautsky (1990) | Germinates in brackish water. |
| *Najas tenuifolia* | Yes | Yes | — | Finlayson et al. (1990) | |
| *Nothosaerva brachiata* | No | Yes | — | Mason (1996) | |
| *Nuphar lutea* | Yes | Yes | — | Heslop-Harrison (1955); Conti and Gunther (1984) | Germinates in light or dark; highest percentage of seeds germinates in dark, submerged conditions. |
| *Nuphar intermedia* | — | Yes | — | Heslop-Harrison (1955) | |
| *Nymphaea nouchali* | Yes | No | — | Mason (1996) | |
| *Nymphaea odorata* | Yes | Yes | — | Conti and Gunther (1984); van der Valk and Rosburg (1997) | Germinates in light or dark. |
| *Nymphaea violacea* | Yes | Yes | — | Finlayson et al. (1990) | |
| *Nymphoides cristatum* | Yes | Yes | — | Mason (1996) | |
| *Nyssa aquatica* | No | Yes | — | DeBell and Auld (1971); Priester (1980); Schneider and Sharitz (1986) | Germinates in full sunlight; seeds remain viable even if submerged in water for months; seeds float. |
| *Nyssa sylvatica* var. *biflora* | — | Yes | — | DeBell and Hook (1969); Conti and Gunther (1984) | Seeds do not float. |
| *Oenanthe aquatica* | — | Yes | — | van der Valk and Verhoeven (1988) | |
| *Oenanthe fistulosa* | — | Yes | — | van der Valk and Verhoeven (1988) | |
| *Oldenlandica uniflora* | — | Yes | — | Schneider and Sharitz (1986) | |
| *Onoclea sensibilis* (spore) | — | Yes | — | Schneider and Sharitz (1986) | |
| *Opuntia compressa* | — | Yes | — | Schneider and Sharitz (1986) | |
| *Orontium aquaticum* | No | Yes | — | Conti and Gunther (1984) | Germinates in light or dark. |
| *Oryza meridionalis* | Yes | Yes | — | Finlayson et al. (1990) | |

| | | | | | |
|---|---|---|---|---|---|
| *Oryza rufipogon* | Yes | Yes | — | Mason (1996) | |
| *Panicum capillare* | — | Yes | — | Shipley and Parent (1991) | |
| *Panicum dichotomiflorum* | — | Yes | — | Baskin and Baskin (1983) | In spring, seed germination at 20/20, 25/15/, 30/15 and 35/20°C diurnal temperture fluctuation; in fall, germination only at 35/20°C; seeds lose ability to germinate at all by late fall; regain ability by January. |
| *Panicum implicatum* | — | Yes | — | Keddy and Reznicek (1982) | |
| *Panicum lanuginosum* | — | Yes | — | Wisheu and Keddy (1991) | |
| *Panicum longifolium* | — | Yes | — | Shipley and Parent (1991) | |
| *Panicum rigidulum* | — | Yes | — | Wisheu and Keddy (1991) | |
| *Panicum spretum* | — | Yes | — | Keddy and Reznicek (1982) | |
| *Paspalidium punctatum* | No | Yes | — | Mason (1996) | |
| *Paspalum distichum* | No | Yes | — | Middleton et al. (1991) | |
| *Paulownia tomentosa* | No | Yes | — | Leck and Graveline (1979) | Depth of flooding not given. |
| *Pedicularis palustris* | — | Yes | — | van der Valk and Verhoeven (1988) | |
| *Peltandra virginica* | Yes | Yes | — | Leck and Graveline (1979) | Depth of flooding not given. |
| *Penthorum sedoides* | — | Yes | — | Schneider and Sharitz (1986); Shipley et al. (1989); Shipley and Parent (1991) | ; |
| *Peplidium maritinum* | No | Yes | — | Mason (1996) | |
| *Persicaria lapathifolia* | No | Yes | — | Haukos and Smith (1994) | |
| *Peucedanum palustre* | — | Yes | — | Meredith (1985); van der Valk and Verhoeven (1988) | |
| *Phalaris arundinacea* | — | Yes | — | Meredith (1985); Shipley et al. (1989); Shipley and Parent (1991) | |
| *Phragmites australis* | — | Yes | Yes | Pederson (1981); Smith and Kadlec (1983); Galinato and van der Valk (1986); van der Valk and Verhoeven (1988) | Germinate in salinities from 0 to 5000 mg $l^{-1}$; higher germination in saline water than in fresh water; limited germination in 2–3 cm of water. |
| *Phytolacca americana* | — | Yes | — | Schneider and Sharitz (1986) | |
| *Pilea pumila* | Yes | Yes | — | Leck and Graveline (1979) | Depth of flooding not given. |

| Species | Germinates Under Water | Germinates in Moist, Unflooded | Salinity | Reference | Comment |
|---|---|---|---|---|---|
| *Plantago major* | — | Yes | — | van der Valk and Verhoeven (1988); Shipley et al. (1989); Shipley and Parent (1991) | |
| *Pluchea camphorata* | — | Yes | — | Schneider and Sharitz (1986) | |
| *Poa trivialis* | — | Yes | — | van der Valk and Verhoeven (1988) | |
| *Polygonum amphibium* | — | Yes | — | Wisheu and Keddy (1991) | |
| *Polygonum arifolium* | Yes | Yes | — | Leck and Graveline (1979) | Depth of flooding not given. |
| *Polygonum aviculare* | — | Yes | — | Hopkins and Parker (1984) | Salt marsh seed bank watered with fresh water. |
| *Polygonum careyi* | — | Yes | — | Keddy and Reznicek (1982) | |
| *Polygonum hydropiperoides* | — | Yes | — | Schneider and Sharitz (1986); Shipley and Parent (1991) | |
| *Polygonum lapathifolium* | — | Yes | — | van der Valk and Davis (1978); Willis and Mitsch (1995) | |
| *Polygonum persicaria* | No | Yes | — | Keddy and Reznicek (1982); Smith and Kadlec (1983) | |
| *Polygonum plebeium* | No | Yes | — | Mason (1996) | |
| *Polygonum punctatum* | No | Yes | — | van der Valk and Davis (1978) | |
| *Polygonum sagittatum* | Yes | Yes | — | Leck and Graveline (1979) | Depth of flooding not given. |
| *Polypogon monspeliensis* | Yes | Yes | — | Smith and Kadlec (1983) | |
| *Pontederia cordata* | Yes | Yes | — | Wisheu and Keddy (1991); Kimber et al. (1995) | |
| *Populus balsamifera* | No | Yes | — | Moss (1938) | Seeds germination and seedling establishment occur only under moist conditions. |
| *Populus deltoides* | Yes | Yes | — | Hosner (1957) | Seeds sink in water and germinate. |
| | No | Yes | — | Pederson (1981) | |
| *Populus deltoides* var. *deltoides* | Yes | Yes | — | Maisenhelder (1958) | Seeds remain viable for less than two weeks. |

| | | | | | |
|---|---|---|---|---|---|
| *Populus deltoides* var. *monilifera* | No | Yes | — | Scott et al. (1993b) | Seeds remain viable for only a few weeks. |
| *Populus fremontii* | — | Yes | — | Fenner et al. (1984) | |
| *Populus heterophylla* | — | Yes | — | Johnson and Beaufait (1965a) | Seeds are short-lived. |
| *Populus petrowskyana* | — | Yes | — | Moss (1938) | Seed germination and seedling establishment occur only when under moist conditions. |
| *Populus sargentii* | — | Yes | — | Moss (1938) | Seeds germination and seedling establishment occur only during moist conditions. |
| *Potamogeton foliosus* | Yes | — | — | Kimber et al. (1995) | |
| *Potamogeton pectinatus* | Yes | No | Yes | Pederson (1981); van Wijk (1983, 1989); van der Valk and Pederson (1989) | Germinated seeds or seedlings rarely observed in the field (Camargue, the Netherlands, and the northern Baltic; van Wijk, 1983, and Kautsky, 1990, respectively); in lab, germination occurs in 0–6‰ $Cl^-$; germination precentage higher after passage through duck (Guppy 1897). |
| *Potentilla palustris* | — | Yes | — | van der Valk and Verhoeven (1988) | |
| *Potentilla supina* | No | Yes | — | Mason (1996) | |
| *Prunus serotina* | No | Yes | — | Hough (1965) | Seeds germinate in bare mineral soil or in leaf litter. |
| *Pseudoraphis spinescens* | Yes | Yes | — | Finlayson et al. (1990) | |
| *Quercus falcata* | — | Yes | — | Larsen (1963) | |
| *Quercus laurifolia* | — | Yes | — | Larsen (1963) | |
| *Quercus lyrata* | — | Yes | — | Larsen (1963) | Soaking delays germination. |
| *Quercus phellos* | — | Yes | — | Larsen (1963) | |
| *Ranunculus lingua* | — | Yes | — | van der Valk and Verhoeven (1988) | |
| *Ranunculus reptans* | — | Yes | — | Wisheu and Keddy (1991) | |

| Species | Germinates Under Water | Germinates in Moist, Unflooded | Salinity | Reference | Comment |
|---|---|---|---|---|---|
| *Ranunculus scleratus* | Yes | Yes | — | Pederson (1981) | |
| *Rhamnus cathartica* | — | Yes | — | Meredith (1985) | |
| *Rhexia virginica* | — | Yes | — | Keddy and Reznicek (1982) | |
| *Rhizophora mangle* | — | — | — | McKee (1995a) | Propagules root only at lower elevations in the intertidal zone. |
| *Rhus copallina* | — | Yes | — | Schneider and Sharitz (1986) | |
| *Rhyncospora capillacea* | — | Yes | — | Shipley and Parent (1991) | |
| *Rhyncospora capitellata* | — | Yes | — | Keddy and Reznicek (1982) | |
| *Rhyncospora fusca* | — | Yes | — | Keddy and Reznicek (1982); Wisheu and Keddy (1991) | |
| *Rorippa islandica* | No | Yes | — | van der Valk and Davis (1978); Pederson (1981) | |
| *Rorippa sinuata* | No | Yes | — | Haukos and Smith (1994) | |
| *Rotala indica* | Yes | Yes | — | Mason (1996) | |
| *Rubus* sp. | — | Yes | — | Wisheu and Keddy (1991) | |
| *Rubus arundelanus* | — | Yes | — | Keddy and Reznicek (1982) | |
| *Rumex acetosa* | No | Yes | — | Voesenek et al. (1992) | Germinates at a wide range of temperatures. |
| *Rumex acetosella* | — | Yes | — | Schneider and Sharitz (1986) | |
| *Rumex crispus* | No | Yes | — | Smith and Kadlec (1983); Voesenek et al. (1992); Haukos and Smith (1994) | Maximum germination at 25/19°–25/13°C diurnal temperature fluctuation. |
| *Rumex dentatus* | No | Yes | — | Mason (1996) | |
| *Rumex hydrolapathum* | — | Yes | — | van der Valk and Verhoeven (1988) | |
| *Rumex maritimus* | No | Yes | — | van der Valk and Davis (1978); Pederson (1981) | Limited germination in 2–3 cm of water. |
| *Rumex verticillatus* | — | Yes | — | Shipley et al. (1989); Shipley and Parent (1991) | |
| *Ruppia spiralis* | — | Yes | Yes | Kautsky (1990) | Germinates in brackish water. |
| *Sabatia kennedyana* | — | Yes | — | Wisheu and Keddy (1991); Shipley and Parent (1991) | |
| *Sagittaria* sp. | Yes | — | — | Kimber et al. (1995) | |

| | | | | | |
|---|---|---|---|---|---|
| *Sagittaria guayanensis* | Yes | No | — | Mason (1996) | |
| *Sagittaria lancifolia* | Yes | Yes | — | van der Valk and Rosburg (1997) | |
| *Sagittaria latifolia* | No | Yes | — | van der Valk and Davis (1978); Schneider and Sharitz (1986); Shipley et al. (1989); Willis and Mitsch (1995) | |
| *Sagittaria longiloba* | Yes | No | — | Haukos and Smith (1994) | |
| *Salicornia rubra* | Yes | Yes | — | Smith and Kadlec (1983) | |
| *Salicornia virginica* | — | Yes | — | Hopkins and Parker (1984) | Salt marsh seed bank watered with fresh water. |
| *Salix nigra* | Yes | Yes | — | Hosner (1957); Hosner and Boyce (1962); McKnight (1980) | Seeds germinate best in very moist soil; seeds must germinate in 24 hours. Seeds sink in water and germinate. |
| *Samollus valerandi* | — | Yes | — | Meredith (1985) | |
| *Saururus cernuus* | — | Yes | — | Schneider and Sharitz (1986) | |
| *Scirpus acutus* | Yes | Yes | — | Pederson (1981); Smith and Kadlec (1983); Shipley et al. (1989) | Limited germination in 2–3 cm of water. |
| *Scirpus americanus* | — | Yes | — | Shipley and Parent (1991) | |
| *Scirpus articulatus* | Yes | No | — | Mason (1996) | |
| *Scirpus atrovirens* | — | Yes | — | Shipley and Parent (1991) | |
| *Scirpus cyperinus* | — | Yes | — | Keddy and Reznicek (1982); Shipley et al. (1989); Wisheu and Keddy (1991); Shipley and Parent (1991) | |
| *Scirpus maritimus* | Yes | Yes | Yes | Kautsky (1990); Smith and Kadlec (1983); van der Valk and Pederson (1989) | Germinates in brackish water. Seed bank samples from a brackish water site in Manitoba, Canada. |
| *Scirpus paludosus* | Yes | Yes | — | Pederson (1981) | |
| *Scirpus robustus* | — | Yes | — | Hopkins and Parker (1984) | Salt marsh seed bank watered with fresh water. |
| *Scirpus supinus* | Yes | No | — | Mason (1996) | |
| *Scirpus torreyi* | — | Yes | — | Shipley and Parent (1991) | |
| *Scirpus validus* | Yes | Yes | — | van der Valk and Davis (1978); Pederson (1981) | |

| Species | Germinates Under Water | Germinates in Moist, Unflooded | Salinity | Reference | Comment |
|---|---|---|---|---|---|
| *Scolochloa festucacea* | No | Yes | Yes | Pederson (1981); Galinato and van der Valk (1986); Neill (1990) | Germinates in salinities from 0 to 5000 mg $l^{-1}$; germination reduced above a salinity of 1000 mg $l^{-1}$; limited germination in 2–3 cm of water (Pederson 1981); germinated in field during drawdown (Neill, 1990). |
| *Scutellaria galericulata* | No | Yes | — | Pederson (1981); van der Valk and Verhoeven (1988) | |
| *Scutellaria laterifolia* | — | Yes | — | Schneider and Sharitz (1986) | |
| *Senecio congestus* | No | Yes | — | Pederson (1981) | Limited germination in 2–3 cm of water. |
| *Sisyrinchium atlanticum* | — | Yes | — | Wisheu and Keddy (1991) | |
| *Sium suave* | No | Yes | — | van der Valk and Davis (1978) | |
| *Solanum americanum* | — | Yes | — | Schneider and Sharitz (1986) | |
| *Solanum nigrum* | No | Yes | — | Mason (1996) | |
| *Solidago canadensis* | No | Yes | — | Pederson (1981) | |
| *Solidago graminifolia* | — | Yes | — | Keddy and Reznicek (1982) | |
| *Solidago rugosa* | — | Yes | — | Keddy and Reznicek (1982) | |
| *Sonchus* spp. | — | Yes | — | Meredith (1985) | |
| *Sonchus arvensis* | No | Yes | — | Pederson (1981) | |
| *Sparganium eurycarpum* | — | Yes | — | Shipley et al. (1989); Shipley and Parent (1991) | |
| *Spartina anglica* | — | Yes | — | Chung (1989) | Germination of seeds fallen from spikelet was lower. Also, germinated in the refrigerator. |
| *Spartina pectinata* | — | Yes | — | Shipley et al. (1989); Shipley and Parent (1991) | |
| *Sphenochloa zeylanica* | Yes | Yes | — | Mason (1996) | |
| *Spiraea latifolia* | — | Yes | — | Keddy and Reznicek (1982) | |
| *Spirodela polyrhiza* | No | Yes | — | van der Valk and Davis (1978) | |
| *Stachys palustris* | No | Yes | — | van der Valk and Davis (1978); Pederson (1981); Meredith (1985) | |

| | | | | | |
|---|---|---|---|---|---|
| *Stellaria media* | — | Yes | — | Meredith (1985) | |
| *Stellaria palustris* | — | Yes | — | van der Valk and Verhoeven (1988) | |
| *Sueda depressa* | No | Yes | — | Pederson (1981) | |
| *Symphytum officinale* | — | Yes | — | Meredith (1985) | |
| *Taxodium ascendens* | No | Yes | — | Conti and Gunther (1984) | Germinates only as a result of alternating temperature. |
| *Taxodium distichum* | No | Yes | — | DuBarry (1963); Williston et al. (1980); McKnight (1980); Schneider and Sharitz (1986) | No germination under water; seeds remain viable in water for up to 30 months. Germination improves if seeds soaked in cold water for 90 days or more. |
| *Teucrium occidentale* | No | Yes | — | Pederson (1981) | Very limited germination in 2–3 cm of water. |
| *Thalictrum flavum* | — | Yes | — | Meredith (1985) | |
| *Triadenum fraseri* | — | Yes | — | Keddy and Reznicek (1982); Shipley and Parent (1991) | |
| *Trianthema portulacastrum* | No | Yes | — | Mason (1996) | |
| *Thelypteris palustris* (spores) | — | Yes | — | van der Valk and Verhoeven (1988) | |
| *Triglochin* sp. | No | Yes | — | Pederson (1981) | |
| *Typha* sp.-1 | Yes | No | — | van der Valk and Pederson (1989) | Seed bank samples from a brcakish water site in Manitoba, Canada. |
| *Typha* sp.-2 | Yes | Yes | — | van der Valk and Rosburg (1997) | |
| *Typha angustata* | No | Yes | — | Mason (1996) | |
| *Typha angustifolia* | — | Yes | — | Shipley et al. (1989) | |
| *Typha glauca* | No | Yes | Yes | van der Valk and Davis (1978); Galinato and van der Valk (1986) | Germinate in salinities from 0 to 5000 mg $l^{-1}$. |
| *Typha latifolia* | — | Yes | — | Schneider and Sharitz (1986) | |
| *Ulmus americana* | No | Yes | — | Hosner (1957); Guilkey (1965); Schneider and Sharitz (1986) | Germination rapid after soaking. Many birds and animals eat the fruit. |
| *Ulmus rubra* | No | Yes | — | Scholz (1965) | Germinate on moist, well-drained soil. |
| *Urtica dioica* | — | Yes | — | Pederson (1981); Meredith (1985) | |

| Species | Germinates Under Water | Germinates in Moist, Unflooded | Salinity | Reference | Comment |
|---|---|---|---|---|---|
| *Utricularia* sp. | Yes | Yes | — | Finlayson et al. (1990) | |
| *Utricularia vulgaris* | Yes | No | — | Pederson (1981) | |
| *Vallisneria spiralis* | Yes | — | — | Kimber et al. (1995); Mason (1996) | |
| *Verbena hastata* | — | Yes | — | Shipley and Parent (1991) | |
| *Veronica peregrina* | No | Yes | — | Leck and Graveline (1979) | Depth of flooding not given. |
| *Viola* sp. | — | Yes | — | Schneider and Sharitz (1986) | |
| *Viola lanceolata* | — | Yes | — | Keddy and Reznicek (1982); Wisheu and Keddy (1991) | |
| *Vitus aestivalis* | — | Yes | — | Schneider and Sharitz (1986) | |
| *Vitus rotundifolia* | — | Yes | — | Schenider and Sharitz (1986) | |
| *Xyris difformis* | — | Yes | — | Keddy and Reznicek (1982); Shipley and Parent (1991) | |
| *Xyris smalliana* | Yes | Yes | — | Conti and Gunther (1984) | |
| *Zannichellia palustris* | Yes | — | Yes | Smith and Kadlec (1983); van der Valk and Pederson (1989); Kautsky (1990) | Germinates in brackish water. |
| *Zizania aquatica* var. *aquatica* | Yes | No | — | Leck and Graveline (1979) | Depth of flooding not given. |
| *Zosterella dubia* | Yes | — | — | Kimber et al. (1995) | |

# *Appendix 3*

# *Internet Guide for Wetland Practitioners, Researchers, and Students of Wetland Ecology*

*This is intended as a broadly based rather than a complete list.*

Africa: Earthwatch South Africa The Nile Crocodile Course.
http://www.earthwatch.org/x/Xleslie.html

Africa: EcoNews Africa.
http://www.web.net/~econews/ena4-17.txt

Africa:South Africa's Threatened Wildlife.
http://www.infoweb.co.za/enviro/ewtbook/page3.htm

Asia Vietnam: World Conservation Monitoring Program.
http://www.wcmc.org.uk/infoserv/countryp/vietnam/app7.html

Australia New South Wales: One Tree Island Research Station.
http://www.bio.usyd.edu.au:80/OTI/

Australia: Australian National Botanic Gardens Library.
http://www.anca.gov.au/library/anbglib.htm

Australia: Cooperative Research Centre for Freshwater Ecology.
http://lake.canberra.edu.au/crcfe/crcfe.html

Australia: Department of the Environment SaTSotE-E.
http://www.environment.gov.au/portfolio/minister/dept/bg27jun_estuar.html

Australia: Environment Australia KNPMP.
http://www.environment.gov.au/portfolio/anca/manplans/kakadu/contents.html

Australia: Environment Australia Online.
http://www.environment.gov.au/index.html

Australia: Environment Australia Online Library Services.
http://www.environment.gov.au/portfolio/library/library.html

Australia: Environment Australia ''Riverland'' Ramsar Sites in Australia.
http://kaos.erin.gov.au/land/wetlands/RAMSAR/site29.html

Australia: Erin Murray-Darling Basin.
http://kaos.erin.gov.au/general/brochures/br_mdb.html

Australia: LIMLOG River and Wetland Ecology Management Database.
http://www.lib.flinders.edu.au/services/databases/local/LIMLOG.html

Australia: Murray-Darling Basin Commission.
http://kaos.erin.gov.au/portfolio/esd/csd95/case6.html

Australia: Murray-Darling Basin Hydrogeology Operating Plan 94–95.
http://www.agso.gov.au/94–95/workplan/project/241.05.html

Australia: Murray-Darling Waterbird Survey.
http://www.vicnet.net.au/vicnet/RAOU/murraydarling.html

Australia: Rivers and Wetlands.
http://www.dwr.csiro.au/rivers/rivers.html

Australia: The National Wetlands Newsletter.
http://kaos.erin.gov.au/land/wetlands/newslett.html

Australia: University of Ballarat Reference Collection on Land Rehabilitation.
http://www.ballarat.edu.au/is/ej/pathfind/pflanreh.htm

Australia: University of Western Australia Centre for Water Research.
http://www.cwr.uwa.edu.au/index2.html

Canada: Environment Canada.
http://www.doe.ca/

Canada: Grand River Watershed Aquatic Resources Research.
http://bordeaux.uwaterloo.ca/aqres/index.html

Canada: Saskatchewan Wetland Conservation Corporation.
http://www.wetland.sk.ca/swccmenu.htm

Canada: The Canada Centre for Inland Waters.
http://www.cciw.ca/Welcome.html

Canada: Wetlands of Eastern Temperate Canada.
http://www.wetlands.ca/wetcentre/wetcanada/regions/temperate/temperate.html

Canada: WetNet.
http://www.wetlands.ca/index.html

Central America: INBIO Manual de las Plantas de Costa Rica via Missouria Botanical Garden.
http://www.mobot.org/manual.plantas/

Europe Italy: Joint Wetland Research at JRC Ispra.
http://willow.sti.jrc.it/iain/birds/birds.html

Europe Sweden: Takern Field Station.
http://www.lysator.liu.se/~ngn/tfhome.html

Europe: International Working Group of the Waterworks in the Rhine River Basin Program.
http://www.chem.vu.nl/0/AAC/RiverBasin/indexEng.html

Europe: Maps of the Rhine.
http://www.u-net.com/maps/germany/rhine.htm

Europe: MedWet and Protected Area Needs for Mediterranean Wetlands.
http://www.ecnc.nl/doc/europe/legislat/medwet.html

Europe: Netherlands Neonet Directory for Water.
http://neonet.nlr.nl/dir_water.html

Europe: Ongoing Activities of the European Centre for Nature Conservation and the Netherlands Ministry of Agriculture NMaF.
http://www.ecnc.nl/doc/europe/activities/activlst.html

Europe: Publications of the Regional Environmental Center for Central and Eastern Europe.
http://www.rec.org/REC/Publications/publications.html

Europe: Rhône River F.
http://limnologie.univ-lyon1.fr/

Europe: ''Sewer of Europe'' Cleans Up Its Act.
http://www.koblenz.fh-rpl.de/koblenz/remstecken/rhine/960327WashingtonPost.html

Europe: The Netherlands Rijkswaterstaat Institute for Inland Water and Waste Water Treatment.
http://www.minvenw.nl/rws/riza/

Europe: The Rhine.
http://www.english.upenn.edu/~jlynch/Frank/Places/rhine.html

General: A Guide to Wetland Restoration on Private Land.
http://ceres.ca.gov/ceres/calweb/DU/Valley_Habitats4.html

General: Neonet ITC Library.
http://neonet.nlr.nl/providers/www.itc.nl/library.html

General: Sierra Club Policy Wetlands.
http://www.sierraclub.org/policy/613.html

General: Society for Environmental Restoration Abstracts.
http://nabalu.flas.ufl.edu/ser/Seattle/Wetlands.html

General: Soil Properties in Wetland Creation and Restoration.
http://weber.u.washington.edu/~robh/Courses/ESC311/1997/Sasha/intro.htm

India: National Parks.
http://www.allindia.com/wild/

Latin America: NOAA Paleoclimatology Program Latin American Pollen Database Publications.
http://julius.ngdc.noaa.gov/paleo/lapdpubs.html

Russian Federation: Greenpeace Russia NABU on Vodlozero National Park.
http://www.sll.fi/mpe/vodla/description.html

Southeast Asia: Malaysia Asian Wetlands Bureau.
http://www.gsf.de/UNEP/maawb.html

Southeast Asia: Wild Bird Society of Japan.
http://www.kt.rim.or.jp/~birdinfo/japan/wbsj.html

Southeast Asia: World Conservation Union IUCN.
http://nabalu.flas.ufl.edu/ser/Seattle/Wetlands.html

United States East: Coastal Marsh Program of the University of Maryland/NASA.
http://www.geog.umd.edu/wetlands/Marsh.html

United States East: Woods Hole Oceanographic Institute.
http://www.whoi.edu/

United States General: Hydric Soils.
http://www.statlab.iastate.edu/soils/osd/
http://www.statlab.iastate.edu/index.html/

United States General: National Wetlands Inventory Program.
http://www.usgs.gov/fact-sheets/national-wetlands-inventory-program/

United States General: United States Department of Agriculture Natural Resources Conservation Service Wetland Science Institute.
http://159.189.24.10/wetsci.htm

United States General: USDA Climate Data.
http://www.wcc.nrcs.usda.gov/water/w_clim.html

United States General: USDA 1996 Farm Bill Provisions.
http://www.nhq.nrcs.usda.gov/OPA/FB96OPA/MitgFact.html

United States Great Lakes: Restoring Wetland Habitat at Duluth's Grassy Point.
http://www.d.umn.edu/~pcollins/grassy.html

United States Great Lakes: Saginaw Bay Watershed; A Strategy for Wetland Restoration.
http://epawww.ciesin.org/glreis/nonpo/nprog/sag_bay/sb/ch1.html

United States Midwest: Illinois Natural History Survey Center for Aquatic Ecology.
http://www.inhs.uiuc.edu/chf/pub/an_report/94_95/CAE.html

United States Midwest: Restoring Iowa's Agricultural Wetlands.
http://www.wetland.sk.ca/wetlinks.htm

United States Midwest: USGS Northern Prairie Science Center.
http://www.npsc.nbs.gov/resource/resource.htm#literat

United States Pacific Northwest: Wetland Ecosystem Team.
http://www.fish.washington.edu/people/asif/WET.html

United States Southeast: Cache River Illinois Corridor Restoration Plan of the SIUC Restoration and Landscape Ecology Class.
http://www.science.siu.edu/plant-biology/plb545/spring-96/landscape.htm

United States Southeast: Everglades Ecosystem Destruction and Rehabilitation.
http://geog.gmu.edu/gess/Everglades/pract.html#top

United States Southeast: Everglades Information Network.
http://everglades.fiu.edu/

United States Southeast: Kennedy Space Center Jurisdictional Wetlands.
http://atlas.ksc.nasa.gov/wetlands/wetlands.html

United States Southeast: LUMCON Louisiana Research Programs.
http://www.lumcon.edu/ResearchPrograms.htm

United States Southeast: The Conservancy of Southwest Florida.
http://www.arch.usf.edu/FICUS/conserve/groups/conservy/overview/overview.htm

United States West: Suisun Marsh Natural History Association.
http://community.net/marsh/index.html

United States/Canada: PLANTS-Database of North American Plants.
http://www.ars-grin.gov/cgi-bin/

United States: National Estuary Program of the U.S.E.P.A.
http://www.epa.gov/nep/nep.html

United States: U.S. Army Corps of Engineers Publications.
http://www.usace.army.mil/inet.usace-docs/

United States: U.S. Army Corps of Engineers Publications Wetland Delineation Manual.
http://www.wes.army.mil/el/wetlands/wlpubs.html

United States: U.S. Army Corps of Engineers Waterways Experiment Station.
http://www.wes.army.mil

United States: USGS National Wetlands Research Center.
http://www.nwrc.gov/

World General: Wetlinks.
http://www.wetland.sk.ca/wetlinks.htm

World: Kew Garden Reference List for Plant Re-introductions RPaRP.
http://www.rbgkew.org.uk/conservation/habitat.html

World: Society for Ecological Restoration.
http://nabalu.flas.ufl.edu/ser/SERhome.html

World: Society for Ecological Restoration Library.
http://nabalu.flas.ufl.edu/ser/Library.html#Main Contents

World: Wetlands International-Africa, Europe, Middle East
http://www.wetlands.agro.nl/

World: World Conservation Monitoring Centre.
http://www.wcmc.org.uk/index.html

World: World Conservation Monitoring Centre Marine Information.
http://www.wcmc.org.uk:80/marine/data/

# *Glossary*

**Active restoration** aggressive manipulation of the restoration site to address altered flow regimes due to dams or channelization and revegetation through replanting or reseeding.

**aerenchyma** cavities in the stems and/or roots of aquatic plants.

**Anaerobic** condition of low oxygen.

**Anemochory** seed dispersal via wind.

**Bank protection** channel stabilization (Brookes, 1985).

**Billabong** in Australia, a depression on the floodplain holding water only during the rainy season or in a dead end channel or oxbow.

**Bog** an acid peat deposit.

**Channelization** channel alterations for the purpose of increasing flow and decreasing retention time, including resectioning, realignment, diversion, embankment, bank protection, channel lining, and culverting by dredging, cutting, and obstruction removal (Brookes, 1985).

**Clementsian succession** the idea that successional stages replace one another by autogenic and allogenic processes until a climatic climax is reached. In this view, species are thought of as highly interlinked (Clements, 1916). Clementsian succession is synonymous with the terms *classical succession, relay floristics*, and *facilitation*.

**Colonizer** species that enters an unoccupied habitat for the first time.

**Constancy** the degree to which the ecosystem can resist change (Holling, 1973).

**Creation** establishing an ecosystem that did not originally occupy the site.

**Cyclostome** a vertebrate including lampreys and hagfishes.

**Design** approach in wetland restoration that views the life history strategy as the important factor in developing vegetation on a restoration site. This view favors engineering and replanting strategies directed to producing a wetland type with no fixed endpoint. This strategy emphasizes interventionist approaches (e.g., reengineering of hydrology to encourage dispersal and germination from the extent seed bank, replanting) based on predictable outcomes. At its heart, it is a restatement of Gleasonian succession theory (van der Valk, in press).

**Desnagging** removal of woody obstructions in a waterway.

**Diaspore** dispersal unit of plants, including sexually and asexually produced propagules.

**Dispersal** the movement of a seed or propagule from its site of production (parent plant) to another spatial location.

**Disturbance** ''any relatively discrete event in time that disrupts ecosystem, community or population structure and changes resources, substrate availability or the physical environment'' (White and Pickett, 1985). Resh et al. (1988) add to this definition the qualifier that the frequency and intensity of the event be outside of a predictable range.

**Bull ditch** water drainage system created to increase surface water drainage, with teams of oxen pulling a drag line through wetland or existing stream channels.

**Ditch plug** earthen dam used to obstruct water through a low-flow stream, ravine, or ditch.

**Diversion** type of channelization in which flow is diverted around an area to be protected (Brookes, 1985).

**Donor seed bank** technique by which wetland seeds are transferred from one wetland to another via the soil seed bank.

**Duration** time interval over which a disturbance occurs.

**Ecocline** describing populations of a species that varies along geographical range.

**Ecotone** transition zone between two community types characterized by structural or functional discontinuity.

**Embankment** type of channelization in which a levee, bund, or dike is used to prevent the flow from overflowing onto the floodplain (Brookes, 1985).

**Enhancement** an improvement in a structural or functional attribute.

**Fen** nonacid peat deposit with surface runoff from mineral soils that may be supplied from nearby springs.

**Flood pulsing** the idea that the physical and biotic functions of the floodplain wetland are dependent on the pulsing dynamics of water discharged from the river channel.

**Frequency** mean number of occurrences of a particular disturbance over a given time interval.

**Functional group** a set of organisms that reacts to a disturbance (or perturbation) similarly (Gitay and Noble, 1997).

**Gleasonian succession** in wetlands, cyclic wetland dynamics in which "once a wetland, always a wetland." Change is brought about by the maturation and fluctuation of species in response to disturbance and environmental change (van der Valk, 1981).

**Hydrarch succession** process by which a lake fills in and becomes dry land; a concept now challenged by most ecologists (*see* hydrosere).

**Hydrochory** aquatic seed dispersal.

**Hydrosere** model of wetland succession that suggests that due to autogenic and allogenic processes, wetlands first form from the filling of shallow lakes in an increasingly dry environment until the formation of an upland forest; see *hydrarch succession.*

**Intensity** the force of a disturbance event over a given time interval (e.g., wind speed of a tornado or flow rate of flood waters).

**Interbasin water transfer** the reengineering of water flow from one distinct river or river catchment to another (Davies et al., 1992).

**Intermediate disturbance hypothesis** idea that sites with moderate amounts of disturbance have higher species richness than those with either low or high levels of disturbance (Connell, 1978).

**Jokulhlaup** dam of glacial ice (Icelandic).

**Keystone species** the dominant predator that controls populations and processes in the ecosystem; in the case of beaver, the dominant herbivore that controls stream processes *sensu* Naiman and Mellilo (1984).

**Landscape element** a unit distinct from others in a landscape; similar to the idea of *community*, but includes other objects of anthropogenic origin that would be visible in an aerial photo (e.g., levee, subdivision).

**Landscape reintegration** restoring functional aspects of a fragmented landscape.

**Lock** along a river, a weir with a collapsible panel to allow boat travel.

**Mitigation** the restoration, creation, or enhancement of wetland to compensate for lost wetland.

**Muskeg** in Canada, a bog.

**Paludification** the process by which dry land changes into bog as the water table rises when drainage is hindered by the accumulation of peat.

**Passive restoration** the cessation of land use activities that halt degradation, such as eliminating cattle grazing and lightning strike fire prevention.

**Persistence** existence despite changes due to disturbances, as in the case of wetlands—for example, water level changes, wind storms, and herbivory by native animals.

**Persistent seed bank** seed bank with seeds most retaining viability for more than one year (Grime, 1981).

**Perturbation** according to White and Pickett (1985), roughly the same thing as *disturbance*. Many ecologists use the terms interchangeably. *Perturbation* is used by some ecologists to distinguish a disturbance that is unfamiliar to the ecosystem, such as some human disturbances.

**Pothole** marshy ponds of glacial origin in the U.S. Midwest and central Canadian provinces.

**Preemption** the limitation of species establishment because another species is already there, also called *preemptive competition*; an important component of species interactions in plant communities.

**Press event** longer events such as a habitat alteration in which recovery can take a long time without intervention.

**Pulse event** events of a limited and defined duration (Niemi et al., 1990; Allan and Flecker, 1993) that are repetitive and part of the natural system upon which the organisms depend (Vogl, 1980) and after which species can usually recover via colonization, depending on their dispersal abilities and the proximity of refugia.

**Realignment** type of channelization in which the stream channel is shortened via an artificial cutoff (Brookes, 1985).

**Reclamation** an alteration in a ecosystem that creates another type of ecosystem of value to humans.

**Recoverability** ability of an ecosystem to return to its predisturbance conditions (Holling, 1973).

**Rehabilitation** less than full restoration of the structural and functional aspects of the predisturbance state.

**Resectioning** type of channelization in which the channel is widened or deepened (Brookes, 1985).

**Resilience** the extent to which an ecosystem's composition and function can be disturbed and still recover these predisturbance attributes afterward (Holling, 1973).

**Restoration** returning a site to approximately its condition before alteration, including its predisturbance function and related physical, chemical, and biological characteristics; full restoration is the complete return of a site to its original state.

**Return interval** mean time interval between disturbance events.

**River Continuum Concept** the idea that stream function shifts progressively from headwater to broad meandering segments downstream.

**Safe site** a zone with conditions under which a seed can germinate, that is, to escape pregermination predation and overcome dormancy (Harper et al., 1965).

**Sapling** a young tree more than 3 feet (91.4 cm) tall.

**Seedling** a plant or tree less than one year old, often with cotyledon leaves attached.

**Self-design** approach in wetland restoration that emphasizes the ability of a wetland, given enough time, to organize itself around engineered components. This strategy deemphasizes interventionist approaches and views wetland development as an ecosystem-level process, a reworking of Odum (1969) succession approaches (Mitsch and Jørensen, 1989).

**Serial Discontinuity Concept** the idea that impoundments shift longitudinal gradients along rivers from that predicted by the River Continuum Concept (Ward and Stanford, 1983).

**Severity** the impact of the disturbance on the individual, community, or landscape (e.g. amount of cover removed).

**Snag** woody obstruction in a waterway.

**Species richness** number of species present in an area.

**Stability** in communities, the ability to resist change (constancy) and return to preexisting conditions after disturbance (Holling, 1973).

**Stop log structure** water control structure that allows the manager to control the water level in a wetland by adding or removing boards within the structure.

**Stratification** in seed technology, the improvement of germination success by cold treatment at (0–5°C).

**String** in a patterned fen, a transverse peat ridge (Glaser, 1987).

**Succession** any change in the quality or quantity of vegetation in a community due to maturation or fluctuation of species (van der Valk, 1981).

**Terrestrialization** the process of infilling of a shallow lake to become a bog (Klinger, 1996b).

**Transient seed bank** seed bank with seeds most of which retain their viability for less than one year (Grime, 1981).

**Turion** overwintering bud.

**Water regime** annual and long-term fluctuations in water levels of wetlands (Cowardin et al., 1979); from a river perspective, the pattern of water discharge or flow in the river (Sparks, 1992).

**Wetland** (1) ''lands transitional between terrestrial and aquatic systems where the water table is usually at or near the surface or the land is covered by shallow water'' (Cowardin et al., 1979); (2) ''areas of marsh, fen, peatland or water, whether natural or artificial, permanent of temporary, with water that is static or flowing, fresh, brackish, or salt, including areas or marine water the depth of which at low tide does not exceed six metres'' (Ramsar Convention, 1971).

# References

Abbe, T. B., and D. R. Montgomery, 1996, Large woody debris jams, channel hydraulics and habitat formation in large rivers, Regulated Rivers: Research and Management **12**:201–221.

Adamoli, J., 1992, Research and management in the Pantanal of Brazil, *In* Proceedings of the Third International Wetlands Conference, Maltby, E., P. J. Dugan, and J. C. Lefeuvre, eds., September 19–23, 1988, Rennes, France, The World Conservation Union (IUCN), Gland, Switzerland, pp. 61–67.

Adams, D. A., M. A. Buford, and D. M. Dumond, 1987, In search of the wetland boundary, Wetlands **7**:59–70.

Adis, J., S. I. Golovatch, and S. Hamann, 1996, Survival strategy of the terricolous millipede *Cutervodesmus adisi* Golovatch (Fuhrmannodesmideae, Polydesmida) in a blackwater inundation forest of central Amazonia (Brazil) in response to the flood pulse, Memoires du Museum National d'Histoire Naturelle **169**:523–532.

Admiraal, A. N., M. J. Morris, T. C. Brooks, J. W. Olson, and M. V. Miller, 1997, Illinois Wetland Restoration and Creation Guide, Illinois Natural History Survey, Champaign, Ill.

Agami, M., and Y. Waisel, 1986, Regeneration of *Najas marina* L. and of *Potamogeton lucens* L. after selective clipping of an established mixed stand, *In* Proceedings EWRS/AAB, 7th Symposium on Aquatic Weeds, September 15–17, 1986, Loughborough, Leicestershire, United Kingdom, European Weed Research Society, Association of Applied Biologists, Loughborough, Leicestershire, U.K., pp. 3–7.

Ahlgren, C. E., and H. L. Hansen, 1957, Some effects of temporary flooding on coniferous trees, Journal of Forestry **55**:647–659.

Aide, T. M., and J. Cavelier, 1994, Barriers to lowland tropical forest restoration in the Sierra Nevada de Santa Marta, Colombia, Restoration Ecology **2**:219–229.

Aide, T. M., J. K. Zimmerman, M. Rosario, and H. Marcano, 1996, Forest recovery in abandoned cattle pastures along an elevational gradient in northeastern Puerto Rico, Biotropica **28**:537–548.

Aiken, S. G., and K. F. Walz, 1979, Turions of *Myriophyllum exalbescens*, Aquatic Botany **6**:357–363.

Aiken, S. G., P. R. Newroth, and I. Wile, 1979, The biology of Canadian weeds. 34. *Myriophyllum spicatum* L., Canadian Journal of Plant Science **59**:201–215.

Akanil, N., and B. A. Middleton, 1997, Leaf litter decomposition along the Porsuk River, Eskisehir, Turkey, Canadian Journal of Botany **75**:1394–1397.

Alexander, T. R., 1971, Sawgrass biology related to the future of the Everglades ecosystem, Soil and Crop Science Society of Florida **31**:72–74.

Alexander, T. R., and A. G. Crook, 1974, Recent vegetational changes in southern Florida, *In* Environments of South Florida: Present and Past, Gleason, P. J., ed., Miami Geological Society, Miami, Florida, pp. 61–72.

Allan, J. D., and A. S. Flecker, 1993, Biodiversity conservation in running waters, BioScience **43**:32–43.

Anderson, J. R., and A. K. Morison, 1988, A Study of the Native Fish Habitat in the Goulburn River, Shepparton—Impact of a Proposal to Operate a Paddle-Steamer, Technical Report Series No. 68 Arthur Rylah Institute for Environmental Research, Kaiela Fisheries Research Station, Shepparton, Victoria, Australia.

Anonymous, 1975, Salt problems in Perth's hills, Ecos (CSIRO Australia) **4**:3–12.

Arber, A., 1920, Water Plants, A Study of Aquatic Angiosperms, Cambridge University Press, Cambridge.

Armstrong, J., and W. Armstrong, 1991. A convective throughflow of gases in *Phragmites australis* (Cav.) Trin ex Steud., Aquatic Botany **39**:75–88.

Armstrong, W., R. Brändle, and M. B. Jackson, 1994, Mechanisms of flood tolerance in plants, Acta Botanica Neerlandica **43**:307–358.

Aronson, J., and E. Le Floc'h, 1996, Vital landscape attributes: missing tools in restoration ecology, Restoration Ecology **4**:377–387.

Arthington, A. H., 1996, The effects of agricultural land use and cotton production on tributaries of the Darling River, Australia, GeoJournal **40**:115–125.

Ashton, P. J., and H. R. Van Vliet, 1997, South African approaches to river water-quality protection, *In* River Quality: Dynamics and Restoration, Laenen, A., and D. A. Dunnette, eds., Lewis Publishers, Boca Raton, Florida, pp. 403–411.

Auld, T. D., and R. A. Bradstock, 1996, Soil temperatures after the passage of a fire: do they influence the germination of buried seeds? Australian Journal of Ecology **21**:106–109.

Bailey, A. D., and T. Bree, 1981, Effect of improved land drainage on river flood flows. Proceedings of a conference, Institution of Civil Engineers, ICE. Flood Studies Report—Five Years On. Thomas Telford Ltd., London, pp. 131–143.

Baker, B. W., and G. L. Wright, 1990, The Murray Valley: Its Hydrologic Regime and the Effects of Water Development on the River, *In* Proceedings of the Royal Society of Victoria 90, Royal Society of Victoria, Melbourne, Victoria, pp. 103–110.

Baker, J. B., 1977, Tolerance of planted hardwoods to spring flooding, Southern Journal of Applied Forestry **1**:23–25.

Baker, V. R., 1988, Overview, *In* Flood Geomorphology, Baker, V. R., R. C. Kochel, and P. C. Patton, eds., John Wiley & Sons, New York, pp. 1–11.

Baldwin, A. H., W. J. Platt, K. L. Gathen, J. M. Lessman, and T. J. Rauch, 1995, Hurricane damage and regeneration in fringe mangrove forests of southeast Florida, U.S.A., Journal of Coastal Research **21**:169–183.

Barbe, J., 1984, Les végétaux aquatiques. Données biologiques et écologiques clés de détermination des macrophtes de France, Bulletin Francais de Pisciculture, Numéro Spécial 42, Paris.

Bardach, J., 1972, Introduction, *In* Proceedings of an International Symposium on River Ecology and the Impact of Man, Oglesby, R. T., C. A. Carlson, and J. A. McCann, eds., June 20–23, 1971, Amherst, Massachusetts. Academic Press, New York, p. 465.

Barrat-Segretain, M. H., and C. Amoros, 1995, Influence of flood timing on the recovery of macrophytes in a former river channel, Hydrobiologia **316**:91–101.

Barrat-Segretain, M. H., and C. Amoros, 1996, Recolonization of cleared riverine macrophyte patches: importance of the border effect, Journal of Vegetation Science **7**:769–776.

Barro, S. C., P. M. Wohlengemuth, and A. G. Campbell, 1989, Post-fire interactions between riparian vegetation and channel morphology and the implications for stream channel rehabilitation choices, *In* Proceedings of the California Riparian Systems Conference, Protection, Management and Restoration for the 1990's, September 22–24, Davis, Calif., General Technical Report PSW-110, U.S.D.A. Forest Service Pacific Southwest Forest and Range Experiment Station, Berkeley, California, pp 51–53.

Bartley, M. R., and D. H. N. Spence, 1987, Dormancy and propagation in helophytes and hydrophytes, Archiv für Hydrobiologie, Beiheft. Ergebnisse der Limnologie **27**:139–155.

Baskin, C. C., J. M. Baskin, and E. W. Chester, 1993, Seed germination ecophysiology of four summer annual mudflat species of Cyperaceae, Aquatic Botany **45**:41–52.

Baskin, J. M., and C. C. Baskin, 1983, Seasonal changes in the germination responses of fall *Panicum* to temperature and light, Canadian Journal of Plant Science **63**:973–979.

Baskin, J. M., and C. C. Baskin, 1985, The annual dormancy cycle in buried weed seeds: a continuum, BioScience **35**:492–498.

Bassett, P. A., 1980, Some effects of grazing on vegetation dynamics in the Camargue, France, Vegetatio **43**:173–184.

Bayley, P. B., 1991, The flood pulse advantage and the restoration of river-floodplain systems, Regulated Rivers: Research and Management **6**:75–86.

Bayley, P. B., 1995, Understanding large river-floodplain ecosystems, BioScience **45**:153–158.

Bazzaz, F. A., 1986, Life history of colonizing plants: some demographic, genetic, and physiological features, *In* Symposium proceedings, Ecology of Biological Invasions of North America and Hawaii, Baker, H. G. H. A. Mooney, and J. A. Drake, eds. October 21–25, 1984, Asilomar, California, Springer-Verlag, New York, pp. 96–110.

Bedford, B. L., 1996, The need to define hydrologic equivalence at the landscape scale for freshwater wetland mitigation, Ecological Applications **6**:57–68.

Bedford, B. L., E. H. Zimmerman, and J. H. Zimmerman, 1974, The Wetlands of Dane County, Wisconsin, Dane County Regional Planning Commission, Madison, Wisconsin.

Bedinger, M. S., 1978, Relation between forest species and flooding, *In* Wetland Functions and Values: The State of Our Understanding; Proceedings of the National Symposium on Wetlands, Greeson, P. E., J. R. Clark, and J. E. Clark, eds., November 7–10, 1978, Lake Buena Vista, Florida, American Water Resources Association, Minneapolis, Minnesota, pp. 427–435.

Beeftink, W. G., 1977, Saltmarshes, *In* The Coastline, Barnes, R. S. K., ed., John Wiley & Sons, Chichester, United Kingdom, pp. 93–121.

Begon, M., J. L. Harper, and C. R. Townsend, 1990, Ecology: Individuals, Populations and Communities, 2d ed., Blackwell Scientific Publications, Boston.

Behnke, R. J., and R. F. Raleigh, 1979, Grazing and the riparian zone: impact and management perspectives, *In* Strategies for Protection and Management of Floodplain Wetlands and Other Riparian Ecosystems, December 11–13, 1978, Proceedings of the Symposium, Callaway Gardens, Georgia, USDA Forest Service, General Technical Report WO-12, Washington, D.C., pp. 263–267.

Bell, D. T., and F. L. Johnson, 1974, Flood caused tree mortality around Illinois reservoirs, Transactions of the Illinois State Academy of Science **67**:28–67.

Bell, S., 1995, New woodlands in the landscape, *In* The Ecology of Woodland Creation, Ferris-Kaan, R., ed., John Wiley & Sons, Chichester, United Kingdom, pp. 27–47.

Bell, S. S., M. S. Fonseca, and L. B. Motten, 1997, Linking restoration and landscape ecology, Restoration Ecology **5**:318–323.

Belsare, D. K., 1994, Vanishing wetland wildlife of Southeast Asia, *In* Global Wetlands: Old World and New, Mitsch, W. J., ed., Elsevier, Amsterdam, pp. 841–856.

Benner, P. A., and J. R. Sedell, 1997, Upper Williamette River landscape: a historic perspective, *In* River Quality: Dynamics and Restoration. Laenen, A., and D. A. Dunnette, eds., Lewis Publishers, Boca Raton, Florida, pp. 23–47.

Bennington, C. C., J. B. McGraw, and M. C. Vavrek, 1991, Ecological genetic variation in seed banks. II. Phenotypic and genetic differences between young and old subpopulations of *Luzula parviflora*, Journal of Ecology **79**:627–643.

Berkmüller, K., S. Mukherjee, and B. Mishra, 1990, Grazing and cutting pressures on Ranthambhore National Park, Rajasthan, India, Environmental Conservation **17**:135–140.

Betz, R., 1986, One decade of research in prairie restoration at the Fermi National Acceleration Laboratory (Fermilab), Batavia, Illinois, The Prairie-Past, Present and Future, Proceedings of the Ninth Midwest Prairie Conference, Clamby, G. K., and R. H. Pemble, eds, Tricollege University, Moorhead, Minnesota. July 29–August 1, 1984, TriCollege University, Moorhead, Minnesota, pp. 179–1851.

Bijlmakers, L. L., and E. O. A. M. de Swart, 1995, Large-scale wetland-restoration of the Ronde Venen, The Netherlands, Water Science and Technology **31**:197–205.

Bishel-Machung, L., R. P. Brooks, S. S. Yates, and K. L. Hoover, 1996, Soil properties of reference wetlands and wetland creation projects in Pennsylvania, Wetlands **16**:532–541.

Björk, S., 1976, The restoration of degraded wetlands. Proceedings International Conference on the Conservation of Wetlands and Waterfowl, December 2–6, 1974, Heiligenhafen, Federal Republic of Germany. International Waterfowl Research Bureau, Slimbridge, United Kingdom, pp. 349–353.

Blanch, S. J., and M. A. Brock, 1994, Effects of grazing and depth on two wetland plant species, Australian Journal of Marine and Freshwater Research **45**:1387–1394.

Bond, W. J., and B. W. van Wilgen, 1996, Fire and Plants, Chapman and Hall, London.

Bontje, M. P., 1988, The application of science and engineering to restore a salt marsh, 1987, *In* Proceedings of a Conference, Increasing Our Wetland Resources, Zelazny, J., and J. S. Feierabend, eds., October 4–7, 1988, Washington, D.C., National Wildlife Federation, Washington, D.C., pp. 267–273.

Bontje, M. P., and S. Stedman, 1991, A successful salt marsh restoration in the New Jersey meadowlands, *In* Proceedings of the Eighteenth Annual Conference on Wetland Restoration and Creation, Webb, F. J., ed., May 16–17, 1991, Hillsborough Community College, Tampa, Florida, pp. 5–16.

Bornette, G., and C. Amoros, 1996, Disturbance regimes and vegetation dynamics: role of floods in riverine wetlands, Journal of Vegetation Science **7**:615–622.

Boutin, C., and P. A. Keddy, 1993, A functional classification of wetland plants, Journal of Vegetation Science **4**:591–600.

Bowers, J. K., 1995, Innovations in tidal marsh restoration: The Kenilworth Marsh account, Restoration and Management Notes **13**:155–161.

Bradshaw, A. D., 1987, Restoration: an acid test for ecology, *In* Restoration Ecology: A Synthetic Approach to Ecological Research, Jordan, W. R., III, ed., Cambridge University Press, Cambridge, pp. 23–45.

Brammer, H., 1990, Floods in Bangladesh, The Geographical Journal **156**:158–165.

Bravard, J.-P., C. Amoros, and G. Pautou, 1986, Impact of civil engineering works on the successions of communities in a fluvial system, Oikos **47**:92–111.

Bren, L. J., 1992, Tree invasion of an intermittent wetland in relation to changes in the flooding frequency of the River Murray, Australia, Australian Journal of Ecology **17**:395–408.

Bren, L. J., I. C. O'Neill, and N. L. Gibbs, 1987, Flooding in the Barmah Forest and its relation to flow in the Murray-Edward River system. Australian Journal of Forestry **17**:127–144.

Brink, V. C., 1954, Survival of plants under flood in the Lower Fraser River Valley, British Columbia, Ecology **35**:94–95.

Brinkman, K. A., 1965, Black walnut (*Juglans nigra* L.), *In* Silvics of Forest Trees of the United States, Fowells, H. A., ed., Agriculture Handbook 271, U.S.D.A. Forest Service, Washington, D.C., pp. 203–207.

Brinson, M. M., H. D. Bradshaw, R. N. Holmes, and J. B. Elkins, Jr., 1980, Litterfall, stemflow, and throughfall nutrient fluxes in an alluvial swamp forest, Ecology **61**:827–835.

Brinson, M. M., A. E. Lugo, and A. E. Brown, 1981a, Primary productivity, decomposition, and consumer activity in freshwater wetlands, Annual Review of Ecology and Systematics **12**:123–161.

Brinson, M. M., B. L. Swift, R. C. Plantico, and J. S. Barclay, 1981b, Riparian Ecosystems: Their Ecology and Status, FWS/OBS-81/17, U.S. Fish and Wildfire Service, Kearneysville, West Virginia.

Britton, R. H., and A. J. Crivelli, 1993, Wetlands of southern Europe and North Africa: Mediterranean wetlands, *In* Wetlands of the World I. Inventory, Ecology and Management. Whigham, D. F., D. Dykyová, and S. Henjny, eds., Kluwer, Dordrecht, the Netherlands, pp. 129–194.

Brix, H., B. K. Sorrell, and P. T. Orr, 1992, Internal pressurization and convective gas flow in some emergent freshwater macrophytes, Limnology and Oceanography **37**:1420–1433.

Broadfoot, W. M., and H. L. Williston, 1973, Flooding effects on southern forests, Journal of Forestry **71**:584–587.

Broadfoot, W. M., 1967, Shallow-water impoundment increases soil moisture and growth of hardwoods. Soil Scientists of America Proceedings **31**:562–564.

Brock, T. C. M., G. van der Velde, and H. M. van de Steeg, 1987, The effects of extreme water level fluctuations on the wetland vegetation of a nymphaeid-dominated oxbow lake in The Netherlands, Archiv für Hydrobiologie, Beiheft. Ergebnisse der Limnologie **27**:57–73.

Brock, T. C. M., H. Mielo, and G. Oostremeijer, 1989, On the life cycle and germination of *Hottonia palustris* L. in a wetland forest, Aquatic Botany **35**:153–166.

Brookes, A., 1985, River channelization: traditional engineering methods, physical consequences and alternative practices, Progress in Physical Geography **9**:44–73.

Brookes, A., 1990, Restoration and enhancement of engineered river channels: some European experiences, Regulated Rivers: Research and Management **5**: 45–56.

Brookes, A., J. Baker, and C. Redmond, 1996, Floodplain restoration and riparian zone management, *In* River Channel Restoration: Guiding Principles for Sustainable Projects, Brookes, A., and F. D. Shields, Jr., eds., John Wiley & Sons, Chichester, United Kingdom, pp. 201–229.

Brookes, A., and F. D. Shields, Jr., 1996, Towards an approach to sustainable river restoration, *In* River Channel Restoration: Guiding Principles for Sustainable Projects, Brookes, A., and F. D. Shields, Jr., eds., John Wiley & Sons, Chichester, United Kingdom, pp. 385–402.

Broome, S. W., 1990, Creation and restoration of tidal wetlands of the southeastern United States, *In* Wetland Creation and Restoration: The Status of the Science, Kusler, J. A., and M. E. Kentula, eds., Island Press, Washington, D.C., pp. 37–72.

Broome, S. W., I. A. Mendelssohn, and K. L. McKee, 1995, Relative growth of *Spartina patens* (Ait.) Muhl. and *Scirpus olneyi* Gray occurring in a mixed stand as affected by salinity and flooding depth, Wetlands **15**:20–30.

Broome, S. W., E. D. Seneca, and W. W. Woodhouse, Jr., 1988, Tidal salt marsh restoration, Aquatic Botany **32**:1–22.

Brown, V. K., 1992, Plant succession and life history strategy, Trends in Ecology and Evolution **7**:143–144.

Brown, S., 1981, A comparison of the structure, primary productivity, and transpiration of cypress ecosystems in Florida, Ecological Monographs **51**:403–427.

Brunk, E. L., A. R. Allmon, and G. P. Dellinger, 1975, Mortality of trees caused by flooding during the growing season at two Midwest reservoirs, Missouri Department of Conservation, Jefferson City, Missouri.

Biological Sciences Curriculum Study (BSCS) 1992, Biological Science: An Ecological Approach. Kendall/Hunt Publishing, Dubuque, Iowa.

Buchanan, R. A., 1989, Bush Regeneration, TAFE, Sydney, Australia.

Budowski, G., 1966, Fire in tropical lowland areas, Tall Timbers Fire Ecology Conference **5**:5–22.

Buell, M. F., H. F. Buell, and W. A. Reiners, 1968, Radial mat growth on Cedar Creek Bog, Minnesota, Ecology **49**:1198–1199.

Bull, L. A., L. J. Davis, and J. B. Furse, 1991, Lake Okeechobee-Kissimmee River-Everglades Resource Evaluation Report, Wallop-Breaux F-52–5, Game and Fresh Water Fish Commission, Okeechobee, Florida.

Burgis, M. J., and J. J. Symoens, eds., 1987, African Wetlands and Shallow Water Bodies, Institut Français de Recherche Scientifique Pour le Développement en Coopération, ORSTOM, Paris, France.

Burrows, F. M., 1975a, Calculation of the primary trajectories of dust seeds, spores and pollen in unsteady winds, New Phytologist **75**:389–403.

Burrows, F. M., 1975b, Wind-borne seed and fruit movement, New Phytologist **75**: 405–418.

Buschbacher, R. J., 1986, Tropical deforestation and pasture development, BioScience **36**:22–28.

Buschbacher, R., C. Hul, and E. A. S. Serrao, 1992, Reforestation of degraded Amazon pasture lands, *In* Ecosystem Rehabilitation Vol. 2. Ecosystem Analysis and Synthesis. Wali, M. K., ed., SPB Academic Publishing, the Hague, the Netherlands, pp. 257–274.

Cairns, J., Jr., 1988, Increasing diversity by restoring damaged ecosystems, *In* Biodiversity, Wilson, E. O., ed., National Academy Press, Washington, D.C., pp. 333–343.

Cairns, J., Jr., and S. E. Palmer, 1993, Senescent reservoirs and ecological restoration: an overdue reality check, Restoration Ecology **1**:212–219.

Cairns, J., Jr., and J. R. Pratt, 1995, Ecological restoration through behavioral change, Restoration Ecology **3**:51–53.

Camargo, J. A., and D. G. de Jalon, 1990, The downstream impacts of the Burgomillodo Reservoir, Spain, Regulated Rivers: Research and Management **5**: 305–317.

Campbell, A. G., and J. F. Franklin, 1979, Riparian Vegetation in Oregon's Western Cascade Mountains: Composition, Biomass, and Autumn Phenology, Coniferous Forest Biome, U.S./International Biological Program, University of Washington, Seattle.

Carangelo, P. D., 1988, Creation of sea grass habitat in Texas: results of research investigations and applied programs, *In* Proceedings of a Conference, Increasing Our Wetland Resources, Zelazny, J., and J. S. Feierabend, eds., October 4–7, 1988, Washington, D.C., National Wildlife Federation, Washington, D.C., pp. 286–300.

Carlquist, S., 1983, Intercontinental dispersal, Sonderbaende des Naturwissenschaftlichen Vereins in Hamburg **7**:37–47.

Carpenter, J. M., and G. T. Farmer, 1981, Peat Mining: An Initial Assessment of Wetland Impacts and Measures to Mitigate Adverse Effects, U.S. Environmental Protection Agency, Washington, D.C.

Chabrek, R. H., 1972, Vegetation, Water and Soil Characteristics of the Louisiana Coastal Region, Louisiana State University Agricultural Experiment Station, Baton Rouge.

Chabrek, R. H., 1976, Management of wetlands for wildlife habitat improvement, *In* Estuarine Processes. Wiley, M., ed., Academic Press, New York, pp. 226–233.

Chabrek, R. H., 1988, Coastal Marshes: Ecology and Wildlife Management, University of Minnesota Press, Minneapolis.

Chabwela, H. W., 1992, Conservation and development: the sustainable use of wetland resources, *In* Proceedings of the Third International Wetlands Conference, Maltby, E., P. J. Dugan, and J. C. Lefeuvre, eds., September 19–23, 1988, Rennes, France, The World Conservation Union (IUCN), Glard, Switzerland, pp. 31–39.

Chabwela, H. N., and G. A. Ellenbroek, 1990, The impact of hydroelectric developments on the lechwe and its feeding grounds at Kafue Flats, Zambia, *In* Wetland Ecology and Management: Case Studies, Whigham, D. F., R. E. Good, and J. Kvet, eds., Kluwer, Dordrecht, the Netherlands, pp. 95–101.

Changnon, S. A., and J. M. Changnon, 1996, History of the Chicago Diversion and future implications, Journal of Great Lakes Research **22**:100–118.

Chesterfield, E. A., 1986, Changes in the vegetation of the river red gum forest at Barmah, Victoria, Australian Forestry **49**:4–15.

Christensen, N., R. Burchell, A. Ligett, and E. Simms, 1981, The structure and development of pocosin vegetation, *In* Pocosin Wetlands. Richardson, C. J., ed., Hutchison Ross Publishing Company, Stroudsburg, Pennsylvania, pp. 43–61.

Christenson, J., 1997, February 4, California floods change thinking on need to tame rivers, The New York Times, C4.

Chung, C.-H., 1982, Low marshes, China, *In* Creation and Restoration of Coastal Plant Communities, Lewis, R. R., III, ed., CRC Press, Boca Raton, Florida, pp. 131–45.

Chung, C.-H., 1989, Ecological engineering of coastlines with salt-marsh plantations, *In* Ecological Engineering: An Introduction to Ecotechnology, Mitsch, W. J., and S. E. Jørgensen, eds., John Wiley & Sons, New York, pp. 255–289.

Clark, J. R., and J. Benforado, eds., 1981, Wetlands of Bottomland Hardwood Forests, Elsevier, New York.

Clarke, P. J., 1993, Dispersal of grey mangrove (*Avicennia marina*) propagules in southeastern Australia, Aquatic Botany **45**:195–204.

Clarkson, B. R., 1997, Vegetation recovery following fire in two Waikato peatlands at Whangamarino and Moanatuatua, New Zealand, New Zealand Journal of Botany **35**:167–179.

Clements, F. E., 1916, Plant Succession: An Analysis of the Development of Vegetation, Carnegie Institute of Washington, Washington, D.C.

Clements, F. E., 1928, Plant Succession and Indicators, Carnegie Institution, Washington, D.C.

Clewell, A. F., and R. Lea, 1990, Creation and restoration of forested wetland vegetation in the southeastern United States, *In* Wetland Creation and Restoration: The Status of the Science, Kusler, J. A., and M. E. Kentula, eds., Island Press, Washington, D.C., pp. 195–231.

Close, A., 1990, The impact of man on the natural flow regime, *In* The Murray, Mackay, N., and D. Eastburn, eds., Murray Darling Basin Commission, Canberra, Australia, pp. 61–74.

Coble, T. A., and B. D. Vance, 1987, Seed germination in *Myriophyllum spicatum* L., Journal of Aquatic Plant Management **25**:8–10.

Cohen, A. D., 1974, Evidence of fires in the ancient Everglades and coastal swamp of southern Florida, *In* Environments of South Florida: Present and Past. Gleason, P. J., ed., Miami Geological Society, Miami, Florida, pp. 213–218.

Cohn, J. P., 1994, Restoring the Everglades, BioScience **44**:579–583.

Coiaccetto, E., 1996, A model for use in the management of coastal wetlands, Landscape and Urban Planning **36**:27–47.

Confer, S. R., and W. A. Niering, 1992, Comparison of created and natural freshwater emergent wetland in Connnecticut (USA), Wetlands Ecology and Management **2**:143–156.

Connell, J. H., 1978, Diversity in tropical rainforest and coral reefs, Science **199**: 1302–1310.

Connell, J. H., and R. O. Slatyer, 1977, Mechanisms of succession in natural communities and their role in community stability and organization, American Naturalist **111**:1119–1144.

Conner, W. H., and J. W. Day, Jr., 1989, Response of coastal wetland forests to human and natural changes in environment with emphasis on hydrology, *In* Proceedings of the Symposium, The Forested Wetlands of the Southern United States, Hook, D. D., and R. Lea, eds., June 12–14, 1988, Orlando, Florida, North Carolina University Press, Asheville, North Carolina, pp. 34–43.

Conner, W. H., and K. Flynn, 1989, Growth and survival of bald cypress (*Taxodium distichum* (L.) Rich.) planted across a flooding gradient in a Louisiana bottomland forest, Wetlands **9**:207–217.

Conner, W. H., J. R. Toliver, and F. H. Sklar, 1986, Natural regeneration of baldcypress (*Taxodium distichum* (L.) Rich) in a Louisiana swamp, Forest Ecology and Management **14**:305–317.

Conner, W. H., and J. W. Day, Jr., 1992, Water level variability and litterfall productivity of forested freshwater wetlands in Louisiana, American Midland Naturalist **128**:237–245.

Conover, M. R., 1985, Alleviating nuisance Canada goose problems through Methiocarb induced aversive conditioning, Journal of Wildlife Management **49**:631–636.

Conrad, S. T., 1997, Reproductive ecology and diaspore bank of the bryophytes of a bald cypress swamp, M.S thesis, Southern Illinois University, Carbondale.

Conti, R. S., and P. P. Gunther, 1984, Relations of phenology and seed germination to the distribution of dominant plants in Okefenokee Swamp, *In* The Okefenokee Swamp: Its Natural History, Geology, and Geochemistry, Cohen, A. D., D. J. Casagrande, M. J. Andrejko, and G. R. Best, eds., Wetland Surveys, Los Alamos, New Mexico, pp. 144–167.

Cook, C. D. K., 1985, Range extensions of aquatic vascular plant species, Journal of Aquatic Plant Management **23**:1–6.

Cook, C. D. K., 1987, Dispersion in aquatic and amphibious vascular plants, *In* Plant Life in Aquatic and Amphibious Habitats, Crawford, R. M. M., ed., Blackwell Scientific Publications, Oxford, pp. 179–190.

Cook, C. D. K., 1990a, Aquatic Plant Book, SPB Academic Publishing, the Hague, the Netherlands.

Cook, C. D. K., 1990b, Seed dispersal of *Nymphoides peltata* (S. G. Gmelin) O. Kuntze (Menyanthaceae), Aquatic Botany **37**:325–340.

Cook, G. D., S. A. Setterfield, and J. P. Maddison, 1996, Shrub invasion of a tropical wetland: implications for weed management, Ecological Applications **6**: 531–537.

Cortner, M., 1993, September 15, Flood victims vote for a new town, The Waterloo Republic-Times, 1–3.

Costa, J. E., 1988, Floods from dam failures, *In* Flood Geomorphology. Baker, V. R., R. C. Kochel, and P. C. Patton, eds., John Wiley & Sons, New York, pp. 439–463.

Coultas, C. L., 1979, Transplanting needlerush (*Juncus roemerianus*), *In* Wetlands Restoration and Creation: Proceedings of the Seventh Annual Conference, Cole, D. P., ed., May 16–17, 1980, Hillsborough Community College, Tampa, Florida, pp. 117–124.

Cowardin, L. M., V. Carter, F. C. Golet, and E. T. LaRoe, 1979, Classification of Wetlands and Deepwater Habitats of the United States, U.S. Fish and Wildlife Service, U.S. Department of the Interior, Washington, D.C.

Craighead, F. C., 1971, The Trees of South Florida, University of Miami Press, Coral Gables, Florida.

Crawford, R. M. M., 1972, Some metabolic aspects of ecology, Botanical Society of Edinburgh **41**:309–321.

Crawford, R. M. M., 1978, Biochemical and ecological similarities in marsh plant and diving animals, Naturwissenschaften **65**:194–201.

Crawford, R. M. M., 1983, Root survival in flooded soils, *In* Ecosystems of the World 4A; Mires: Swamp, Bog, Fen and Moor; General Studies, Gore, A. J. P., ed., Elsevier, Amsterdam, pp. 257–283.

Crum, H., 1988, A Focus on Peatlands and Peat Mosses, University of Michigan Press, Ann Arbor, Michigan.

Cummins, K. W., 1979, The natural stream ecosystem, *In* The Ecology of Regulated Streams, Proceedings of the First International Symposium on Regulated Streams, Ward, J. V., and J. A., Stanford, eds., April 18–20, 1979, Erie, Pennsylvania, Plenum Press New York, pp. 7–24.

Cummins, K. W., C. E. Cushing, and G. W. Minshall, 1995, Introduction: an overview of stream ecosystems, *In* River and Stream Ecosystems, Cushing, C. E., K. W. Cummins, and G. W. Minshall, eds., Elsevier, Amsterdam, pp. 1–8.

Cummins, K. W., G. W. Minshall, J. R. Sedell, C. E. Cushing, and R. C. Petersen, 1984, Stream ecosystem theory, Verhandlungen Internationale Vereinigung für Theoretische und Angewandte Limnologie **22**:1818–1827.

Cunningham, W. P., and B. W. Saigo, 1995, Environmental Science: A Global Concern, 3d ed., William C. Brown, Dubuque, Iowa.

Cunningham, W. P., and B. W. Saigo, 1997, Environmental Science: A Global Concern., 4th ed., Wm. C. Brown, Dubuque, Iowa.

Cypert, E., 1961, The effects of fires in the Okefenokee Swamp in 1954 and 1955, American Midland Naturalist **66**:485–503.

Dacey, J. W. H., 1980, Internal wind in water lilies: an adaptation for life in anaerobic sediments, Science **210**:1017–1019.

Dacey, J. W. H., 1981, Pressurized ventilation in the yellow water lily, Ecology **62**:1127–1147.

Dalby, R., 1957, Problems of land reclamation. 5. Saltmarsh in the Wash, Agricultural Review **2**:31–37.

Dalrymple, N. K., G. H. Dalrymple, and K. A. Fanning, 1993, Vegetation of restored rock—plowed wetlands of the East Everglades, Restoration Ecology **1**: 220–225.

Damman, A. W. H., and T. W. French, 1987, The Ecology of Peat Bogs of the Glaciated Northeastern United States: A Community Profile, Biological Report 85 (7.16), July 1987 U.S. Fish and Wildlife Service, Washington, D.C.

Danvind, M., and C. Nilsson, 1997, Seed floating ability and distribution of alpine plants along a northern Swedish river, Journal of Vegetation Science **8**:271–276.

Darnell, R. M., W. E. Pequegnat, B. M. James, F. J. Benson, and R. A. Defenbaugh, 1976, Impacts of Construction Activities in Wetlands of the United States, EPA-60013-76-045, April 1976, U.S. Environmental Protection Agency, Corvallis, Oregon.

Davies, B. R., and K. F. Walker, 1986, River systems as ecological units. An introduction to the ecology of river systems, *In* The Ecology of River Systems, Davies, B. R., and K. F. Walker, eds., Dr. W. Junk Publishers, Dordrecht, the Netherlands, pp. 1–23.

Davies, B. R., M. Thoms, and M. Meador, 1992, An assessment of the ecological impacts of interbasin water transfers, and their threats to river basin integrity and conservation, Aquatic Conservation: Marine and Freshwater Ecosystems **2**: 325–349.

Davies, R. J., 1987, Trees and Weeds, Forestry Commission Handbook 2, Her Majesty's Stationery Office, London.

Davis, F. W., E. A. Keller, A. Parikh, and J. Florsheim, 1989, Recovery of the chaparral riparian zone after wildfire, *In* Proceedings of the California Riparian Systems Conference, Protection, Management and Restoration for the 1990s, September 22–24, 1988, Davis, California, U.S.D.A. Technical Report PSW-110, U.S.D.A. Forest Service, Pacific Southwest Forest and Range Experiment Station, Berkeley, California, pp. 194–203.

Davis, J. H., Jr., 1943, The Natural Features of Southern Florida. Florida Geological Survey Bulletin No. 25, Tallahassee, Florida.

Davis, S. M., and Ogden, J. C. 1997, Toward ecosystem restoration, *In* Everglades: The Ecosystem and Its Restoration. *Edited by* S. M. Davis and J. C. Ogden, St. Lucie Press, Boca Raton, Florida. pp. 769–796.

DeBell, D. S., and D. D. Hook, 1969, Seeding habits of swamp tupelo, Research Report Paper SE-47, U.S.D.A. Forest Service, Washington, D.C.

DeBell, D. S., and I. D. Auld, 1971, September, Establishment of swamp tupelo seedlings after regeneration cuts, U.S.D.A. Forest Service, Washington, D.C. Research Note SE-164:1–7.

Décamps, H., A. M. Planty-Tabacchi, and E. Tabacchi, 1995, Changes in the hydrological regime and invasion by plant species along riparian systems of the Adour River, France, Regulated Rivers: Research and Management **11**:23–33.

Décamps, H., and E. Tabacchi, 1994, Species richness in vegetation along river margins, *In* Giller, P. S., A. G. Hildrew, and D. G. Raffaelli, eds., Aquatic Ecology: Scale Pattern and Process, The 34th Symposium of the British Ecological Society, University College, Cork. Blackwell Scientific Publications, Oxford, pp. 1–20.

de la Cruz, A. A., and C. T. Hackney, 1980, The Effects of Winter Fire and Harvest on the Vegetational Structure and Productivity of Two Tidal Marsh Communities in Mississippi, Mississippi-Alabama Sea Grant Consortium, Ocean Springs, Mississippi.

DeLaune, R. D., R. J. Buresh, and W. H. Patrick, Jr., 1979, Relationship of soil properties to standing crop biomass of *Spartina alterniflora* in a Louisiana marsh, Estuarine and Coastal Marine Science **8**:477–487.

Dellinger, G. P., E. L. Brunk, and A. D. Allmon, 1976, Tree mortality caused by flooding at two midwestern reservoirs, *In* Proceedings of the Thirtieth Annual Conference of the Southeastern Association of Fish and Wildlife Agencies, October 24–27, 1976, Jackson, Mississippi, pp. 645–648.

Demaree, D., 1932, Submerging experiments with *Taxodium*, Ecology **13**:258–262.

de Mars, H., M. J. Wassen, and H. Olde Venterink, 1997, Flooding and groundwater dynamics in fens in eastern Poland, Journal of Vegetation Science **8**:319–328.

Demissie, M., and A. Khan, 1991, Wetland drainage and streamflow trends in Illinois, *In* Hydraulic Engineering, Proceedings of the 1991 National Conference, Shane, R. M., ed., Nashville, Tennessee, American Society of Civil Engineers, New York, pp. 1050–1054.

Dennis, J. V., 1988, The Great Cypress Swamps, Louisiana State University Press, Baton Rouge, Louisiana.

Denyer, W. B. G., and C. G. Riley, 1964, Dieback and mortality of tamarack caused by high water, Forestry Chronicle **40**:334–338.

DeShield, M. A., Jr., M. R. Reddy, S. Leonard, N. H. Assar, and W. T. Brown, 1994, Inundation tolerance of riparian plant species, Wetland Journal **6**:20–21.

de Vlaming, V., and V. W. Proctor, 1968, Dispersal of aquatic organisms: viability of seeds recovered from the droppings of captive killdeer and mallard ducks, American Journal of Botany **55**:20–26.

Dexter, B. D., H. J. Rose, and N. Davies, 1986, River regulation and associated forest managment problems in the River Murray red gum forests, Australian Forestry **49**:16–27.

Diamond, J. M., 1975, The island dilemma: lessons for modern biogeographic studies for the design of natural reserves, Biological Conservation **7**:129–146.

Dickson, R. E., J. F. Hosner, and N. W. Hosley, 1965, The effects of four water regimes upon the growth of four bottomland tree species, Forest Science **11**: 299–305.

Dinerstein, E., 1989, The foliage-as-fruit hypothesis and the feeding behavior of South Asian ungulates, Biotropica **21**:214–218.

Dister, E., D. Gomer, P. Obrdlik, P. Petermann, and E. Schneider, 1990, Water management and ecological perspectives of the Upper Rhine's floodplains, Regulated Rivers: Research and Management **5**:1–15.

Dolbeer, R. A., N. R. Holler, and D. W. Hawthorne, 1994, Prevention and Control of Wildlife Damage, University of Nebraska, Lincoln, Nebraska.

Doody, J. P., 1984, The conservation of British saltmarshes, *In* Saltmarshes: Morphodynamics, Conservation and Engineering Significance. Allen, J. R. L., and K. Pye, eds., Cambridge University Press, Cambridge, pp. 80–114.

Dorge, C. L., W. J. Mitsch, and J. R. Wiemhoff, 1984, Cypress wetlands in southern Illinois. *In* Cypress Swamps. Ewel, K. C., and H. T. Odum, eds. University of Florida Presses, Gainesville, Florida, pp. 393–404.

Douglas, M. S., 1947, The Everglades: River of Grass, Hurricane House, Miami, Florida.

Downs, P. W., 1994, Characterization of river channel adjustments in the Thames Basin, south-east England, Regulated Rivers: Research and Management **9**:151–175.

Drozdiak, W., 1996, March 27, Sewer of Europe, The Washington Post on the Net, 3.

DuBarry, A. P., 1963, Germination of bottomland tree seed while immersed in water, Journal of Forestry **61**:225–226.

Dudgeon, D., 1996, Anthropogenic influences on Hong Kong streams, GeoJournal **40**:53–61.

Duever, M. J., 1984, Environmental factors controlling plant communities of the Big Cypress Swamp, *In* Environments of South Florida: Past and Present. Gleason, P. J., ed., Miami Geological Society, Miami, Florida, pp. 127–137.

Duever, M. J., J. E. Carlson, J. F. Meeder, L. C. Duever, L. H. Gunderson, L. A. Riopelle, T. R. Alexander, R. L. Myers, and D. P. Spangler, 1986, The Big Cypress National Preserve, National Audubon Society, New York.

Duever, M. J., and J. M. McCollom, 1987, Cypress tree-ring analysis in relation to wetlands and hydrology, *In* Proceedings of the International Symposium on Ecological Aspects of Tree-Ring Analysis, August 17–21, 1986, Marymount College, Tarrytown, New York, U.S. Department of Energy, Washington, D.C., pp. 249–260.

Dumortier, M., A. Verlinden, H. Beeckman, and K. van der Mijnsbrugge, 1996, Effects of harvesting dates and frequencies on above and below-ground dynamics in Belgian wet grasslands, Ecoscience **3**:190–198.

Dunn, W. J., 1989, Wetland succession—What is the appropriate paradigm?, *In* Wetlands Concerns and Successes, Symposium Proceedings, Fisk, D. W., ed., Tampa, Florida, September 17–22, 1989, American Water Resources Association, Bethesda, Maryland, pp. 473–488.

Dwyer, R. L., and M. A. Turner, 1983, Simulation of river flow and DO dynamics affected by peaking discharges from a hydroelectric dam, *In* Developments in Environmental Modelling, 5, Analysis of Ecological Systems: State-of-the-Art in Ecological Modelling, Proceedings of a Symposium at Colorado State University, May 24–28, 1982, Laurenroth, W. K., G. V. Skogerboe, and M. Flug, eds., Fort Collins, Colorado, Elsevier, Amsterdam, pp. 933–949.

Ebersole, J. L., W. J. Liss, and C. A. Frissell, 1997, Restoration of stream habitats in the western United States: restoration as reexpresssion of habitat capacity, Environmental Management **21**:1–14.

Echeverria, J. D., P. Barrow, and R. Roos-Collins, 1989, Rivers at Risk, Island Press, Washington, D.C.

Eckenfelder, G. V., 1995, Dams through the decades: observing a broadening knowledge base, Hydroreview (April):34–38.

Eggler, W. A., and W. G. Moore, 1961, The vegetation of Lake Chicot, Louisiana, after eighteen years of impoundment, The Southwestern Naturalist **6**:175–183.

Egler, F. E., 1977, The Nature of Vegetation: Its Management and Mismanagement, Aton Forest, Norfolk, Connecticut.

Ehrenfeld, J. G., and L. A. Toth, 1997, Restoration ecology and the ecosystem perspective. Restoration Ecology **5**:307–317.

Eleuterius, L. N., 1976, Transplanting maritime plants to dredged material in Mississippi waters, *In* Proceedings of the Specialty Conference on Dredging and Its Environmental Effects, Krenkel, P. A., J. Harrison, and J. C. Burdick III, eds., January 26–28, 1976, Mobile, Alabama, American Society of Civil Engineers New York, pp. 900–918.

Ellenberg, H., 1988, Vegetation Ecology of Central Europe, Strutt, G. K., translator, Cambridge University Press, Cambridge.

Ellery, W. N., and T. S. McCarthy, 1994, Principles for the sustainable utilization of the Okavango Delta ecosystem, Botswana, Biological Conservation **70**:159–168.

Elmore, W., and R. L. Beschta, 1987, Riparian areas: perceptions in management, Rangelands **9**:260–265.

Environmental Defense Fund and World Wildlife Fund, 1992, How Wet is a Wetland?, Environmental Defense Fund and World Wildlife Fund, New York and Washington, D.C.

Erlandson, C. S., 1987, The potential role of seed banks in the restoration of drained prairie wetlands, M.S. thesis, Department of Botany, Iowa State University, Ames, Iowa.

Erwin, K. L., 1991, An Evaluation of Wetland Mitigation in the South Florida Water Management District, C89-0082-A1, South Florida Water Management District, Volume 1, West Palm Beach, Florida.

Evans, J., 1988, Natural Regeneration of Broadleaves, Forestry Commission Bulletin 78, Her Majesty's Stationery Office, London.

Evink, G. L., 1980, Studies of Causeways in the Indian River, Florida, FL-ER-7-80, Florida Department of Transportation, Tallahassee, Florida.

Ewel, J. J., 1986, Invasibility: lessons from South Florida, *In* Ecology of Biological Invasions of North America and Hawaii, Symposium Proceedings, October 21–25, 1984, Asilomar, California, Baker, H. G. H. A., Mooney, and J. A. Drake, eds., Springer-Verlag, New York, pp. 214–230.

Ewel, K. C., 1990, Swamps, *In* Ecosystems of Florida. Myers, R. L., and J. J. Ewel, eds., University of Central Florida Press, Orlando, Florida, pp. 281–323.

Ewel, K. C., and W. J. Mitsch, 1978, The effects of fire on species composition in cypress dome ecosystems, Florida Scientist **41**:25–32.

Faber, P. M., E. Keller, A. Sands, and B. M. Massey, 1989, The Ecology of Riparian Habitats of the Southern California Coastal Region: A Community Profile, Biological Report 85 (727), U.S. Fish and Wildlife Service, Portland, Oregon.

Faber-Langendoen, D., and P. F. Maycock, 1989, Community patterns and environmental gradients of buttonbush, *Cephalanthus occidentalis*, ponds in lowland forests of southern Ontario, The Canadian Field-Naturalist **103**:479–485.

Fagerstedt, K. V., 1992, Development of aerenchyma in roots and rhizomes of *Carex rostrata* (Cyperaceae), Nordic Journal of Botany **12**:115–120.

Farrelly, D., 1984, The Book of Bamboo, Thames and Hudson, London.

Faulkner, S. P., and W. H. Patrick, Jr., 1992, Redox processes and diagnostic wetland soil indicators in bottomland hardwood forests, Soil Science Society of America Journal **56**:856–865.

Federal Emergency Management Agency, 1993, Water Control Infrastructure: National Inventory of Dams 1992, Federal Emergency Management Agency and U.S. Army Corps of Engineers, Washington, D.C.

Feller, I. C., and K. L. McKee, in press, Light-gap creation in a Belizian red mangrove forest by a wood-boring insect, Biotropica.

Fenner, M., 1985, Seed Ecology, Chapman and Hall, London.

Fenner, P., W. W. Brady, and D. R. Patton, 1984, Observations on seeds and seedling of Fremont cottonwood, Desert Plants **6**:55–58.

Fenner, P., W. W. Brady, and D. R. Patton, 1985, Effects of regulated water flows on regeneration of Fremont cottonwood, Journal of Range Management **38**:135–138.

Finlayson, C. M., 1991, Plant ecology and management of an internationally important wetland in monsoonal Australia, *In* Proceedings of an International Symposium on Wetlands and River Corridor Management, Kusler, J. A., and S. Daly, eds., July 5–9, 1989, Charleston, South Carolina, Association of State Wetland Managers, Berne, New York, pp. 90–98.

Finlayson, C. M., 1991b, Production and major nutrient composition of three grass species on the Magela floodplain, Northern Territory, Australia, Aquatic Botany **41**:263–280.

Finlayson, C. M. 1993, Vegetation changes and biomass on an Australian monsoonal floodplain. *In* Wetlands and Ecotones: Studies on Land-Water Interactions, Gopal B., A. Hillbricht-Ilkowska and R. G. Wetzel, eds., National Institute of Ecology, New Delhi. pp. 157–171.

Finlayson, C. M., B. J. Bailey, and I. D. Cowie, 1989, Macrophyte Vegetation of the Magela Creek Floodplain, Alligator Rivers Region, Northern Territory, Supervising Scientist for the Alligator Rivers Region, Australian Government Publishing Service, Canberra.

Finlayson, C. M., I. D. Cowie, and B. J. Bailey, 1990, Sediment seedbanks in grassland on the Magela Creek floodplain, northern Australia, Aquatic Botany **38**: 163–176.

Finlayson, C. M., and I. Von Oertzen, 1993, Wetlands of Australia: northern (tropical) Australia, *In* Wetlands of the World I: Inventory, Ecology and Management, Whigham, D., D. Dykyjová, and S. Hejny, eds., Kluwer, Dordrecht, the Netherlands, pp. 195–304.

Firth, P. L., and K. L. Hooker, 1989, Plant community structure in disturbed and undisturbed forested wetlands, *In* Wetland: Concerns and Successes, Fisk, D. W., ed., September 17–22, 1989, Tampa, Florida, American Water Resources Association, Bethesda, Maryland, pp. 101–113.

Fisher, S. G., L. J. Gray, N. B. Grimm, and D. E. Busch, 1982, Temporal succession in a desert stream ecosystem following flash flooding, Ecological Monographs **52**:93–110.

Flowers, T. J., M. A. Hajibagheri, and N. J. W. Clipson, 1986, Halophytes, Quarterly Review of Biology **61**:313–337.

Fojt, W. J., 1994, Dehydration and the threat to East Anglian fens, England, Biological Conservation **69**:163–175.

Fonseca, M. S., 1990, Regional analysis of the creation and restoration of seagrass systems, *In* Wetland Creation and Restoration: The Status of the Science, Kusler, J. A., and M. E. Kentula, eds., Island Press, Washington, D.C., pp. 171–193.

Fonseca, M. S., W. J. Kenworthy, and F. X. Courtney, 1996, Development of planted seagrass beds in Tampa Bay, Florida, USA. I. plant components, Marine Ecology Progress Series **132**:127–139.

Fonseca, M. S., W. J. Kenworthy, F. X. Courtney, and M. O. Hall, 1994, Seagrass planting in the southeastern United States: methods for accelerating habitat development, Restoration Ecology **2**:198–212.

Fowell, H. A., 1965, Silvics of Forest Trees of the United States, USDA, Washington, D.C.

Fredrickson, L. H., 1979, Lowland hardwood wetlands: current status and habitat values for wildlife, *In* Greeson, P. E., J. R. Clark, and J. E. Clark, eds., November 7–10, 1978, Lake Buena Vista, Florida, American Water Resources Association, Minneapolis, Minnesota, pp. 296–306.

Fredrickson, L. H., and F. A. Reid, 1990, Impacts of hydrologic alteration on management of freshwater wetlands, *In* Management of Dynamic Ecosystems, Proceedings of a Symposium in the 51st Midwest Fish and Wildlife Conference, Sweeney, J. M., ed., December 5, 1989, Springfield, Illinois, North Central Section, The Wildlife Society, West Lafayette, Indiana, pp. 72–90.

Friedrich, G., and D. Müller, 1984, Rhine, *In* Ecology of European Rivers, Whitton, B. A., ed., Blackwell Scientific Publications, Oxford, pp. 265–315.

Frissell, C. A., W. J. Liss, C. E. Warren, and M. D. Hurley, 1986, A hierarchical framework for stream habitat classification: viewing streams in a watershed context, Environmental Management **10**:199–214.

Fritzell, E. K., 1989, Mammals in prairie wetlands, *In* Northern Prairie Wetlands. van der Valk, A. G., ed., Iowa State University Press, Ames, Iowa, pp. 268–301.

Funk, D. F., 1965, Honey locust (*Gleditsia triacanthos* L.), *In* Silvics of Forest Trees of the United States, Fowells, H. A., ed, Agriculture Handbook 271, U.S.D.A. Forest Service, Washington, D.C. pp. 198–201.

Galatowitsch, S. M., and A. G. van der Valk, 1994, Restoring Prairie Potholes: An Ecological Approach, Iowa State University Press, Ames, Iowa.

Galatowitsch, S. M., and A. G. van der Valk, 1995, Natural revegetation during restoration of wetlands in the southern Prairie Pothole Region of North America. *In* Restoration of Temperate Wetlands. Wheeler, B. D., S. C. Shaw, W. J. Fojt, and R. A. Robertson, eds. John Wiley and Sons, Chichester, United Kingdom, pp. 129–142.

Galatowitsch, S. M., and A. G. van der Valk, 1996, The vegetation of restored and natural prairie wetlands, Ecological Applications **6**:102–112.

Galinato, M. I., and A. G. van der Valk, 1986, Seed germination traits of annuals and emergents recruited during drawdowns in the Delta Marsh, Manitoba, Canada, Aquatic Botany **26**:89–102.

Garbisch, E. W., 1991, The do's and dont's of wetland planning, Wetland Journal **7**:14–16.

Garbisch, E. W., Jr., P. B. Woller, and R. J. McCallum, 1975, Salt Marsh Establishment and Development, TM-52, U.S. Army Corps of Engineers Coastal Engineering Research Center, Fort Belvoir, Virginia.

Gates, F. C., 1926, Plant succession about Douglas Lake, Cheboygan County, Michigan, Botanical Gazette **82**:170–182.

Gentry, R. C., 1974, Hurricanes in South Florida, *In* Environments of South Florida: Present and Past. Gleason, P. J., ed., Miami Geological Society, Miami, Florida, pp. 73–81.

Gibson, D. J., 1996, Textbook misconceptions: the climax concept of succession, American Biology Teacher **55**:135–140.

Gill, A. M., J. R. L. Hoare, and N. P. Cheney, 1990, Fires and their effects in the wet-dry tropics of Australia, *In* Fire in the Tropical Biota, Goldammer, J. C., ed., Springer-Verlag, Berlin, pp. 159–178.

Gill, A. M., P. H. R. Moore, and R. J. Williams, 1996, Fire weather in the wet-dry tropics of the World Heritage Kakadu National Park, Australia, Australian Journal of Ecology **21**:302–308.

Gill, R. M. A., J. Gurnell, and R. C. Trout, 1995, Do woodland mammals threaten the development of new woods?, *In* The Ecology of Woodland Creation, Ferris-Khan, R., ed., John Wiley & Sons, Chichester, United Kingdom, pp. 201–224.

Gilman, K., 1982, Nature conservation in wetlands; two small fen basins in western Britain, *In* Proceedings of the International Scientific Workshop on Ecosystem in Freshwater Wetlands and Shallow Water Bodies, July 12–26, 1981, Minsk-Pinsk-Tskhaltoubo, USSR, Centre of International Projects GKNT, Moscow, pp. 290–310.

Gippel, C. J., 1995, Environmental hydraulics of large woody debris in streams and rivers, Journal of Environmental Engineering **121**:388–395.

Gippel, C. J., B. S. Finlayson, and I. C. O'Neill, 1996a, Distribution and hydraulic significance of large woody debris in a lowland Australian river, Hydrobiologia **318**:179–194.

Gippel, C. J., I. C. O'Neill, and B. L. Finlayson, 1992, The Hydraulic Basis of Snag Management, Centre for Environmental Applied Hydrology, Department of Civil and Agricultural Engineering, The University of Melbourne, Parkville, Australia.

Gippel, C. J., I. C. O'Neill, B. L. Finlayson, and I. Schnatz, 1996b, Hydraulic guidelines for the reintroduction and management of large woody debris in lowland rivers, Regulated Rivers: Research and Management **12**:233–236.

Gitay, H., and I. R. Noble, 1997, What are functional types and how should we seek them?, *In* Plant Functional Types, Smith, T. M., H. H. Shugart, and F. I. Woodward, eds., Cambridge University Press, Cambridge, pp. 3–19.

Glaser, P. H., and G. A. Wheeler, 1980, The development of surface patterns in the Red Lake peatland, northern Minnesota, *In* Proceedings of the 6th International Peat Congress, August 17–23, 1980, Duluth, Minnesota, International Peat Congress, Duluth, Minnesota, pp. 31–35.

Glass, S., 1987, Rebirth of a river, Restoration and Management Notes **5**:6–14.

Gleason, H. A., 1917, The structure and development of the plant association, Bulletin of the Torrey Botanical Club **43**:463–481.

Gleason, H. A., 1926, The individualistic concept of the plant association, Bulletin of the Torrey Botanical Club **53**:7–26.

Gleason, H. A., 1927, Further views on the succession concept, Ecology **8**:299–326.

Glenn, E., T. L. Thompson, R. Frye, J. Riley, and D. Baumgartner, 1995, Effects of salinity on growth and evapotranspiration of *Typha domingensis* Pers., Aquatic Botany **52**:75–91.

Godshalk, G. L., and R. G. Wetzel, 1979, Decomposition of aquatic angiosperms. III. *Zostera marina* L. and a conceptual model of decomposition, Aquatic Botany **5**:301–364.

Godwin, H., 1923, Dispersal of pond floras, Journal of Ecology **11**:160–164.

Godwin, H., 1968, Evidence for the longevity of seeds, Nature **220**:708–709.

Godwin, H., 1978, Fenland: Its Ancient Past and Uncertain Future, Cambridge University Press, Cambridge.

Golden, T., 1997, February 2, Flood breach deadlock over Yosemite's future, New York Times, On the Web.

Gole, P., 1990, The greening of our hills, Journal of the Ecological Society **3**:13–25.

Good, B., 1993, Louisiana's wetlands: combatting erosion and revitalizing native ecosystems, Restoration and Management Notes **11**:125–133.

Good, B. J., and W. H. Patrick, Jr., 1987, Gas composition and respiration of water oak (*Quercus nigra* L.) and green ash (*Fraxinus pennsylvanica* Marsh.) roots after prolonged flooding, Plant and Soil **97**:419–427.

Gopal, B., 1992, Tropical wetlands: degradation and need for rehabilitation, *In* Ecosystem Rehabilitation, Volume 2: Ecosystem Analysis and Synthesis, Wali, M. K., ed., SPB Academic Publishing, the Hague, the Netherlands, pp. 277–296.

Gopal, B., and K. Krishnamurthy, 1993, Wetlands of South Asia, *In* Wetlands of the World I.: Inventory, Ecology and Management. Whigham, D., D. Dykyjová, and S. Hejny, eds., Kluwer, Dordrecht, the Netherlands, pp. 345–414.

Gopal, B., and M. Sah, 1995, Inventory and classification of wetlands in India, *In* Classification and Inventory of the World's Wetlands. Finlayson, C. M., and A. G. van der Valk, eds., Kluwer, Dordrecht, the Netherlands, pp. 39–56.

Gore, A., 1992, Earth in the Balance, Houghton Mifflin, Boston.

Gosling, P. G., 1987, Dormant tree seeds can exhibit similar properties to seed of low vigour, *In* Advances in Practical Arboriculture, Forestry Commission Bulletin 65, Patch, D., ed., Her Majesty's Stationery Office, London, pp. 28–31.

Gosselink, J. G., 1984, The Ecology of Delta Marshes of Coastal Louisiana: A Community Profile, U.S. Fish and Wildlife Service, Washington, D.C.

Gosselink, J. G., S. E. Bayley, W. H. Conner, and R. E. Turner, 1981, Ecological factors in the determination of riparian wetland boundaries, *In* Wetlands of Bottomland Hardwood Forests, Proceedings of a Workshop on Bottomland Hardwood Forest Wetlands of the Southeastern United States, Clark, J. R., and J. Benforado, eds., June 1–5, 1980, Lake Lanier, Georgia, Elsevier, Amsterdam, pp. 197–219.

Gosz, J. R., and P. J. H. Sharpe, 1989, Broad-scale concepts for interactions of climate, topography, and biota at biome transitions, Landscape Ecology **3**:229–243.

Goulding, M., 1993, The role of fishes in seed dispersal and plant distribution in Amazonian floodplain ecosystems, Sonderbaende des Naturwissenschlaftlichen Vereins in Hamburg **7**:271–283.

Graham, A., R. Begg, P. Graham, and S. Raskin, 1982, Buffalo in the Northern Territory, Conservation Commission of the Northern Territory, Darwin, Australia.

Green, W. E., 1947, Effect of water impoundment on tree mortality and growth, Journal of Forestry **45**:118–120.

Greenway, H., and A. Munns, 1980, Mechanism of salt tolerance in nonhalophytes, Anuual Review of Plant Physiology **31**:149–190.

Gregory, K. J., 1977, The context of river channel changes, *In* River Channel Changes, Gregory, K. J., ed., John Wiley & Sons, Chichester, United Kingdom, pp. 1–12.

Gregory, K. J., 1992, Vegetation and river channel process interactions. *In* River Conservation and Management. Boon, P. J., P. Calow, and G. E. Petts, eds. Wiley, Chichester, U.K., pp. 255–269.

Gregory, K. J., and R. J. Davis, 1992, Coarse woody debris in stream channels in relation to river channel management in woodland areas, Regulated Rivers: Research and Management **7**:117–136.

Grillas, P., 1990, Distribution of submerged macrophytes in the Camargue in relation to environmental factors, Journal of Vegetation Science **1**:393–402.

Grime, J. P., 1977, Evidence for the existence of three primary strategies in plants and its relevance to ecological and evolutionary theory, American Naturalist **111**:1169–1184.

Grime, J. P., 1979, Plant Strategies and Vegetation Processes, John Wiley & Sons, New York.

Grime, J. P., 1981, The role of seed dormancy in vegetation dynamics, Annals of Applied Biology **98**:555–558.

Grosse, W., J. Frye, and S. Latterman, 1992, Root aeration in wetland trees by pressurized gas transport, Tree Physiology **10**:285–295.

Grubaugh, J. W., and R. V. Anderson, 1988, Spatial and temporal availability of floodplain habitat: long-term changes at Pool 19, Mississippi River, American Midland Naturalist **119**:402–411.

Guilkey, P. C., 1965, American elm (*Ulmus americana* L.), *In* Silvics of Forest Trees of the United States, Fowells, H. A., ed., Agriculture Handbook 271, U.S.D.A. Forest Service, Washington, D.C., pp. 725–731.

Gunderson, L. H., and W. F. Loftus, 1993, The Everglades, *In* Biodiversity of the Southeastern United States: Lowland Terrestrial Communities, Martin, W. H., S. G. Boyce, and A. C. Echternacht, eds., John Wiley & Sons, New York, pp. 199–255.

Gunther, P. P., D. J. Casagrande, and R. R. Cherney, 1984, The viability and fate of seeds as a function of depth in the peats of Okefenokee Swamp, *In* The Okefenokee Swamp: Its Natural History, Geology, and Geochemistry, Cohen, A. D., D. J. Casagrande, M. J. Andrejko, and G. R. Best, eds., Wetland Surveys, Los Alamos, New Mexico, pp. 168–179.

Guppy, H. B., 1897, On the postponement of the germination of seeds of aquatic plants, Proceedings of the Royal Physicians Society of Edinburgh, **13**:344–360.

Haag, R. W., 1983, Emergence of seedlings of aquatic macrophytes from lake sediments, Canadian Journal of Botany **61**:148–156.

Hackney, C. T., and S. M. Adams, 1992, Aquatic communities of the southeastern United States: past, present, and future, *In* Biodiversity of the Southeastern United States: Aquatic Communities, Hackney, C. T., S. M. Adams, and W. H. Martin, eds., John Wiley & Sons, New York, pp. 747–760.

Hadley, R. F., M. R. Karlinger, A. W. Burns, and T. R. Eschner, 1987, Water development and associated hydrologic changes in the Platte River, Nebraska, USA., Regulated Rivers: Research and Management **1**:331–341.

Hair, J. D., G. T. Hepp, L. M. Luckett, K. P. Reese, and D. K. Woodward, 1979, Beaver pond ecosystems and their relationships to multi-use natural resource management, *In* Strategies for Protection and Management of Floodplain Wetlands and Other Riparian Ecosystems, Proceedings of the Symposium, Callaway Gardens, Georgia, General Technical Report WO-12, Washington, D.C., pp. 80–92.

Hall, T. F., W. T. Penfound, and A. D. Hess, 1946, Water level relationships of plants in the Tennessee Valley with particular reference to malaria control, Journal of the Tennessee Academy of Science **21**:18–59.

Hall, T. F., and G. E. Smith, 1955, Effects of flooding on woody plants, West Sandy Dewatering Project, Kentucky Reservoir, Journal of Forestry **53**:281–285.

Halls, L. K., ed., 1977, Southern fruit-producing woody plants used by wildlife, Forest Service General Technical Report SO-16, USDA, Washington, D.C.

Haltiner, J., J. B. Zedler, K. E. Boyer, G. D. Williams, and J. C. Callaway, 1997, Influence of physical processes on the design, functioning and evolution of restored tidal wetlands in California, Wetlands Ecology and Management **4**:73–91.

Hammer, D. A., 1997, Creating Freshwater Wetlands, *2d ed.,* Lewis Publishers, Boca Raton, Florida.

Hammer, U. T., 1986, Saline Lake Ecosystems of the World, Dr. W. Junk Publishers, Dordrecht, the Netherlands.

Hanna, M., 1997, February 5, 1997, Namibia's water shortage threatens African oasis, CNN Interactive, 4 pages.

Hansen, A. J., P. G. Risser, and F. di Castri, 1992, Epilogue: biodiversity and ecological flows across ecotones, *In* Landscape Boundaries: Consequences for Biotic Diversity and Ecological Flows, Hansen, A. J., and F. di Castri, eds., Springer-Verlag, New York, pp. 423–438.

Harker, D., S. Evans, M. Evans, and K. Harker, 1993, Landscape Restoration Handbook, Lewis Publishers, Boca Raton, Florida.

Harmer, R., and G. Kerr, 1995, Creating woodlands: to plant trees or not?, *In* The Ecology of Woodland Creation, Ferris-Kaan, R., ed., John Wiley & Sons, Chichester, United Kingdom, pp. 113–128.

Harmer, R., G. Kerr, and R. Boswell, 1997, Characteristics of lowland broadleaved woodland being restocked by natural revegetation, Forestry **70**:199–210.

Harmon, M. E., J. F. Franklin, F. J. Swanson, P. Sollins, S. V. Gregory, J. D. Lattin, N. H. Anderson, S. P. Cline, N. G. Aumen, J. R. Sedell, G. W. Lienkaemper, K.

Cromack, Jr., and K. W. Cummins, 1986, Ecology of coarse woody debris in temperate ecosystems, Advances in Ecological Research **15**:133–302.

Harms, W. R., H. T. Schreuder, D. D. Hook, C. L. Brown, and F. W. Shropshire, 1980, The effects of flooding on the swamp forest in Lake Ocklawaha, Florida, Ecology **61**:1412–1421.

Harms, W. R., 1973, Some effects of soil type and water regime on growth of tupelo seedlings, Ecology **54**:188–193.

Harper, J. L., 1977, Population Biology of Plants, Academic Press, London.

Harper, J. L., J. T. Williams, and G. R. Sagar, 1965, The behaviour of seeds in the soil: I. The heterogeneity of soil surfaces and its role in determining the establishment of plants from seed, Journal of Ecology **53**:273–286.

Harris, R. W., A. T. Leiser, and R. E. Fissell, 1975, Plant Tolerance to Flooding, RWH-200-7/1/75, Department of Environmental Horticulture, Davis, California.

Harris, S. C., T. H. Martin, and K. W. Cummins, 1995, A model for aquatic invertebrate response to Kissimmee River restoration, Restoration Ecology **3**:181–194.

Harris, S. W., and W. H. Marshall, 1963, Ecology of water-level manipulations on a northern marsh, Ecology **44**:331–343.

Hart, J. H., 1990, Nothing is permenent except change, *In* Management of Dynamic Ecosystems, Proceedings of a Symposium on the 51st Midwest Fish and Wildlife Conference, Sweeney, J. M., ed. Springfield, Illinois, Dec. 5, 1989, The Wildlife Society, North Central Chapter, West Lafayette, Indiana, pp. 1–17.

Hartman, G., 1996, Habitat selection by European beaver (*Castor fiber*) colonizing a boreal landscape, Journal of Zoology, London **240**:317–325.

Hartman, J. M., 1983, Effects of wrack accumulation on four vegetation zones in a New England salt marsh, August 7–11, 1983 [program paper G561, Session No. 54], University of North Dakota. American Institute of Biological Sciences, Washington, D.C.

Haufler, J. B., 1990, Static management of forest ecosystems, *In* Management of Dynamic Ecosystems, Proceedings of a Symposium on the 51st Midwest Fish and Wildlife Conference, Sweeney, J. M., ed. Springfield, Illinois, Dec. 5, 1989, West The Wildlife Society, North Central Section, West West Lafayette, Indiana, pp. 123–130.

Haukos, D. A., and L. M. Smith, 1994, Composition of seed banks along an elevational gradient in playa wetlands, Wetlands **14**:301–307.

Hayden, A., 1939, Notes on *Typha angustifolia* in Iowa, Iowa State Journal of Science **13**:341–351.

Hecht, S. B., 1993, The logic of livestock and deforestation in Amazonia, BioScience **43**:687–695.

Heinselman, M. L., 1970, Landscape evolution, peatland types, and the environment in the Lake Agassiz Peatlands Natural Area, Minnesota, Ecological Monographs **40**:235–261.

Heinselman, M. L., 1975, Boreal peatlands in relation to environment, *In* Coupling of Land and Water Systems, Hasler, A. D., ed., Springer-Verlag, New York, pp. 93–104.

Hellings, S. E., and J. L. Gallagher, 1992, The effects of salinity and flooding on *Phragmites australis*, Journal of Applied Ecology **29**:41–49.

Helliwell, D. R., 1989, Soil transfer as a method of moving grassland and marshland vegetation, *In* Biological Habitat Reconstruction, Buckley, G. P., ed., Belhaven Press, London, pp. 258–263.

Herbst, G. N., 1980, Effects of burial on food value and consumption of leaf detritus by aquatic invertebrates in a lowland forest stream, Oikos **35**:411–424.

Hesslop-Harrison, Y., 1955, *Nuphar*, Journal of Ecology **43**:342–364.

Hewitt, N., and K. Miyanishi, 1997, The role of mammals in maintaining plant species richness in a floating *Typha* marsh in southern Ontario, Biodiversity and Conservation **6**:1085–1102.

Hey, D. L., and N. S. Philippi, 1995, Flood reduction through wetland restoration: the upper Mississippi River Basin as a case history, Restoration Ecology **3**:4–17.

Hickin, E. J., 1983, River channel changes: retrospect and prospect, *In* Modern and Ancient Fluvial Systems, Collinson, J. D., and J. Lewin, eds., Blackwell Scientific Publications, Oxford, pp. 61–83.

Hillman, W. S., 1961, The Lemnaceae, or duckweeds, a review of the descriptive and experimental literature, The Botanical Review **27**:221–287.

Hills, J. M., K. J. Murphy, I. D. Pulford, and T. H. Flowers, 1994, A method for classifying European riverine wetland ecosystems using functional vegetation groups, Functional Ecology **8**:242–252.

Hobbs, R. J., D. A. Saunders, L. A. Lobry de Bruyn, and A. R. Main, 1993, Changes in biota, *In* Reintegrating Fragmented Landscapes: Towards Sustainable Production and Nature Conservation, Hobbs, R. J. and D. A. Saunders, ed., Springer-Verlag, New York, pp. 65–106.

Hobbs, R. J., and D. A. Norton, 1996, Towards a conceptual framework for restoration ecology, Restoration Ecology **4**:93–110.

Hodkinson, D. J., and K. Thompson, 1997, Plant dispersal: the role of man, Journal of Applied Ecology **34**:1484–1496.

Hofius, K., 1991, Co-operation in hydrology of the Rhine basin countries, *In* Proceedings of an International Symposium of the XXth General Assembly of the International Union of Geodesy and Geophysics, van de Ven, F. H. M., D. Gutknecht, D. P. Loucks, and K. A. Salewicz, eds., August 11–24, 1991, Vienna, Austria, International Association of Hydrological Sciences via Galliard Ltd., Great Yarmouth, United Kingdom, pp. 25–35.

Hofstetter, R. H., 1974, The effect of fire on the pineland and sawgrass communities of southern Florida, *In* Environments of South Florida: Present and Past. Gleason, P. J., ed., Miami Geological Society, Miami, Florida, pp. 201–209.

Holling, C. S., 1973, Resilience and stability of ecological systems, Annual Review of Ecology and Systematics **4**:1–23.

Holtz, S., 1986, Tropical seagrass restoration, Restoration and Management Notes **4**:5–11.

Hook, D. D., 1984a, Adaptations to flooding with fresh water, *In* Flooding and Plant Growth, Kozlowski, T. T., ed., Academic Press, Orlando, Florida, pp. 365–294.

Hook, D. D., 1984b, Waterlogging tolerance of lowland tree species of the South, Southern Journal of Applied Forestry **8**:136–149.

Hook, D. D., C. L. Brown, and P. P. Kormanik, 1970, Lenticel and water root development of swamp tupelo under various flooding conditions, Botanical Gazette **131**:217–224.

Hooke, J. M., and C. E. Redmond, 1989, River-channel changes in England and Wales, Journal of the Institute of Water and Environmental Management **3**:329–335.

Hopkins, B., 1983, Successional process, *In* Tropical Savannas: Successional Processes. Bourlière, F., ed., Elsevier, New York, pp. 605–616.

Hopkins, D. R., and V. T. Parker, 1984, A study of the seed bank of a salt marsh in northern San Francisco Bay, American Journal of Botany **71**:348–355.

Hosner, J. F., 1957, Effects of water upon the seed germination of bottomland trees, Forest Science **3**:67–70.

Hosner, J. F., 1958, The effects of complete inundation upon seedlings of six bottomland tree species, Ecology **39**:371–374.

Hosner, J. F., 1960, Relative tolerance to complete inundation of fourteen bottomland tree species, Forest Science **6**:246–251.

Hosner, J. F., and S. G. Boyce, 1962, Tolerance to water saturated soil of various bottomland hardwoods, Forest Science **8**:180–186.

Hough, A. F., 1965, Black cherry (*Prunus serotina* Ehrh.), *In* Silvics of Forest Trees of the United States, Fowells, H. A., ed., Agriculture Handbook 271, USDA Forest Service, Washington, D.C., pp. 539–545.

House, R., 1996, An evaluation of stream restoration structures in a coastal Oregon stream, 1981–1993, North American Journal of Fisheries Management **16**:272–281.

Howard-Williams, C., 1977, Swamp ecosystems, The Malayan Nature Journal **31**: 113–125.

Hughes, J. D., 1983, American Indian Ecology, Texas Western Press, El Paso, Texas.

Hügin, G., 1981, Die Auenwälder des südlichen Oberrheintals—ihre Veränderung und Gefährdung durch den Rheinausbau, Landschaft und Stadt **13**:78–91

Hutchings, P. A., and P. Saenger, 1987, Ecology of Mangroves, University of Queensland Press, St. Lucia, Queensland, Australia.

Hutnik, R. J., and H. W. Yauney, 1965, Red maple (*Acer rubrum* L.), *In* Silvics of Forest Trees of the United States, Fowells, H. A., ed., Agriculture Handbook 271, USDA Forest Service, Washington, D.C. pp. 57–62.

Ibàñez, C., N. Prat and A. Canicio, 1996, Changes in the hydrology and sediment transport produced by large dams on the lower Ebro River and its estuary, Regulated Rivers: Research and Management **12**:51–62.

Interagency Floodplain Management Review Committee, 1994, Sharing the Challenge: Floodplain Management into the 21st Century, Administration Floodplain Management Task Force, Washington, D.C.

Iowa Department of Natural Resources, 1989, Details of C.M.P. Water Control Structures, Red Rock WMA—Impoundments #1 and #2; Project Number 82. Warren County, Iowa, unpublished.

Ives, J. D., 1989, Deforestation in the Himalayas: the cause of increased flooding in Bangladesh and northern India?, Land Use Policy (July); 187–193.

Ives, J. D., 1991, Floods in Bangladesh: who is to blame?, New Scientist (April): 34–37.

Ives, R. L., 1942, The beaver-meadow complex, Journal of Geomorphology **5**:191–203.

Jackson, L. L., N. Lopoukhine, and D. Hillyard, 1995, Ecological restoration: a definition and comments, Restoration Ecology **3**:71–75.

Jackson, M. B., 1982, Ethylene as a growth promoting hormone under flooded conditions, *In* Plant Growth Substances 1982, Wareing, P. F., ed., Academic Press, London, pp. 291–301.

Jacobs, T. A., 1990, Regulation and management of the River Murray, *In* The Murray, Mackey, N. and D. Eastburn, eds., Murray Darling Basin Commission, Canberra, Australia, pp. 35–58.

Jakobsson, K., 1981, Impact of beaver (*Castor fiber* L.) on riverside vegetation at Pålböleån, N Sweden, Wahlenbergia **7**:89–98.

Jansen, A. J. M., M. C. C. de Graaf, and J. G. M. Roelofs, 1996, The restoration of species-rich heathland in the Netherlands, Vegetatio **126**:73–88.

Janzen, D. H., 1984, Dispersal of small seeds by big herbivores: foliage is the fruit, American Naturalist **123**:338–353.

Janzen, D. H., 1986, Guanacaste National Park: Tropical Ecological and Biocultural Restoration, Editorial Universidad Estatal A Distancia, San José, Costa Rica.

Jenkins, S. H., and P. E. Busher, 1979, *Castor canadensis*. Mammalian Species **120**:1–8.

Joglekar, D. V., and G. T. Wadekar, 1951, The effects of weirs and dams on the regime of rivers. International Association for Hydraulic Research, Bombay, Colorado State University, Fort Collins, Colorado, pp. 349–363.

Johansson, M. E., and C. Nilsson, 1993, Hydrochory, population dynamics and distribution of the clonal aquatic plant *Ranunculus lingua*, Journal of Ecology **81**:81–91.

Johnson, R. L., and W. R. Beaufait, 1965a, Swamp cottonwood (*Populus heterophylla* L.), *In* Silvics of Forest Trees of the United States, Fowells, H. A., ed.,

Agriculture Handbook 271, U.S.D.A. Forest Service, Washington, D.C., pp. 535–537.

Johnson, R. L. and W. R. Beaufait, 1965b, water hickory (Carya aquatica (M: chx f.) Natt.), In Silvics of Forest Trees of the United States, Fowells, H. A., ed., Agriculte Handbook 271, U.S.D.A. Forest Service, Washington, D.C., pp. 136–138.

Johnson, W. C., 1994, Woodland expansion in the Platte River, Nebraska: patterns and causes, Ecological Monographs **64**:45–84.

Johnston, C. A., 1994, Ecological engineering by beavers, *In* Global Wetlands: Old World and New, Mitsch, W. J., ed., Elsevier, Amsterdam, pp. 379–384.

Johnston, C. A., and R. J. Naiman, 1987, Boundary dynamics at the aquatic-terrestrial interface: the influence of beaver and geomorphology, Landscape Ecology **1**:47–57.

Johnston, C. A., and R. J. Naiman, 1990, Aquatic patch creation in relation to beaver population trends, Ecology **71**:1617–1621.

Johnstone, I. M., 1986, Plant invasion windows: a time-based classification of invasion potential, Biological Review **61**:369–394.

Jongman, R. H., 1992, Vegetation, river management and land use in the Dutch Rhine floodplains, Regulated Rivers: Research and Management **7**:279–289.

Jorgensen, E. E., and L. E. Nauman, 1994, Disturbance in wetlands associated with commercial cranberry (*Vaccinium macrocarpon*) production, American Midland Naturalist **132**:152–158.

Junk, W. J., 1982, Amazonian floodplains: their ecology, present and potential use, *In* Proceedings of the International Scientific Workshop on Ecosystem Dynamics in Freshwater Wetlands and Shallow Water Bodies, Minsk-Pinsk-Tskhaltoubo, USSR, July 12–26, 1982, Scientific Committee on Problems of the Environment (SCOPE) United Nations Environment Program (UNEP), New York, pp. 98–126.

Junk, W. J., 1983, Ecology of swamps on the middle Amazon, *In* Ecosystems of the World 4B, Mires: Swamp, Bog, Fen and Moor. Gore, A. P., ed., Elsevier, Amsterdam, pp. 269–294.

Junk, W. J., 1986, Aquatic plants of the Amazon system, *In* The Ecology of River Systems, Davies, B. R., and K. F. Walker, eds., Dr. W. Junk Publishers, Dordrecht, the Netherlands, pp. 319–337.

Junk, W. J., 1993, Wetlands of tropical South America, *In* Wetlands of the World I: Inventory, Ecology and Management. Whigham, D., D. Dykyjová, and S. Hejny, eds., Kluwer, Dordrecht, the Netherlands, pp. 679–739.

Junk, W. J., P. B. Bayley, and R. E. Sparks, 1989, The flood pulse concept in river-floodplain systems, *In* Proceedings of the International Large River Symposium (LARS), Dodge, D. P., ed., 1989, Honey Harbour, Ontario, Canada, Canadian Special Publication of Fisheries and Aquatic Sciences 106, Department of Fisheries and Oceans, Ottawa, pp. 110–127.

Junk, W. J., and K. Furch, 1991, Nutrient dynamics in Amazonian floodplains: decomposition of herbaceous plants in aquatic and terrestrial environments, Verhandlungen Internationale Vereinigung für Theoretische und Angewandte Limnologie **24**:2080–2084.

Junk, W. J., and C. Howard-Williams, 1984, Ecology of aquatic macrophytes in Amazonia. *In* The Amazon. Limnology and Landscape Ecology of a Mighty Tropical River and Its Basin, Sioli, H., ed., Dr. W. Junk Publishers, Dordrecht, the Netherlands, pp. 269–309.

Kadlec, J. A., and L. M. Smith, 1984, Marsh plant establishment on newly flooded salt flats, Wildlife Society Bulletin **12**:388–394.

Kamat, N. D., 1967, Dispersal of charophytes by the pintail, Current Science **36**: 134.

Kantrud, H. A., 1986, Effects of vegetation manipulation on breeding waterfowl in prairie wetlands—a literature review, Fish and Wildlife Service Technical Report 3, U.S. Department of the Interior, Fish and Wildlife Service, Washington, D.C.

Kantrud, H. A., J. B. Millar, and A. G. van der Valk, 1989, Vegetation of the wetlands of the Prairie Pothole Region, *In* Northern Prairie Wetlands, van der Valk, A. G., ed., Iowa State University Press, Ames, Iowa, pp. 132–187.

Karr, J. R., K. D. Fausch, P. L. Angermeier, P. R. Yant, and I. J. Schlosser, 1986, Assessing the Biological Integrity of Running Waters: A Method and Its Rationale, Illinois Natural History Survey, Champaign, Illinois.

Katz, N. J., 1926, *Sphagnum* bogs of central Russia: phytosociology, ecology and succession, Journal of Ecology **14**:177–202.

Kauffman, J. B., R. L. Case, D. Lytjen, N. Otting, and D. L. Cummings, 1995, Ecological approaches to riparian restoration in northeast Oregon, Restoration and Management Notes **13**:12–15.

Kaufmann, P. R., 1987, Channel morphology and hydraulic characteristics of torrent-impacted forest streams in the Oregon Coast Range, USA, Ph.D. Dissertation, Oregon State University, Corvallis, Oregon.

Kaul, R. B., 1978, Morphology of germination and establishment of aquatic seedings in Alismataceae and Hydrocharitaceae, Aquatic Botany **5**:139–147.

Kautsky, L., 1988, Life strategies of aquatic soft bottom macrophytes, Oikos **53**: 126–135.

Kautsky, L., 1990, Seed and tuber banks of aquatic macrophytes in the Askö area, northern Baltic proper, Holarctic Ecology **13**:143–148.

Keddy, P. A., 1992, Water level fluctuations and wetland conservation, *In* Proceedings of an International Symposium: Wetland of the Great Lakes: Protection and Restoration Policies, Status of the Science, May 16–18, 1990, Niagara Falls, New York, The Association of State Wetland Managers, Berne, New York, pp. 70–91.

Keddy, P. A., and A. A. Reznicek, 1982, The role of seed banks in the persistence of Ontario's coastal plant flora, American Journal of Botany **69**:13–22.

Keddy, P. A., and P. Constabel, 1986, Germination of ten shoreline plants in relation to seed size, soil particle size and water level: an experimental study, Journal of Ecology **74**:133–141.

Keddy, P. A., and A. A. Reznicek, 1986a, Great Lakes vegetation dynamics: the role of fluctuating water levels and buried seeds, Journal of Great Lakes Research **12**:25–36.

Keeland, B. D., and R. R. Sharitz, 1995, Seasonal growth patterns of *Nyssa sylvatica* var. *biflora, Nyssa aquatica*, and *Taxodium distichum* as affected by hydrologic regime, Canadian Journal of Forest Research **25**:1084–1096.

Keller, E. A., and F. J. Swanson, 1979, Effects of large organic material on channel form and fluvial processes, Earth Surface Processes **4**:361–380.

Kellerhals, R., and M. Church, 1989, The morphology of large rivers: characterization and management, *In* Proceedings of the International Large River Symposium, Dodge, D. P., ed., September 14–21, 1986, Honey Harbour, Ontario, Canada, Canadian Special Publication of Fisheries and Oceans 106, Department of Fisheries and Oceans, Ottawa, pp. 31–48.

Kemp, W. M., Lewis, M. R., and Jones, T. W., 1986, Comparison of methods for measuring production by the submerged macrophyte *Potamogeton perfoliatus* L. Limnology and Oceanography **31**:1322–1334.

Kennedy, H. E., Jr., 1970, Growth of newly planted water tupelo seedlings after flooding and siltation, Forest Science **16**:250–256.

Kennedy, H. E., and R. M. Krinard, 1974, 1973 Mississippi River Flood's Impact On Natural Hardwood Forests and Plantations. U.S. Forest Service, Report SO-177, New Orleans, Louisiana.

Kennedy, R. A., M. E. Pumpho, and T. C. Fox, 1987, Germination physiology of rice and rice weeds: metabolic adaptation to anoxia, *In* Plant Life in Aquatic and Amphibious Habitats, Crawford, R. M. M., ed., Blackwell Scientific Publications, Oxford, pp. 193–203.

Kenworthy, W. J., M. S. Fonseca, and J. Homziak, 1980, Development of a transplanted seagrass (*Zostera marina* L.) meadow in Back Sound, Carteret County, North Carolina, *In* Wetlands Restoration and Creation: Proceedings of the Seventh Annual Conference, Cole, D. P., ed., May 16–17, 1980, Tampa, Florida, Hillsborough Community College, Tampa, Florida, pp. 175–193.

Kern, K., 1992a, Rehabilitation of streams in south-west Germany, *In* River Conservation and Management, Boon, P. J., P. Calow, and G. E. Petts, eds., John Wiley & Sons, Chichester, United Kingdom, pp. 321–335.

Kern, K., 1992b, Restoration of lowland rivers: the German experience, *In* Lowland Floodplain Rivers, Carling, P. A., and G. E. Petts, eds., John Wiley & Sons, Chichester, United Kingdom, pp. 279–297.

Khodachek, E. A., 1997, Seed reproduction in arctic environments, Opera Botanica **132**:129–135.

Kimber, A., C. E. Korschgen, and A. G. van der Valk, 1995, The distribution of *Vallisneria americana* seeds and seedling light requirements in the Upper Mississippi River, Canadian Journal of Botany **73**:1966–1973.

Kirby, J. M., J. R. Webster, and E. F. Benfield, 1983, The role of shredders in detrital dynamics of permanent and temporary streams, *In* Dynamics of Lotic Ecosystems. Fontaine, T. D., III, and S. M. Bartell, eds., Ann Arbor Science Publishers, Ann Arbor, Michigan, pp. 425–435.

Kirchner, J. E., and M. R. Karlinger, 1983, Effects of Water Development on Surface-water Hydrology, Platte River Basin in Colorado, Wyoming, and Nebraska Upstream from Duncan, Nebraska. Geologic Survey Paper 1277, Hydrologic and Geomorphic Studies of the Platte River Basin, Washington, D.C.

Klein, J.-P., J. M. Sanchez-Perez, and M. Tremolieres, 1996, Conservation and management of the Rhine nature reserves in France, Archiv für Hydrobiologie Supplement **113**:345–352.

Klimas, C. V., 1987, Baldcypress response to increased water levels Caddo Lake, Louisiana-Texas, Wetlands **7**:25–37.

Klimas, C. V., 1988, River regulation effects on floodplain hydrology and ecology, *In* The Ecology and Management of Wetlands, Volume I: Ecology of Wetlands, Hook, D. D., W. H. McKee, Jr., H. K. Smith, J. Gregory, V. G. Burrell, Jr., M. R. DeVoe, R. E. Sojka, S. Gilbert, R. Banks, L. H. Stolzy, C. Brooks, T. D. Matthews, and T. H. Shear, eds., Timber Press, Portland, Oregon, pp. 40–49.

Klinger, L. F., 1996a, Coupling of soils and vegetation in peatland succession, Arctic and Alpine Research **28**:380–387.

Klinger, L. F., 1996b, The myth of the classic hydrosere model of bog succession, Arctic and Alpine Research, **28**:1–9.

Klinger, L. F., S. A. Elias, V. M. Behan-Pelletier, and N. E. Williams, 1990, The bog climax hypothesis: fossil arthropod and stratigraphic evidence in peat sections from southeast Alaska, USA, Holartic Ecology **13**:72–80.

Knopf, F. L., and M. L. Scott, 1990, Altered flows and created landscapes in the Platte River headwaters, 1840–1990, *In* Management of Dynamic Ecosystems, Proceedings of a Symposium at the 51st Midwest Fish and Wildlife Conference, Sweeney, J. M., ed., December 5, 1989, Springfield, Illinois, The Wildlife Society, North Central Section, West Lafayette, Indiana, pp. 47–69.

Knutson, P. L., and W. W. Woodhouse, Jr., 1982, Pacific coastal marshes, *In* Creation and Restoration of Coastal Plant Communities, Lewis, R. R., III, ed., CRC Press, Boca Raton, Florida, pp. 111–130.

Koebel, J. W., Jr., 1995, An historical perspective on the Kissimmee River Restoration Project, Restoration Ecology, **3**:149–159.

Kohlhepp, G., 1984, Development planning and practices of economic exploitation in Amazonia, *In* The Amazon. Limnology and Landscape Ecology of a Mighty Tropical River and Its Basin. Sioli, H., ed., Kluwer, Dordrecht, the Netherlands, pp. 649–674.

Komarek, E. V., Sr., 1976, Fire ecology review, Proceedings of the Tall Timbers Fire Ecology Conference, October 8–10, Missoula, Montana, The Tall Timbers Fire Ecology Station, Tallahassee, Florida.

Kondolf, G. M., 1993, Lag in stream channel adjustment to livestock exclosure, White Mountains, California, Restoration Ecology **1**:226–230.

Kondolf, G. M., and M. Larson, 1995, Historical channel analysis and its application to riparian and aquatic habitat restoration, Aquatic Conservation: Marine and Freshwater Ecosystems **5**:109–126.

Koonce, A. L., and A. González-Cabán, 1990, Social and ecological aspects of fire in Central America, *In* Fire in the Tropical Biota, Goldammer, J. G., ed., Springer-Verlag, Berlin, pp. 135–158.

Kozlowski, T. T., 1984, Responses of woody plants to flooding, *In* Flooding and Plant Growth, Kozlowski, T. T., ed., Academic Press, Orlando, Florida, pp. 129–163.

Kozlowski, T. T., and C. E. Ahlgren, eds., 1974, Fire and Ecosystems, Academic Press, New York.

Kunii, H., and K. Maeda, 1982, Seasonal and long-term changes in surface cover of aquatic plants in a shallow pond, Ojaga-ike, Chiba, Japan, Hydrobiologia **87**: 45–55.

Kushlan, J. A., 1991, The Everglades, *In* The Rivers of Florida, Livingston, R. J., ed., Springer-Verlag, New York, pp. 121–142.

Kusler, J., and R. Smardon, 1992, Introduction: Key recommendations, *In* Proceedings of an International Symposium, Wetlands of the Great Lakes: Protection and Restoration Policies, Status of the Science, May 16–18, 1990, Niagara Falls, New York, The Association of State Wetland Managers, Berne, New York, pp. 2–5.

LaBaugh, J. W., 1989, Chemical characteristics of water in northern prairie wetlands, *In* Northern Prairie Wetlands, van der Valk, A. G., ed., Iowa State University Press, Ames, Iowa, pp. 56–90.

Langeland, K. A., and D. L. Sutton, 1980, Regrowth of *Hydrilla* from axillary buds, Journal of Aquatic Plant Management **18**:27–29.

Large, A. R. G., K. Prach, M. A. Bickerton, and P. M. Wade, 1994, Alteration of patch boundaries on the floodplain of the regulated River Trent, U.K., Regulated Rivers: Research and Management **9**:71–78.

Larsen, H. S., 1963, Effects of soaking in water on acorn germination of four southern oaks, Forest Science **9**:236–241.

Larsen, J. A., 1982, Ecology of the Northern Lowland Bogs and Conifer Forests, Academic Press, New York.

Larson, J. S., A. J. Mueller, and W. P. MacConnell, 1980, A model of natural and man-induced changes in open freshwater wetlands on the Massachusetts coastal plain, Journal of Applied Ecology **17**:667–673.

Larson, M., 1995, Developments in river and stream restoration in Germany, Restoration and Management Notes, **13**:77–83.

Lathbury, M. E., 1996, Toward natural flood control: floodplain wetlands, Wetland Journal **8**:10–13.

Leck, M. A., 1989, Wetland seed banks, *In* Ecology of Seed Banks, Leck, M. A., V. T. Parker, and R. L. Simpson, eds., Academic Press, San Diego, California, pp. 283–305.

Leck, M. A., 1996, Germination of macrophytes from a Delaware River tidal freshwater wetland, Bulletin of the Torrey Botanical Club **123**:48–67.

Leck, M. A., and K. J. Graveline, 1979, The seed bank of a freshwater tidal marsh, American Journal of Botany **66**:1006–1015.

Leck, M. A., and R. L. Simpson, 1987, Seed bank of a freshwater tidal wetland: turnover and relationship to vegetion change, American Journal of Botany, **74**: 360–370.

Leck, M. A., and R. L. Simpson, 1995, Ten-year seed bank and vegetation dynamics of a tidal freshwater marsh, American Journal of Botany **82**:1547–1557.

Lehmkuhl, J. F., 1994, A classification of subtropical riverine grassland and forest in Chitwan National Park, Nepal, Vegetatio **111**:29–43.

Leigh, E. G., Jr., S. J. Wright, E. A. Herre, and F. E. Putz, 1993, The decline of tree diversity on newly isolated tropical islands: a test of a null hypothesis and some implications, Evolutionary Ecology **7**:76–102.

Lelek, A., 1989, The Rhine River and some its tributaries under human impact in the last two centuries, *In* Proceedings of the International Large River Symposium (LARS), Dodge, D. P., ed., September 14–21, 1986, Honey Harbour, Ontario, Canada, Canadian Special Publication of Fisheries and Aquatic Sciences 106, Department of Fisheries and Oceans, Ottawa, pp. 469–486.

Lelek, A., and G. Buhse, 1992, Fische des Rheins—Früher und Heute, Springer-Verlag, Berlin.

Leopold, L. B., and T. Maddock, Jr., 1954, The Flood Control Controversy: Big Dams, Little Dams, and Land Management, Ronald Press, New York.

Lessmann, J. M., I. A. Mendelssohn, M. W. Hester, and K. L. McKee, 1997, Population variation in growth response to flooding of three marsh grasses, Ecological Engineering **8**:31–47.

Levine, D. A., and D. E. Willard, 1989, Regional analysis of fringe wetlands, in the Midwest: creation and restoration, *In* Wetland Creation and Restoration: The Status of the Science, Kusler, J. A., and M. E. Kentula, eds., Island Press, Washington, D.C., pp. 305–332.

Lewis, R. R., III., 1982, Mangrove forests, *In* Creation and Restoration of Coastal Plant Communities, Lewis, R. R., III, ed., CRC Press, Boca Raton, Florida, pp. 153–171.

Lewis, R. R., III, 1989, Wetlands Restoration/Creation/Enhancement Technology: Suggestions for Standardization, In Wetland Creation and Restoration: The Status of the Science, Volume II, Kusler, J. A., and M. E. Kentula, eds., U.S. Environmental Protection Agency, Washington, D.C.

Lewis, R. R., III., 1990, Creation and restoration of coastal plain wetland in Florida, *In* Wetland Creation and Restoration: The Status of the Science, Kusler, J. A., and M. E. Kentula, eds., Island Press, Washington, D.C., pp. 73–101.

Lewis, R. R., III., and R. C. Phillips, 1979, Experimental sea grass mitigation in the Florida Keys, *In* Wetlands Restoration and Creation: Proceedings of the Seventh Annual Conference, Cole, D. P., ed., May 16–17, 1979 Tampa, Florida, Hillsborough Community College, Tampa, Florida, pp. 155–173.

Linsley, R. K., M. A. Kohler, and J. L. H. Paulhus, 1975, Hydrology for Engineers, 2d ed. McGraw-Hill Book Company, New York.

Lienkaemper, G. W., and F. J. Swanson, 1987, Dynamics of large woody debris in streams in old-growth Douglas-fir forests, Canadian Journal of Forest Research **17**:150–156.

Ligon, F. K., W. E. Dietrich, and W. J. Trush, 1995, Downstream ecological effects of dams, BioScience **45**:183–192.

Little, S., 1965, Atlantic white-cedar (*Chamaecyparis thyoides*) (L.) B.S.P.), *In* Silvics of Forest Trees of the United States, Fowells, H. A., ed., Agriculture Handbook 271. U.S.D.A. Forest Service, Washington, D.C. pp. 151–156.

Lloyd, L. N., K. F. Walker, and T. J. Hillman, 1991, Environmental Significance of Snags in the River Murray, Department of Primary Industries and Energy, Land and Water Resources Research and Development Corporation, Australian Water Research Advisory Council Completion Report, Canberra, Australia.

Loftin, M. K., 1990, The Kissimmee River—yesterday and today, *In* Proceedings of the Kissimmee River Restoration Symposium, Loftin, M. K., L. A. Toth, and J. Obeysekera, eds., October 1988, Orlando, Florida, South Florida Water Management District, West Palm Beach, Florida, pp. 5–8.

Loftin, M. K., L. A. Toth, and J. T. B. Obeysekera, 1990, Kissimmee River Restoration: Alternative Plan Evaluation and Preliminary Design Report, South Florida Water Management District, West Palm Beach, Florida.

Loftus, T. T., 1994, Status and assessment of *Taxodium distichum* (L.) Rich. and *Nyssa aquatica* L. in Horseshoe Lake, Alexander County, Illinois: Phase One—Baseline Study, M.S. thesis, Southern Illinois University, Carbondale, Illinois.

Lonsdale, W. M., and A. M. Lane, 1994, Tourist vehicles as vectors of weed seeds in Kakadu National Park, northern Australia, Biological Conservation **69**:277–283.

Loope, L., M. Duever, A. Herndon, J. Snyder, and D. Jansen, 1984, Hurricane impact on uplands and freshwater swamp forest, BioScience **44**:238–246.

Lotti, T., 1965a, Cherrybark oak (*Quercus falcata* var. *pagodaefolia*), *In* Silvics of Forest Trees of the United States, Fowells, H. A., ed., Agriculture Handbook 271, U.S.D.A. Forest Service, Washington, D.C. p. 569–572.

Lotti, T., 1965b, Swamp chestnut oak (*Quercus michauxii* Nutt.), *In* Silvics of Forest Trees of the United States, Fowells, H. A., ed., Agriculture Handbook 271, U.S.D.A. Forest Service, Washington, D.C. pp. 622–624.

Loucks, O. L., 1970, Evolution of diversity, efficiency, and community stability, American Zoologist **10**:17–25.

Loucks, O. L., 1990, Restoration of the pulse control function of wetland and its relationship to water quality objectives, *In* Wetland Creation and Restoration:

The Status of the Science, Kusler, J. A., and M. E. Kentula, eds., Island Press, Washington, D.C., pp. 467–477.

Lowry, M. M., and R. L. Beschta, 1994, Effect of a beaver pond on groundwater elevation and temperatures in a recovering stream system, *In* Proceedings of the Annual Summer Symposium of the American Water Resources Association, Effects of Human-induced Changes on Hydrologic Systems, Marston, R. A., and V. R. Hasfurther, eds., June 26–29, 1994, Jackson Hole, Wyoming, American Water Resources Association, Bethesda, Maryland, pp. 503–513.

Lugo, A. E., and F. N. Scatena, 1995, Ecosystem-level properties of the Luquillo Experimental Forest with emphasis on the Tabonuco forest, *In* Tropical Forests: Management and Ecology, Lugo, A. E., and C. Lowe, eds., Springer-Verlag, New York, pp. 59–108.

Lunt, I. D., 1996, A transient soil seed bank for the yam-daisy *Microseris scapigera*, The Victorian Naturalist **113**:16–19.

Lyon, J. G., 1993, Practical Handbook for Wetland Identification and Delineation, Lewis Publishers, Boca Raton, Florida.

MacArthur, R. H., and E. O. Wilson, 1967, The Theory of Island Biogeography, Princeton University Press, Princeton, New Jersey.

Mackay, N., 1990, Understanding the Murray, *In* The Murray, Mackay, N., and D. Eastburn, eds., Murray Darling Basin Commission, Canberra, Australia, pp. ix–xix.

Magilligan, F. J., and P. F. McDowell, 1997, Stream channel adjustments following elimination of cattle grazing, Journal of the American Water Resources Association **33**:867–878.

Magnuson, J. J., H. A. Regier, W. J. Christies, and W. C. Sonzogni, 1980, To reinhabit and restore Great Lakes ecosystems, *In* The Recovery Process in Damaged Ecosystems. Cairns, J., Jr., ed., Ann Arbor Science Publishers, Ann Arbor, Michigan, pp. 95–112.

Maheshwari, B. L., K. F. Walker, and T. A. McMahon, 1995, Effects of regulation on the flow regime of the River Murray, Australia, Regulated Rivers: Research and Management **10**:15–38.

Mahler, D., 1988, New device speed seed harvest, Restoration and Management Notes **6**:23.

Main, A. R., 1993, Landscape reintegration: problem definition, *In* Reintegrating Fragmented Landscapes: Towards Sustainable Production and Nature Conservation, Hobbs, R. J., and D. A. Saunders, eds., Springer-Verlag, New York, pp. 189–208.

Maisenhelder, L. C., 1958, Understory plants of bottomland forests, Occasional Paper 165, USDA Forest Service, Southern Forest Experiment Station, New Orleans, Louisiana.

Malakoff, D., 1998, Restored wetlands flunk real-world test, Science **280**:371–372.

Mallik, A. U., 1990a, Microscale succession and vegetation management by fire in a freshwater marsh of Atlantic Canada, *In* Wetland Ecology and Management:

Case Studies, Whigham, D. F., R. E. Good, and J. Kvet, eds., Kluwer, Dordrecht, the Netherlands, pp. 19–29.

Mallik, A. U., 1990b, Smoldering combustion, thermal decomposition and nutrient content following controlled burning of *Typha* dominated organic mat, *In* Wetland Ecology and Management: Case Studies, Whigham, D. F., R. E. Good, and J. Kvet, eds., Kluwer, Dordrecht, the Netherlands, pp. 7–17.

Maltby, E., 1986, Waterlogged Wealth, International Institute for Environment and Development, London.

Maltby, E., 1988, Global wetlands—history, current status and future, *In* The Ecology and Management of Wetlands, Volume 1: Ecology of Wetlands, Hook, D. D., W. H. McKee, Jr., H. K. Smith, J. Gregory, V. G. Burrell, Jr., M. R. DeVoe, R. E. Sojka, S. Gilbert, R. Banks, L. H. Stolzy, C. Brooks, T. D. Matthews, and T. H. Shear, eds., Timber Press, Portland, Oregon, pp. 3–14.

Marburger, J. E., 1993, Biology and management of *Sagittaria latifolia* Willd. (broad-leaf arrow-head) for wetland restoration and creation, Restoration Ecology **1**:248–255.

Marburger, J. E., 1992, Wetland Plants: Plant Material Technology Needs and Development for Wetland Enhancement, Restoration, and Creation in Cool Temperate Regions of the United States, U.S.D.A. Soil Conservation Service, Washington, D.C.

Margalef, R., 1963, On certain unifying principles in ecology, The American Naturalist **97**:357–374.

Maser, C., and J. R. Sedell, 1994, From the Forest to the Sea: The Ecology of Wood in Streams, River, Estuaries, and Oceans, St. Lucie Press, Delray Beach, Florida.

Mason, D. H., 1996, Coexistence of two floating-leaved species, *Nymphoides indica* and *Nymphoides cristata*, and the role of seed banks in vegetation dynamics at the Keoladeo National Park wetlands, Bharatpur, India, Ph.D. dissertation, Iowa State University, Ames, Iowa.

Mathis, M., 1996, Herbivory and vegetation dynamics in coal slurry ponds reclaimed as wetlands, M.S. thesis, Southern Illinois University, Carbondale, Illinois.

Matlack, G. R., 1987, Diaspore size, shape, and fall behavior in wind-dispersed plant species, American Journal of Botany **74**:1150–1160.

Mattoon, W. R., 1915, The Southern Cypress, Bulletin No. 272, USDA, Washington, D.C.

Mattoon, W. R., 1916, Water requirements and growth of young cypress, *In* Proceedings of the Society of American Foresters, Washington, D.C., Society of American Foresters, pp. 192–197.

McAtee, W. L., 1925, Notes on drift, vegetable balls, and aquatic insects as a food product of inland waters, Ecology **6**:288–302.

McCarthy, K. A., 1987, Spatial and temporal distributions of species in two intermittent ponds in Atlantic County, New Jersey, M.S. thesis, Rutgers University, New Brunswick, New Jersey.

McClain, W. E., 1986, Illinois Prairie: Past and Future, Illinois Department of Natural Conservation, Champaign, Illinois.

McClure, F. A., 1967, The Bamboos: A Fresh Perspective, Harvard University Press, Cambridge, Massachusetts.

McCoy, M. B., and J. M. Rodríguez, 1994, Cattail (*Typha domingensis*) eradication methods in the restoration of a tropical, seasonal, freshwater marsh, *In* Global Wetlands: Old World and New, Mitsch, W. J., ed., Elsevier, Amsterdam, pp. 469–482.

McDermott, R. E., 1954, Effects of saturated soil on seedling growth of some bottomland hardwood species, Ecology **35**:36–41.

McGinley, M. A., and T. G. Whitham, 1985, Central place foraging by beavers (*Castor canadensis*): test of foraging predictions and the impact of selective feeding on the growth form of cottonwoods (*Populus fremontii*), Oecologia **66**: 558–562.

McGraw, J. B., M. C. Vavrek, and C. C. Bennington, 1991, Ecological genetic variation in seed banks I. establishment of a time transect, Journal of Ecology **79:**617–625.

McGuire, J. M., 1997, February 23, The high ground, St. Louis Post-Dispatch, 1–11.

McInerney, J. D., 1996, Textbook misconceptions at issue, American Biology Teacher **58**:328.

McIninch, S. M., and E. W. Garbisch, 1991, Oxygen requirements in dormant wetland plants, Wildflower **4**:6–12.

McKee, K. L., 1993, Soil physicochemical patterns and mangrove species distribution—reciprocal effects?, Journal of Ecology **81**:477–487.

McKee, K. L., 1995a, Mangrove species distribution and propagule predation in Belize: an exception to the dominance-predation hypothesis, Biotropica **27**:334–345.

McKee, K. L., 1995b, Seedling recruitment patterns in a Belizean mangrove forest: effects of establishment ability and physico-chemical factors, Oecologia **101**: 448–460.

McKee, K. L., and A. H. Baldwin, in press, Disturbance regimes in North American wetlands, *In* Ecosystems of Disturbed Ground, Walker, L. R., ed., Ecosystems of the World Series. Elsevier, Amsterdam.

McKee, K. L., and I. A. Mendelssohn, 1989, Response of a freshwater marsh plant community to increased salinity and increased water level, Aquatic Botany **34**: 301–316.

McKevlin, M. R., D. D. Hook, and A. A. Rozelle, 1998, Adaptations of plants to flooding and soil waterlogging. *In* Southern Forested Wetlands Ecology and Management. Messina, M. G., and W. H. Conner, eds. Lewis Publishers, Boca Raton, Florida, pp. 173–203.

McKinley, C. E., and F. P. Day, Jr., 1979, Herbaceous production in cut-burned, uncut-burned and control areas of a *Chamaecyparis thyoides* (L.) BSP (Cupressaceae) stand in the Great Dismal Swamp, Bulletin of the Torrey Botanical Club **106**:20–28.

McKnight, J. S., 1965, Black willow (*Salix nigra* Marsh.), *In* Silvics of Forest Trees of the United States, Fowells, H. A., ed., Agriculture Handbook 271, U.S.D.A. Forest Service, Washington, D.C. pp. 650–652.

McKnight, J. S., D. D. Hook, O. G. Langdon, and R. L. Johnson, 1981. Flood tolerance and related characteristics of trees of the bottomland forests of the southern United States, *In* Wetlands of Bottomland Hardwood Forests, Proceedings of a Workshop on Bottomland Hardwood Forest Wetlands of the Southeastern United States, Clark, J. R., and J. Benforado, eds., June 1–5, 1980, Lake Lanier, Georgia, Elsevier, Amsterdam, pp. 29–43.

McMannon, M., and R. M. M. Crawford, 1971, A metabolic theory of flooding tolerance: the significance of enzyme distribution and behaviour, New Phytologist **70**:299–306.

Meehan, W. R., J. R. Sedell, and W. S. Duval, 1985, Influence of Forest and Rangeland Management on Anadromous Fish Habitat in Western North America; 5. Water Transportation and Storage of Logs, General Technical Report PNW-186, USDA Forest Service, Pacific Northwest Forest and Range Experiment Station, Portland, Oregon.

Menges, E. S., and D. M. Waller, 1983, Plant strategies in relation to elevation and light in floodplain herbs, The American Naturalist **122**:454–473.

Meredith, T. C., 1985, Factors affecting recruitment from the seed bank of sedge (*Cladium mariscus*) dominated communities at Wicken Fen, Cambridgeshire, England, Journal of Biogeography **12**:463–472.

Merriam, G., and D. A. Saunders, 1993, Corridors in restoration of fragmented landscapes, *In* Nature Conservation 3: Reconstruction of Fragmented Ecosystems Global and Regional Perspectives, Saunders, D. A., R. J. Hobbs, and P. R. Ehrlich, eds., Surrey Beatty and Sons, Chipping Norton, New South Wales, Australia, pp. 71–87.

Merz, R. W., 1965, Shellbark hickory (*Carya lacinosa* (Michx. f.) Loud.), *In* Silvics of Forest Trees of the United States, Fowells, H.A., ed., Agriculture Handbook 271, U.S.D.A Forest Service, Washington, D.C., pp. 132–135.

Metzler, G. M., and L. A. Smock, 1990, Storage and dynamics of subsurface detritus in a sand-bottomed stream, Canadian Journal of Fisheries and Aquatic Science **47**:588–594.

Middleton, B. A., 1978, Vegetational response to grazing in the Lodi Wildlife Area, senior thesis, unpublished, University of Wisconsin, Madison.

Middleton, B. A., 1989, Succession and goose herbivory in monsoonal wetlands of the Keoladeo National Park, Bharatpur, India, Ph.D. dissertation, Iowa State University, Ames, Iowa.

Middleton, B. A., 1990, Effect of water depth and clipping frequency on the growth and survival of four wetland plant species, Aquatic Botany **37**:189–196.

Middleton, B. A., 1992, Habitat and food preferences of Greylag and Barheaded Geese wintering in the Keoladeo National Park, India, Journal of Tropical Ecology **8**:181–193.

Middleton, B. A., 1994a, Decomposition and litter production in a northern bald cypress swamp, Journal of Vegetation Science **5**:271–274.

Middleton, B. A., 1994b, Management of monsoonal wetland for Greylag (*Anser anser* L.) and Barheaded Geese (*Anser indicus* L.) in the Keoladeo National Park, India, International Journal of Ecology and Environmental Sciences **26**: 163–171.

Middleton, B. A., 1995a, The Role of Flooding in Seed Dispersal: Restoration of Cypress Swamps along the Cache River, Illinois, U.S. Geological Survey and Water Resources Center, Champaign, Illinois.

Middleton, B. A., 1995b, Sampling devices for the measurement of seed rain and hydrochory in rivers, Bulletin of the Torrey Botanical Club, **122**:152–155.

Middleton, B. A., 1995c, Seed banks and species richness potential of coal slurry ponds reclaimed as wetlands, Restoration Ecology **3**:311–318.

Middleton, B. A., 1996, Characteristics of plants in forested wetlands. *In* Symposium Proceedings, Management of Forested Wetlands in the Central Hardwood Region, Roberts, S. D., and R. A. Rathfon, eds. October 11–13, 1994, Evansville, Indiana, Department of Forestry and Natural Resources Purdue University, West Lafayette, Indiana.

Middleton, B. A., 1998, The water buffalo controversy in the Keoladeo National Park, India, Ecological Modelling **106**:93–98.

Middleton, B. A., accepted, Herbivory and succession in monsoonal wetlands. Wetlands Ecology and Management.

Middleton, B. in review, Hydrochory, seed banks, and regeneration across landscape boundaries in cypress swamps.

Middleton, B. A., and K. Fessel, unpublished, Seed germination and seedling survivorship under various water and sedimentation levels of swamp species, Southern Illinois University, Carbondale, Illinois.

Middleton, B. A., and D. H. Mason, 1992, Seed herbivory by nilgai, feral cattle, and wild boar in the Keoladeo National Park, India, Biotropica **24**:538–543.

Middleton, B. A., A. G. van der Valk, D. H. Mason, R. L. Williams, and C. B. Davis, 1991, Vegetation dynamics and seed banks of a monsoonal wetland overgrown with *Paspalum distichum* L. in northern India, Aquatic Botany **40**:239–259.

Middleton, B. A., A. G. van der Valk, R. L. Williams, D. H. Mason, and C. B. Davis, 1992, Litter decomposition in an Indian monsoonal wetland overgrown with *Paspalum distichum,* Wetlands **12**:37–44.

Middleton, B. A., E. Sanchez-Rojas, B. Suedmeyer, and A. Michels, 1997, Fire in a tropical dry forest of Central America: A natural part of the disturbance regime? Biotropica **29**:515–517.

Milhous, R. T., 1994, Instream flows and cottonwood establishment in the Bosque del Apache reach of the Rio Grande, *In* Proceedings of the Annual Summer Symposium of the American Water Resources Association, Effects of Human-induced Changes on Hydrologic Systems, Marston, R. A., and V. R. Hasfurther, eds., Jackson Hole, Wyoming, American Water Resources Association, Minneapolis, Minnesota, pp. 535–544.

Miller, T. S., 1987, Techniques used to enhance, restore, or create freshwater wetlands in the Pacific Northwest, *In* Proceedings of the Eighth Annual Meeting, Mutz, K. M., and L. C. Lee, eds., May 26–29, 1987, Seattle, Washington, Society of Wetland Scientists, Lawrence, Kansas, pp. 116–121.

Milleson, J. F., R. L. Goodrick, and J. A. Van Arman, 1980, Plant communities of the Kissimmee River valley, Technical Publication 80–7, South Florida Water Management District, West Palm Beach, Florida.

Minckler, L. S., 1965, White oak (*Quercus alba* L.), *In* Silvics of Forest Trees of the United States, Fowells, H. A., ed, Agriculture Handbook 271, U.S.D.A. Forest Service, Washington, D.C., pp. 631–637.

Ministère d l'Environnement et Agences de L'Eau, 1985, L'Entretien des Cours d'Eau, Cahier Technique de la Direction de la Prévention des Pollutions, Paris.

Mitchell, J. K., 1974, Natural hazards research, *In* Perspectives on Environment. Manners, I. R., and M. M. Mikesell, eds., Association of American Geographers, Washington, D.C., pp. 311–341.

Mitsch, W. J., 1979, Interaction between a riparian swamp and a river in southern Illinois, *In* Strategies for the Protection and Management of Floodplain Wetlands and Other Riparian Ecosystems, Proceedings of the Symposium, December 11–13, 1978, Callaway Gardens Georgia, U.S. Forest Service General Technical Report WO-12, Washington, D.C., pp. 63–72.

Mitsch, W. J., 1989, Wetlands of Ohio's Coastal Lake Erie: A Hierarchy of Systems, R/ER-13-PD, Ohio Sea Grant Program, Columbus, Ohio.

Mitsch, W. J., 1992, Combining ecosystem and landscape approaches to Great Lakes wetlands, Journal of Great Lakes Research **18**:552–570.

Mitsch, W. J., 1993, Ecological engineering: a cooperative role with the planetary life-support system, Environmental Science and Technology **27**:438–445.

Mitsch, W. J., 1996, Ecological engineering: a new paradigm for engineers and ecologists, *In* Engineering within Ecological Constraints, Schultze, P. C., ed., National Academy of Engineering, Washington, D.C., pp. 111–128.

Mitsch, W. J., in press, Self-design and wetland creation: early results of a freshwater marsh experiment, *In* Proceedings of INTECOL's V International Wetlands Conference, McComb, A. J., and J. A. Davis, eds., September 22–28, 1996, Perth, Australia, Gleneagles Press, Adelaide.

Mitsch, W. J., and J. G. Gosselink, 1993, Wetlands, Van Nostrand Reinhold, New York.

Mitsch, W. J., and S. E. Jørgensen, 1989, Introduction to ecological engineering, *In* Ecological Engineering: An Introduction to Ecotechnology, Mitsch, W. J., and S. E. Jørgensen, eds., John Wiley & Sons, New York, pp. 3–12.

Mitsch, W. J., R. H. Mitsch, and R. H. Turner, 1994, Wetlands of the Old and New Worlds: ecology and management, *In* Global Wetlands: Old World and New, Mitsch, W. J., ed., Elsevier, Amsterdam, pp. 3–56.

Mitsch, W. J., J. R. Taylor, and K. B. Benson, 1991, Estimating primary productivity of forested wetland communities in different hydrologic landscapes, Landscape Ecology **5**:75–92.

Mitsch, W. J., and R. F. Wilson, 1996, Improving the success of wetland creation and restoration with know-how, time, and self-design, Ecological Applications **6**:77–83.

Molles, M. C., Jr., C. S. Crawford, and L. M. Ellis, 1995, Effects of an experimental flood on litter dynamics in the Middle Rio Grande riparian ecosystem, Regulated Rivers: Research and Management **11**:275–281.

Morlan, J. C., and R. E. Frenkel, 1992, The Salmon River Estuary, Restoration and Management Notes **10**:21–23.

Morris, R. C., 1965, Overcup oak (*Quercus lyrata Walt.*), *In* Silvics of Forest Trees of the United States, Fowells, H. A., ed., Agriculture Handbook 271, U.S.D.A. Forest Service, Washington, D.C., pp. 600–602.

Moss, B., 1988, Ecology of Fresh Waters: Man and Medium, Blackwell Scientific Publications, Oxford.

Moss, E. H., 1938, Longevity of seed and establishment of seedlings in species of *Populus*, Botanical Gazette **99**:529–542.

Msangi, J. P., and G. A. Ellenbroek, 1990, Should there be man-made lakes in Africa?, *In* Wetland Ecology and Management: Case Studies. Whigham, D. F., R. E., Good, and J. Kvet, eds., Kluwer, Dordrecht, the Netherlands, pp. 103–116.

Mulamoottil, G., B. G. Warner, and E. A. McBean, 1996, Introduction, *In* Wetlands: Environmental Gradients, Boundaries, and Buffers, Mulamoottil, G., B. G. Warner, and E. A. McBean, eds., Lewis Publishers, Boca Raton, Florida, pp. 1–17.

Muñoz, I., and N. Prat, 1989, Effects of river regulation on the lower Ebro River (NE Spain), Regulated Rivers: Research and Management **3**:345–354.

Murphy, K. J., B. Rørslett, and I. Springuel, 1990, Strategy analysis of submerged lake macrophyte communities: an international example, Aquatic Botany **36**: 303–323.

Murphy, P. G., and A. E. Lugo, 1986, Ecology of a dry tropical forest, Annual Review of Ecology and Systematics **17**:67–88.

Muzika, R. M., J. B. Gladden, and J. D. Haddock, 1987, Structural and functional aspects of succession in southeastern floodplain forests, following a major disturbance, The American Midland Naturalist **117**:1–9.

Myers, R. L., 1983, Site susceptibility to invasion by the exotic tree *Melaleuca quinquenervia* in southern Florida, Journal of Applied Ecology **20**:645–658.

Myers, R. L., 1984, Ecological compression of *Taxodium distichum* var *nutans* by *Melaleuca quinquenervia* in southern Florida, *In* Cypress Swamps, Ewel, K. C., and H. T. Odum, eds., University of Florida Press, Gainesville, Florida, pp. 358–364.

Myers, R. L., 1990, Palm swamps, *In* Forested Wetlands. Lugo, A. E., M. M. Brinson, and S. L. Brown, eds., Elsevier, Amsterdam, pp. 267–286.

Naiman, R. J., 1988, Animal influences on ecosystem dynamics, BioScience **38**: 750–752.

Naiman, R. J., and J. M. Melillo, 1984, Nitrogen budget of a subarctic stream altered by beaver (*Castor canadensis*), Oecologia **62**:150–155.

Naiman, R. J., C. A. Johnston, and J. C. Kelley, 1988a, Alteration of North American streams by beaver, BioScience **38**:753–762.

Naiman, R. J., H. Décamps, J. Pastor, and C. A. Johnston, 1988b, The potential importance of boundaries to fluvial ecosystems, Journal of the North American Benthological Society **7**:289–306.

Naiman, R. J., T. Manning, and C. A. Johnston, 1991, Beaver population fluctuations and tropospheric methane emissions in boreal wetlands, Biogeochemistry **12**:1–15.

Naiman, R. J., D. M. McDowell, and B. S. Farr, 1984, The influence of beaver (*Castor canadensis*) on the production dynamics of aquatic insects, Verhandlungen Internationale Vereinigung für Theoretische und Angewandte Limnologie **22**:1801–1810.

Naiman, R. J., J. M. Melillo, and J. E. Hobbie, 1986, Ecosystem alteration of boreal forest streams by beaver (*Castor canadensis*), Ecology **67**:1254–1269.

Naiman, R. J., and H. Décamps, 1997, The ecology of interfaces: riparian zones, Annual Review of Ecology and Systematics **28**:621–658.

National Research Council, 1982, Ecological Aspects of Development in the Humid Tropics, National Academy Press, Washington, D.C.

National Research Council, 1992, Restoration of Aquatic Ecosystems, National Academy Press, Washington, D.C.

National Research Council, 1995, Wetlands: Characteristics and Boundaries. National Academy Press, Washington, D.C.

Naveh, Z., and A. S. Lieberman, 1984, Landscape Ecology: Theory and Application, Springer-Verlag, New York.

Neill, C., 1990, Effects of nutrients and water levels on species composition in prairie whitetop (*Scolochloa festucacea*) marshes, Canadian Journal of Botany **68**:1015–1020.

Nelson, J. E., and P. Pajak, 1990, Fish habitat restoration following dam removal on a warmwater river, *In* Rivers and Streams Technical Committee, The Restoration of Midwestern Stream Habitat, Proceedings of a Symposium held at the 52nd Midwest Fish and Wildlife Conference, December 4–5, 1989 Minneapolis,

Minnesota, Minnesota Department of Natural Resources, St. Paul, Minnesota, pp. 53–63.

Neuhold, J. M., 1981, Strategy of stream ecosystem recovery, *In* Stress Effects on Natural Ecosystems. Barrett, G. W., and R. Rosenberg, eds., John Wiley & Sons, Chichester, United Kingdom, pp. 261–265.

Newling, C. J., 1993, Restoration of bottomland hardwood forests in the Lower Mississippi Valley, Restoration and Management Notes **8**:23–28.

Newman, S., J. B. Grace, and J. W. Koebel, 1996, Effects of nutrients and hydroperiod on *Typha, Cladium*, and *Eleocharis*: implications for Everglades restoration, Ecological Applications **6**:774–783.

Newson, M. D., 1986, River basin engineering-fluvial geomorphology, Journal of the Institution of Water Engineers and Scientists **40**:307–324.

Newson, D. M., 1994, Hydrology and the River Environment, Clarendon Press, Oxford.

Nichols, S. A., and B. H. Shaw, 1986, Ecological life histories of the three aquatic nuisance plants, *Myriopyllum spicatum, Potamogeton crispus* and *Elodea canadensis*, Hydrobiologia **131**:3–21.

Nielsen, M. B., 1996, River restoration: report of a major EU Life demonstration project, Aquatic Conservation: Marine and Freshwater Ecosystems **6**:187–190.

Niemi, G. J., P. DeVore, N. Detenbeck, D. Taylor, A. Lima, J. Pastor, J. D. Yount, and R. J. Naiman, 1990, Overview of case studies on recovery of aquatic ecosystems from disturbance, Environmental Management **14**:571–588.

Niering, W., 1994, Wetland vegetation change: a dynamic process, Wetland Journal **6**:6–15.

Niering, W. A., 1987, Vegetation dynamics (succession and climax) in relation to plant community management, Conservation Biology **1**:287–295.

Niering, W. A., 1988, Wetlands hydrology and vegetation dynamics, *In* Proceedings of the National Wetland Symposium: Mitigation of Impacts and Losses, Kusler, J. A., ed., October 8–10, 1986, New Orleans, Louisiana, Association of State Wetland Managers, Berne, New York, pp. 320–322.

Nilsson, C., and R. Jansson, 1995, Floristic differences between riparian corridors of regulated and free-flowing boreal rivers, Regulated Rivers: Research and Management **11**:55–66.

Nixon, S. W., 1982, The Ecology of New England High Salt Marshes: A Community Profile, FWS/OBS-81/55, March 1992, U.S. Fish and Wildlife Service, Washington, D.C.

Noble, I. R., and H. Gitay, 1996, A functional classification for predicting the dynamics of landscapes, Journal of Vegetation Science **7**:329–336.

Noble, R. E., and P. K. Murphy, 1975, Short term effects of prolonged backwater flooding on understory vegetation, Castanea **40**:22–28.

Noble, I. R., and R. O. Slatyer, 1977, The effect of disturbance on plant succession, *In* Proceedings of the Conference, Exotic Species in Australia—Their Establishment and Success, Anderson, D., ed., May 19–20, 1977. Adelaide, Australia. Ecological Society of Australia, Canberra, pp. 135–145.

Noble, I. R., and R. O. Slatyer, 1980. The use of vital attributes to predict successional changes in plant communities subject to recurrent disturbances, Vegetatio **43**:5–21.

Nohara, and M. Kimura, 1997, Growth characteristics of *Nelumbo nucifera* Gaertn, in response to water depth and flooding. Ecological Research **12**:11–20.

Noss, R. F., 1993, Wildlife corridors, *In* Ecology of Greenways: Design and Function of Linear Conservation Areas, Smith, D. S., and P. C. Hellmund, eds., University of Minnesota Press, Minneapolis, Minnesota, pp. 43–68.

Nutter, W. L., and J. W. Gaskin, 1989, Role of streamside management zones in controlling discharges to wetlands, *In* Proceedings of the Symposium, The Forested Wetlands of the Southern United States, Hook, D. D., and R. Lea, eds., July 12–14, 1988, Orlando, Florida, U.S.D. A Forest Service, Asheville, North Carolina, pp. 81–84.

Nyman, J. A., and R. H. Chabreck, 1995, Fire in coastal marshes: history and recent concerns, *In* Fire in Wetlands: a Management Perspective; Proceedings of the Tall Timbers Fire Ecology Conference 19, Fire in Wetlands: A Management Perspective, Cerulean, S. I., and R. T. Engstrom, eds., Tall Timbers Research Station, Tallahassee, Florida, pp. 134–141.

O'Connor, N. A., 1991, The effects of habitat complexity on the macroinvertebrates colonising wood substrates in a lowland stream, Oecologia **85**:504–512.

Odum, E. P., 1969, The strategy of ecosystem development, Science **164**:262–270.

Odum, E. P., 1971, Fundamentals of Ecology. W. B. Saunders, Philadelphia.

Odum, H. T., 1989, Experimental study of self-organization in estuarine ponds, *In* Ecological Engineering: An Introduction to Ecotechnology, Mitsch, W. J., and S. E. Jørgensen, eds., John Wiley & Sons, New York, pp. 291–340.

Odum, W. E., 1988, Predicting ecosystem development following creation and restoration of wetlands, *In* Proceedings of a Conference, Increasing Our Wetland Resources, Zelazny, J., and J. S. Feierabend, eds., October 4–7, 1987, Washington, D.C., National Wildlife Federation, Washington D.C., pp. 67–70.

Ogaard, L., and J. A. Leitch, 1981, The Fauna of the Prairie Wetlands: Research Methods and Annotated Bibliography, Agricultural Experiment Station, North Dakota State University, Fargo, North Dakota.

Ohly, J. J., 1987, Untersuchungen über die Eignum der nattürlichen Plflanzenbestände auf den Überschwemmungsgebieten (Várzea) am mittleren Amazonas, Brasilien, als Weide für den Wasserbüffel (Bubalus bubalis) während der terrestrischen Phase des Ökosystems, Göttingen Beiträge zur Tierhygiene in den Tropen und Subtropen Heft 24, Institute für Planzenbau und Tierhygiene in den Tropen und Subtropen, Göttingen, Germany.

O'Keeffe, J. H., R. W. Palmer, B. A. Byren, and B. R. Davies, 1990, The effects of impoundment on the physicochemistry of two contrasting southern African river systems, Regulated Rivers: Research and Management **5**:97–110.

Orians, G. H., and N. E. Pearson, 1979, On the theory of central place foraging, *In* Analysis of Ecological System, Horn, D. J., G. R. Stairs, and R. D. Mitchell, eds., Ohio State University Press, Columbus, Ohio, pp. 155–177.

Osborne, D. J., 1982, The ethylene regulation of cell growth in specific target tissues of plants, *In* Plant Growth Substances 1982, Wareing, P. F., ed., Academic Press, London, pp. 279–290.

Overpeck, J. T., R. S. Webb, and T. Webb, III., 1992, Mapping eastern North American vegetation change of the past 18 ka: no-analogs and the future, Geology **20**:1071–1074.

Pakarinen, P., 1994, Impacts of drainage on Finnish peatlands and their vegetation, International Journal of Ecology and Environmental Sciences **20**:173–183.

Palmer, M. A., R. F. Ambrose, and N. L. Poff, 1997. Ecological theory and community restoration ecology. Restoration Ecology **5**:291–300.

Park, C. C., 1977, Man-induced changes in stream channel capacity, *In* River Channel Changes, Gregory, K. J., ed., John Wiley & Sons, Chichester, United Kingdom, pp. 121–144.

Parker, V. T., 1997. The scale of successional models and restoration objectives. Restoration Ecology **5**:301–306.

Parker, V. T., and M. A. Leck, 1985, Relationships of seed banks to plant distribution patterns in a freshwater tidal wetland, American Journal of Botany **72**: 161–174.

Parker, V. T., R. L. Simpson, and M. A. Leck, 1989, Pattern and process in the dynamics of seed banks. *In* Ecology of Soil Seed Banks. Leck, M. A., V. T. Parker, and R. L. Simpson, eds. Academic Press, San Diego, California, pp. 367–384.

Paterson, L., 1976, An introduction to the ecology and zoo-geography of the Okavango Delta, *In* Proceedings of the Symposium on the Okavango Delta and its Future Utilisation, August 20–September 2, 1976, Gaborone, Botswana, Botswana Society, Gaborone, Botswana, pp. 55–60.

Patrick, W. H., Jr., G. Dissmeyer, D. D. Hook, V. W. Lambou, H. M. Leitman, and C. H. Wharton, 1981, Characteristics of wetlands ecosystems of southeastern bottomland hardwood forests, *In* Proceedings of the Workshop on Bottomland Hardwood Forest Wetlands of Southeastern U.S., Clark, J. R., and J. Benforado, eds., Lake Lanier, Georgia, Elsevier Amsterdam, pp. 276–300.

Payne, N. F., 1992, Techniques for Wildlife Habitat Management of Wetlands, McGraw-Hill, New York.

Pearce, C. M., and L. D. Cordes, 1988, The distribution and ecology of water horsetail (*Equisetum fluviatile*) in northern wetlands, Journal of Freshwater Ecology **4**:383–394.

Pearce, F., 1991, The rivers that won't be tamed, New Scientist (April):38–41.

Pearsall, W. H., 1920, The aquatic vegetation of English lakes, Journal of Ecology **8**:163–201.

Pedersen, A., 1975, Growth measurements of five *Sphagnum* species in South Norway, Norwegian Journal of Botany **22**:277–284.

Pederson, R. L., 1981, Seed bank characteristics of the Delta Marsh, Manitoba: applications for wetland management, *In* Selected Proceedings of the Midwest Conference on Wetland Values and Management, Richardson, B., ed., June 17–19, 1981, St. Paul, Minnesota, Freshwater Society, St. Paul, Minnesota, pp. 61–69.

Penfound, W. T., 1949, Vegetation of Lake Chicot, Louisiana, in relation to wildlife resources. Proceedings of the Louisiana Academy of Science **12**:47–56.

Peterson, R. C., B. L. Madsen, M. A. Wilzbach, C. H. D. Magadza, A. Paarlberg, A. Kullberg and K. W. Cummins et al., 1987, Stream management, emerging global similarities, Ambio **16**:166–179.

Peterson, R. C., L. B.-M. Petersen, and J. Lacoursiére, 1992, A building-block model for stream restoration, *In* River Conservation and Management, Boon, P. J., P. Calow, and G. E. Petts, eds., John Wiley & Sons, Chichester, United Kingdom, pp. 293–309.

Petts, G. E., 1977, Channel response to flow regulation: the case of the River Derwent, Derbyshire, *In* River Channel Changes, Gregory, K. J., ed., John Wiley & Sons, Chichester, United Kingdom, pp. 145–164.

Petts, G. E., 1984, Impounded Rivers: Perspectives for Ecological Management, John Wiley & Sons, Chichester, United Kingdom.

Petts, G. E., and J. Lewin, 1979, Physical effects of reservoirs on river systems, *In* Man's Impact on the Hydrological Cycle in the United Kingdom, Hollis, G. E., ed., Geo Abstracts Ltd., Norwich, United Kingdom, pp. 79–92.

Pezeshki, S. R., 1990, A comparative study of the response of *Taxodium distichum* and *Nyssa aquatica* seedlings to soil anaerobiosis and salinity, Forest Ecology and Management **33/34**:531–541.

Pezeshki, S. R., 1992, Response of *Pinus taeda* L. to soil flooding and salinity, **49**: 149–159.

Pezeshki, S. R., 1994, Plant response to flooding, *In* Plant-Environment Interactions, Wilkinson, R. E., ed., Marcel Dekker, New York, pp. 289–321.

Pezeshki, S. R., R. D. DeLaune, and W. H. Patrick, Jr., 1990a, Differential response of selected mangroves to soil flooding and salinity: gas exchange and biomass partitioning, Canadian Journal of Forest Research **20**:869–874.

Pezeshki, S. R., R. D. Delaune, and W. H. Patrick, Jr., 1990b, Flooding and salt water intrusion: potential effects on survival and productivity of wetland forests along the U.S. Gulf Coast, Forest Ecology and Management **33/34**:287–301.

Pezeshki, S. R., W. H. Patrick, Jr., R. D. Delaune, and E. D. Moser, 1989, Effects of waterlogging and salinity interaction on *Nyssa aquatica* seedllings, Forest Ecology and Management **27**:41–51.

Phillips, R. C., 1982, Seagrass meadows, *In* Creation and Restoration of Coastal Plant Communities, Lewis, R. R., III, ed., CRC Press, Boca Raton, Florida, pp. 173–201.

Pickett, S. T. A., and M. L. Cadenasso, 1995, Landscape ecology: spatial heterogeneity in ecological systems, Science **269**:331–334.

Pickett, S. T. A., and J. N. Thompson, 1978, Patch dynamics and the size of nature reserves, Biological Conservation **13**:27–37.

Pickett, S. T. A., and P. S. White, eds., 1985, The Ecology of Natural Disturbance and Patch Dynamics, Academic Press, San Diego, California.

Piegay, H., 1993, Nature, mass and preferential sites of coarse woody debris deposits in the Lower Ain Valley (Mollon Reach), France, Regulated Rivers: Research and Management **8**:359–372.

Pierce, G. J., A. B. Amerson, and L. R. Becker, Jr., 1982, Pre-1960 floodplain vegetation of the lower Kissimmee River valley, Florida, Biological Service Report 8–23, Environmental Consultant, Inc., Dallas, Texas.

Pimm, S. L., G. E. Davis, L. Loope, C. T. Roman, T. J. Smith, III, and J. T. Tilmant, 1994, Hurricane Andrew, BioScience **44**:224–229.

Pinay, G., A. Fabre, P. Vervier, and F. Gazell, 1992, Control of C, N, P distribution in soils of riparian forests, Landscape Ecology **6**:121–132.

Platt, S. G., and C. G. Brantley, 1993, Switchcane: propagation and establishment in the southeastern United States, Restoration and Management Notes **11**:134–137.

Poels, C. L. M., M. A. van der Gaag, and J. F. J. van de Kerkhoff, 1980, An investigation into the long-term effects of Rhine water on rainbow trout, Water Research **14**:1029–1035.

Poff, N. L., and J. V. Ward, 1990, Physical habitat template of lotic systems: recovery in the context of historical pattern of spatiotemporal heterogeneity, Environmental Management **14**:629–645.

Pollett, F. C., 1979, Report on wetland activities across Canada, *In* Proceedings of a Workshop on Canadian Wetlands, Rubec, C. D. A., and F. C. Pollett, eds., June 11–13, 1979, Saskatoon, Saskatchewan, Ecological Land Classification Series, No. 12, Environment Canada Lands Directorate, Ottawa, pp. 69–76.

Por, F. D., 1995, The Pantanal of Mato Grosso (Brazil) World's Largest Wetlands, Monographiae Biologicae, Volume 73, Dumont, H. J., and M. J.A. Werger, eds., Kluwer, Dordrecht, the Netherlands.

Porsild, A. E., C. R. Harington, and G. A. Mulligan, 1967, *Lupinus arcticus* Wats. grown from seed of Pleistocene age, Science **158**:113–114.

Post, R. A., 1996, Functional profile of black spruce wetlands in Alaska, EPA 910/R-96-006, U.S. Environmental Protection Agency, Region 10, Seattle, Washington.

Power, M. E., W. E. Dietrich, and J. C. Finlay, 1996, Dams and downstream aquatic biodiversity: potential food web consequences of hydrologic and geomorphic change, Environmental Management **20**:887–895.

Prach, K., and P. Pyšek, 1994, Spontaneous establishment of woody plants in central European derelict sites and their potential for reclamation, Restoration Ecology **2**:190–197.

Prat, N., and C. Ibañez, 1995, Effects of water transfers projected in the Spanish National Hydrological Plan on the ecology of the Lower River Ebro (N. E. Spain) and its delta, Water Science and Technology **31**:79–86.

Pressey, R. L., 1990, Wetlands, *In* The Murray, Mackay, N., and D. Eastburn, eds., Murray Darling Basin Commission, Canberra, Australia, pp. 167–181.

Pressey, R. L., and M. J. Middleton, 1982, Impacts of flood mitigation works on coastal wetlands, Wetlands **2**:27–45.

Priester, D. S., 1980, Stump sprouts of swamp and water tupelo produce viable seeds, Southern Journal of Applied Forestry **3**:149–151.

Priestly, D. A., and M. A. Posthumus, 1982, Extreme longevity of lotus seeds from Pulantien, Nature **299**:148–149.

Prigogine, I., 1980, From Being to Becoming, W. H. Freeman, San Francisco, California.

Primack, R. B., 1996, Lessons from ecological theory; dispersal, establishment, and population structure, *In* Restoring Diversity: Strategies for Reintroduction of Endangered Plants, Falk, D. A., C. I. Millar, and M. Olwell, eds., Island Press, Washington, D.C., pp. 209–233.

Proctor, V. W., 1962, Viability of *Chara* oospores taken from migratory water birds, Ecology **45**:565–568.

Prowse, T. D., B. Aitken, M. N. Demuth, and M. Peterson, 1996, Strategies for resorting spring flooding to a drying northern delta, Regulated Rivers: Research and Management **12**:237–250.

Putnam, J. A., G. M. Furnival, and J. S. McKnight, 1960, Management and Inventory of Southern Hardwoods. USDA Agricultural Handbook 181, Washington, D.C.

Pyle, W. H., 1995, Riparian habitat restoration at Hart Mountain National Antelope Refuge, Restoration and Management Notes **13**:40–44.

Pyšek, P., and K. Prach, 1994, How important are rivers for supporting plant invasions? *In* Ecology and Management of Invasive Riverside Plants, de Waal, L. C., L. E. Child, P. M. Wade, and J. H. Brock, eds., John Wiley & Sons, Chichester, United Kingdom, pp. 19–26.

Raffaele, E., 1996, Relationship between seed and spore banks and vegetation of a mountain flood meadow (mallín) in Patagonia, Argentina, Wetlands **16**:1–9.

Ramsar Convention, 1971, Proceedings, International Wildfowl Research Bureau, Slimbridge, United Kingdom.

Raskin, I., and H. Kende, 1985, Mechanism of aeration in rice, Science **228**:327–329.

Raup, H. M., 1957, Vegetation adjustment to the instability of the site, Proceedings, Papers of the Union for the Conservation of Nature and Natural Resources, The Union for the Conservation of Nature and Natural Resources, London, pp. 36–48.

Raven, P., R. F. Evert, and S. E. Eichhorn, 1992, Biology of Plants, 5th ed., Worth, New York.

Reddy, K. N., and M. Singh, 1992, Germination and emergence of hairy beggar-ticks (*Bidens pilosa*), Weed Science **40**:195–199.

Reed, P. B., 1988, National List of Plant Species That Occur in Wetlands: National Summary, U.S. Fish and Wildlife Service, Washington, D.C.

Regier, H. A., R. L. Welcomme, R. J. Steedman, and H. F. Henderson, 1989, Rehabilitation of degraded river ecosystems, *In* Proceedings of the International Large River Symposium (LARS), Dodge, D. P., ed., Honey Harbour, Ontario, Canadian Special Publication of Fisheries and Aquatic Sciences 106 Department of Fisheries and Oceans, Ottawa, pp. 86–97.

Reimold, R. J., 1976, Grazing on wetland meadows, *In* Estuarine Processes. Wiley, M., ed., Academic Press, New York, pp. 219–225.

Reinartz, J. A., and E. L. Warne, 1993, Development of vegetation in small created wetland in southeastern Wisconsin, Wetlands **13**:153–164.

Renshaw, J. F., and W. T. Doolittle, 1965, Yellow-poplar (*Liriodendron tulipifera L.), In* Silvics of Forest Trees of the United States, Fowells, H. A., ed., Agriculture Handbook 271, U.S.D.A. Forest Service, Washington, D.C., pp. 256–265.

Resh, V. H., A. V. Brown, A. P. Covich, M. E. Gurtz, H. W. Li, G. W. Minshall, S. R. Reice, A. L. Sheldon, J. B. Wallace, and R. C. Wissmar, 1988, The role of disturbance in stream ecology, Journal of the American Benthological Society **7**:433–455.

Revkin, A. C., 1997, February 22, PCB's in Hudson are found to persist and to enter the air, The New York Times Online, 4.

Richardson, J. L., and A. E. Richardson, 1972, History of an African rift lake and its climatic implications, Ecological Monographs **42**:499–535.

Ridley, J. E., and J. A. Steel, 1975, Ecological aspects of river impoundments, *In* River Ecology, Whitton, B. A., ed., University of California Press, Berkeley, pp. 565–587.

Risser, P. G., 1992, Landscape ecology approach to ecosystem rehabilitation, *In* Ecosystem Rehabilitation: Preamble to Sustainable Development, Wali, M. K., ed., SPB Academic Publishing, the Hague, the Netherlands, pp. 37–46.

Risser, P. G., 1995, The status of the science of examining ecotones, BioScience **45**:318–325.

Roberts, B. A., and A. Robertson, 1986, Salt marshes of Atlantic Canada: their ecology and distribution, Canadian Journal of Botany **64**:455–467.

Roberts, C. R., 1989, Flood frequency and urban-induced channel change: some British examples, *In* Floods: Hydrological, Sedimentological and Geomorphological Implications, Beven, K., and P. Carling, eds., John Wiley & Sons, Chichester, United Kingdom, pp. 57–82.

Robertson, P. A., 1992, Factors affecting tree growth on three lowland sites in southern Illinios, American Midland Naturalist **128**:218–236.

Robinson, J. V., and J. E. Dickerson, Jr., 1987, Does invasion sequence affect community structure?, Ecology **68**:587–595.

Roche, V., 1993, Peruca Dam sabotaged: Serbia denies responsibility, World Rivers Review **8**:1–8.

Rogers, P., P. Lydon, and D. Seckler, 1989, Eastern Waters Study: Strategies to Manage Flood and Drought in the Ganges-Brahmaputra Basin, April 1989, Irrigation Support Project for Asia and the Near East, Washington, D.C.

Roggeri, H., 1995, Tropical Freshwater Wetlands, Kluwer, Dordrecht, the Netherlands.

Rood, S. B., and J. M. Mahoney, 1990, Collapse of riparian poplar forests downstream from dams in western prairies; probable causes and prospects for mitigation, Environmental Management **14**:451–464.

Roth, L. C., 1992, Hurricanes and mangrove regeneration: effects of Hurricane Joan, October 1988, on the vegetation of Isla del Venado, Bluefields, Nicaragua, Biotropica **24**:375–384.

Rudemann, R., and W. J. Schoonmaker, 1938, Beaver dams as geologic agents, Science **88**:523–525.

Rumburg, C. B., and W. A. Sawyer, 1965, Response of wet meadow vegetation to length and depth of surface water from wild-flood irrigation, Agronomy Journal **57**:245–247.

Rumrill, S. S., and C. E. Cornu, 1995, South Slough coastal watershed restoration: a case study in integrated ecosystem restoration, Restoration and Management Notes **13**:53–57.

Rushmore, F. M., 1965, American beech (*Fagus grandifolia* Ehrh.), *In* Silvics of Forest Trees of the United States, Fowells, H. A., ed., Agriculture Handbook 271, U.S.D.A. Forest Service, Washington, D.C., pp. 172–180.

Russell, H. S., 1982, A Long Deep Furrow: Three Centuries of Farming in New England, University Press of New England, Hanover, New Hampshire.

Rydin, H., and S. Borgegård, 1988, Plant species richness on islands over a century of primary succession: Lake Hjälmaren, Ecology **69**:916–927.

Rydin, H., and S. Borgegård, 1991, Plant characteristics over a century of primary succession on islands: Lake Hjälmaren, Ecology **72**:1089–1101.

Rykiel, E. J., 1979, Ecological disturbances, *In* The Mitigation Symposium: A National Workshop on Mitigation Losses of Fish and Wildlife Habitats, Swanson, G. A., ed., USDA Forest Service General Technical Report RM-65, Fort Collins, Colorado, Fort Collins, Colorado, pp. 624–626.

Rzóska, J., 1974, The Upper Nile swamps, a tropical wetland study, Freshwater Biology **4**:1–30.

Sacco, J. N., F. L. Booker, and E. D. Seneca, 1988, Comparison of the macrofaunal communities of a human-initiated salt marsh at two and fifteen years of age, *In* Proceedings of a Conference, Increasing Our Wetland Resources, Zelazny, J., and J. S. Feierabend, eds., October 4–7, 1988, Washington, D.C., National Wildlife Federation, Washington, D.C. pp. 282–285.

Sale, P. J. M., and R. G. Wetzel, 1983, Growth and metabolism of *Typha* species in relation to cutting treatments, Aquatic Botany **15**:321–334.

Sastroutomo, S. S., I. Ikusima, M. Numata, and S. lizumi, 1979, The importance of turions in the propagation of pondweed (*Potamogeton crispus* L.), Ecological Review (Seitaigaku Kenky U.) **19**:75–88.

Scatena, F. N., S. Moya, C. Estrada, and J. D. Chinea, 1996, The first five years in the reorganization of aboveground biomass and nutrient use following Hurricane Hugo in the Bisley Experimental Watersheds, Luquillo Experimental Forest, Puerto Rico, Biotropica **28**:424–440.

Schlesinger, W. H., 1978, Community structure, dynamics and nutrient cycling in the Okefenokee cypress swamp-forest, Ecological Monographs **48**:43–65.

Schneider, R. L., and R. R. Sharitz, 1986, Seed bank dynamics in a southeastern riverine swamp, American Journal of Botany **73**:1022–1030.

Schneider, R. L., and R. R. Sharitz, 1988, Hydrochory and regeneration in a bald cypress-water tupelo swamp forest, Ecology **69**:1055–1063.

Schnitter, J. J., 1994, A History of Dams: The Useful Pyramids, A. A. Balkema, Rotterdam.

Scholz, H. F., 1965, Slippery elm (*Ulmus rubra* L.), *In* Silvics of Forest Trees of the United States, Fowells, H. A., ed., Agriculture Handbook 271, U.S.D.A. Forest Service, Washington, D.C., pp. 736–739.

Schramm, P., 1992, Prairie restoration: a twenty-five year perspective on establishment, Proceedings of the Twelfth North American Prairie Conference, August 5–9, 1990, Cedar Falls, Iowa, University of Northern Iowa, Cedar Falls, Iowa, pp. 169–178.

Schulte-Wülwer-Leidig, A., 1995, Ecological master plan for the Rhine catchment, *In* The Ecological Basis for River Management, Harper, D. M., and A. J. D. Fergusen, eds., John Wiley & Sons, Chichester, United Kingdom, pp. 505–514.

Schumm, S. A., 1969, River metamorphosis, Journal of the Hydraulics Division, Proceedings on the American Society of Civil Engineers, **441**:255–273.

Schumm, S. A., and R. W. Lichty, 1963, Channel widening and flood-plain construction along Cimarron River in Southwestern Kansas, Geological Survey Professional Paper 352-D:69–88.

Scott, D. A., 1993, Wetlands of West Asia—a regional overview, *In* Wetland and Waterfowl Conservation in South and West Asia, Proceedings of an International Symposium, Moser, M., and J. van Vessem, eds., December 14–20, 1991, Karachi, Pakistan, International Waterfowl and Wetlands Research Bureau Slimbridge, Gloucester, pp. 9–22.

Scott, D. F., and W. Lesch, 1996, The effects of riparian clearing and clearfelling of an indigenous forest on streamflow, stormflow and water quality, Suid-Afrikaanse Bosboutydskrif **175**:1–14.

Scott, M. L., G. T. Auble, J. M. Friedman, L. S. Ishchinger, E. D. Eggleston, M. A. Wondzell, P. B. Shafroth, J. T. Back, and M. S. Jordan, 1993a, Flow Recommendations for Maintaining Riparian Vegetation along the Upper Missouri River, Montana, National Biological Survey, Fort Collins, Colorado.

Scott, M. L., M. A. Wondzell, and G. T. Auble, 1993b, Hydrograph characteristics relevant to the establishment and growth of western riparian vegetation, *In* Proceedings of the Thirteenth Annual American Geophysical Union Hydrology Days, Morel-Seytoux, H. J., ed., Hydrology Days Publications, Atherton, California, pp. 237–245.

Scott, M. L., G. T. Auble, and J. M. Friedman, 1997, Flood dependency of cottonwood establishment along the Missouri River, Montana, USA, Ecological Applications **7**:677–690.

Sculthorpe, C. D., 1967, The Biology of Aquatic Vascular Plants, Edward Arnold, London.

Sedell, J. R., F. H. Everest, and F. J. Swanson, 1982, Fish habitat and streamside management: past and present, *In* Proceedings of the Society of American Foresters National Convention, 1982, Bethesda, Maryland, Society of American Foresters, Washington, D.C., pp. 244–255.

Sedell, J. R., and J. L. Froggatt, 1984, Importance of streamside forest to large rivers: the isolation of the Willamette River, Oregon, U.S.A., from its floodplain by snagging and streamside forest removal, Verhandlungen Internationale Vereinigung für Theoretische und Angewandte Limnologie **22**:1828–1834.

Sedell, J. R., J. E. Richey, and F. J. Swanson, 1989, The River Continuum Concept: a basis for the expected ecosystem behavior of very large rivers?, Canadian Special Publications of Fisheries and Aquatic Science **106**:49–55.

Sedell, J. R., R. J. Steedman, H. A. Regier, and S. V. Gregory, 1991, Restoration of human impacted land-water ecotones, *In* Ecotones: The Role of Landscape Boundaries in the Management and Restoration of Changing Environments, Holland, M. M., P. G. Risser, and R. J. Naiman, eds., Chapman and Hall, London, pp. 110–129.

Segelquist, C. A., M. L. Scott, and G. T. Auble, 1993, Establishment of *Populus deltoides* under simulated alluvial groundwater declines, American Midland Naturalist **130**:274–285.

Sengupta, R., 1995, Fluvial sedimentology of Cypress Creek, Union, Johnson, and Pulaski Counties, southern Illinois, M.S. thesis, Southern Illinois University, Carbondale, Illinois.

Seton, E. T., 1953, Lives of Game Animals, Charles T. Branford, New York.

Shaffer, G. P., C. E. Sasser, J. G. Gosselink, and M. Rejmánek, 1992, Vegetation dynamics in the emerging Atchafalaya Delta, Louisiana, USA., Journal of Ecology **80**:677–687.

Shankman, D., 1991, Forest regeneration on abandoned meanders of a Coastal Plain river in western Tennessee, Castanea **56**:157–167.

Shankman, D., 1993, Channel migration and vegetation patterns in the southeastern coastal plain, Conservation Biology **7**:176–183.

Shankman, D., and L. G. Drake, 1990, Channel migration and regeneration of bald cypress in western Tennessee, Physical Geography **11**:343–352.

Sharitz, R. R., J. E. Irwin, and E. J. Christy, 1974, Vegetation of swamps receiving reactor effluents, Oikos **25**:7–13.

Sharitz, R. R., R. L. Schneider, and L. C. Lee, 1990, Composition and regeneration of a disturbed river floodplain forest in South Carolina, *In* Ecological Processes and Cumulative Impacts: Illustrated By Bottomland Hardwood Wetland Ecosystems, Gosselink, J. G., L. C. Lee, and T. A. Muir, eds., Lewis Publishers, Chelsea, Michigan, pp. 195–218.

Sharma, R. K., and L. A. K. Singh, 1986, Wetland Birds in National Chambal Sanctuary, Crocodile Research Centre of Wildlife Institute of India, Hyderabad.

Shaw, R., 1992, 'Nature', 'culture' and disasters: floods and gender in Bangladesh. Routledge, London.

Sheail, J., and T. C. E. Wells, 1983, The fenlands of Huntingdonshire, England: a case study in catastrophic changes, *In* Ecosystems of the World 4B; Mires: Swamp, Bog, Fen and Moor, Gore, A. J. P., ed., Elsevier, Amsterdam, pp. 375–393.

Shelford, V. E., 1954, Some lower Mississippi Valley flood plain biotic communities: their age and elevation, Ecology **35**:126–142.

Shiel, R. J., 1996, Human population growth and over-utilization of the biotic resources of the Murray-Darling River System, Australia, GeoJournal **40**:101–113.

Shields, F. D., Jr., and J. J. Hoover, 1991, Effects of channel restabilization on habitat diversity, Twentymile Creek, Mississippi, Regulated Rivers: Research and Management **6**:163–181.

Shields, F. D., Jr., and R. H. Smith, 1991, Large woody debris effect on channel friction factor, *In* Hydraulic Engineering, Proceedings of the 1991 National Conference, Shane, R. M. ed., July 29–August 2, 1991, Nashville, Tennessee, American Society of Civil Engineers, New York, pp. 757–762.

Shields, F. D., Jr., and R. H. Smith, 1992, Effects of large woody debris removal on physical characteristics of a sand-bed river, Aquatic Conservation of Marine and Freshwater Ecosystems **2**:145–163.

Shields, F. D., Jr., S. S. Knight, and C. M. Cooper, 1995, Incised stream physical habitat restoration with stone weirs, Regulated Rivers: Research and Management **10**:181–198.

Shipley, B., P. A. Keddy, D. R. J. Moore, and K. Lemky, 1989, Regeneration and establishment strategies of emergent macrophytes, Journal of Ecology **77**:1093–1110.

Shipley, B., and M. Parent, 1991, Germination responses of 64 wetland species in relation to seed size, minimum time to reproduction and seedling relative growth rate, Functional Ecology **5**:111–118.

Shirley, M. A., 1992, Recolonization of a restored red mangrove habitat by fish and macroinvertebrates, *In* Proceedings of the 19th Annual Conference on Wetland Restoration and Creation, May 14–15, 1992, Plant City, Florida, Webb, F., ed., Hillsborough Community College, Tampa, Florida, pp. 159–173.

Shukla, J. B., and B. Dubey, 1996, Effect of changing habitat on species: application to Keoladeo National Park, India, Ecological Modelling **86**:91–99.

Shuman, J. R., 1995, Environmental considerations for assessing dam removal alternatives for river restoration, Regulated Rivers: Research and Management **11**: 249–261.

Shure, D. J., M. R. Gottschalk, and K. A. Parsons, 1986, Litter decomposition processes in a floodplain forest, American Midland Naturalist **15**:314–327.

Seibert, P., 1958, Die pflanzengesellschaften im Naturschurtzgebiet ''Pupplinger Au,'' Landschaftspflege und Vegetationskunde (München) **1**:79 S.

Simberloff, D. S., 1974, Equilibrium theory of island biogeography and ecology, Annual Review of Ecology and Systematics **5**:161–182.

Simberloff, D. S., 1990, Community effects of biological introductions and their implications for restoration, *In* Conservation Sciences Publication No. 2, Towns, D. R., C. H. Daugherty, and I. A. E. Atkinson, eds., Department of Conservation, Wellington, New Zealand, pp. 128–136.

Simberloff, D. S., and E. O. Wilson, 1969, Experimental zoogeography of islands: the colonization of empty islands, Ecology **50**:278–296.

Simons, M. 1997, October 19, Big, bold effort helps bring the Danube back to life. The New York Times on the Web, 4.

Simons, R. K., and D. B. Simons, 1991, Sediment problems associated with dam removal, Muskegon River, Michigan, *In* Hydraulic Engineering, Proceedings of the 1991 National Conference, Shane, R. M., ed., Nashville, Tennessee, American Society of Civil Engineers, New York, pp. 680–691.

Simons, R. K., and D. B. Simons, 1994, An analysis of Platte River channel changes, *In* The Variability of Large Alluvial Rivers, Schumm, S. A., and B. R. Winkley, eds., American Society of Civil Engineers, New York, pp. 341–361.

Singh, S. P., S. S. Pahuja, and M. K. Moolani, 1976, Cultural control of *Typha angustata* at different stages of growth, *In* Aquatic Weeds in Southeast Asia, Varshney, C. K., and J. Rzoska, eds., December 12–17, 1973, New Delhi, India, Dr. W. Junk Publishers, the Hague, the Netherlands, pp. 245–247.

Sisk, T. D., and C. R. Margules, 1993, Habitat edges and restoration: methods for quantifying edge effects and predicting the results of restoration efforts, *In* Nature Conservation 3: Reconstruction of Fragmented Ecosystems Global and Regional Perspectives, Saunders, D. A., R. J. Hobbs, and P. R. Ehrlich, eds., Surrey Beatty and Sons, Chipping Norton, New South Wales, Australia, pp. 57–69.

Sjörs, H., 1963, Bogs and Fens on Attawapiskat River, Northern Ontario, Department of Northern Affairs and National Resources, Ottawa, Canada.

Sklar, L., 1992, The dams come tumbling down, World Rivers Review **8**:9–15.

Sklar, L., 1993, Bangladesh: flood action plan flooded with criticism, World Rivers Review **8**:6–15.

Slootweg, R., and M. L. F. van Schooten, 1995, Partial restoration of floodplain functions at the village level: the experience of Gounougou, Benue Valley, Cameroon, *In* Tropical Freshwater Wetlands, Roggeri, H., ed., Kluwer, Dordrecht, the Netherlands, pp. 159–166.

Smith, J. J., 1989, Recovery of riparian vegetation on an intermittent stream following removal of cattle, *In* Proceedings of the California Riparian Systems Conference Protection, Management for the 1990's, September 22–24, 1988, Davis, California, USDA Forest Service Berkeley, California, pp. 217–221.

Smith, J. J., and P. S. Lake, 1993, The breakdown of buried and surface-placed leaf litter in an upland stream, Hydrobiologia **271**:141–148.

Smith, L. M., and J. A. Kadlec, 1983, Seed banks and their role during drawdown of a North American marsh, Journal of Applied Ecology **20**:673–684.

Smith, M., T. Brandt, and J. Stone, 1995, Effect of soil texture and microtopography on germination and seedling growth in *Boltonia decurrens* (Asteraceae), a threated floodplain species, Wetlands **15**:392–396.

Smith, M., T. Keevin, P. Mettler-McClure, and R. Barkau, 1998, Effect of the flood of 1993 on *Boltonia decurrens*, a rare floodplain plant, Regulated Rivers: Research and Management 14:191–202.

Smith, M., and J. S. Moss, 1998, An experimental investigation using stomatal conductance and fluorescence of the flood sensitivity of *Boltonia decurrens*, and its competitors, Functional Ecology **35**: in press.

Smith, M., Y. Wu, and O. Green, 1993, Effect of light and water-stress on photosynthesis and biomass production in *Boltonia decurrens* (Asteraceae), a threatened species, American Journal of Botany **80**:859–864.

Smith, N. A. F., 1971, A History of Dams, Peter Davies, London.

Smith, P., and J. Smith, 1990, Floodplain vegetation, *In* The Murray, Mackay, N., and D. Eastburn, eds., Murray Darling Basin Commission, Canberra, Australia, pp. 215–228.

Smith, T. J., III, M. B. Robblee, H. R. Wanless, and T. W. Doyle, 1994, Mangroves, hurricanes, and lightning strikes, BioScience **44**:256–262.

Smits, A. J. M., R. van Ruremonde, and G. van der Velde, 1989, Seed dispersal of three nymphaeid macrophytes, Aquatic Botany **35**:167–180.

Smock, L. A., and E. Gilinsky, 1992, Coastal plain blackwater rivers, *In* Biodiversity of the Southeastern United States: Aquatic Communities, Hackney, C. T., S. M. Adams, and W. H. Martin, eds., John Wiley & Sons, New York, pp. 271–311.

Snowden, R. J., 1995, Increased groundwater abstraction in Zanzibar: a threat to mangroves? The Environmentalist **15**:27–40.

Solmsdorf, H., W. Lohmeyer, and W. Mrass, 1975, Ermittlung und Untersuchung des schutzwürdigen und naturnahen Bereiche entlang des Rheins (Schutzwürdigen Bereiche im Rheintal), Schriftenreihe für Landschaftspflege und Naturschutz **11**:1–186.

Sorrell, B. K., and W. Armstrong, 1994, On the difficulties of measuring oxygen release by root systems of wetland plants, Journal of Ecology **82**:177–183.

Sparks, R. E., 1992, Risks of altering the hydrologic regime of large rivers, *In* Predicting Ecosystem Risk. Advances in Modern Environmental Toxicology, Volume XX, Cairns, J., Jr., B. R. Niederlehner, and D. R. Orvos, eds., Princeton Scientific Publishing Company, Princeton, New Jersey, pp. 119–152.

Sparks, R. E., P. B. Bayley, S. L. Kohler, and L. L. Osborne, 1990, Disturbance and recovery of large floodplain rivers, Environmental Management **14**:699–709.

Spencer, J. W., 1995, To what extent can we recreate woodland? *In* The Ecology of Woodland Creation, Ferris-Kaan, R., ed., John Wiley & Sons, Chichester, United Kingdom, pp. 1–16.

Spira, T. P., and L. K. Wagner, 1983, Viability of seeds up to 211 years old extracted from adobe brick buildings of California and northern Mexico, American Journal of Botany **70**:303–307.

Stahle, D. W., and M. K. Cleaveland, 1992, Reconstruction and analysis of spring rainfall over the Southeast U.S. for the past 1000 years, Bulletin of the American Meteorological Society **73**:1947–1961.

Staniforth, R. J., and P. B. Cavers, 1976, An experimental study of water dispersal in *Polygonum* spp., Canadian Journal of Botany **54**:2587–2596.

Steedman, R. J., T. H. Whillans, A. P. Behm, K. E. Bray, K. I. Cullis, M. M. Holland, S. J. Stoddart, and R. J. White, 1996, Use of historical information for conservation and restoration of Great Lakes aquatic habitat, Canadian Journal of Fisheries and Aquatic Sciences **53**:415–423.

Stein, J., ed., 1971, Random House Dictionary of the English Language, Random House, New York.

Stern, W. L., and G. K. Voigt, 1959, Effect of salt concentration on growth of red mangrove in culture, Botanical Gazette **121**:36–39.

Stevens, L. E., J. C. Schmidt, T. J. Ayers, and B. T. Brown, 1995, Flow regulation, geomorphology, and Colorado River Marsh development in the Grand Canyon, Arizona, Ecological Applications **5**:1025–1039.

Stevens, W. K., 1997, February 25, Grand Canyon roars again as ecologic clock is turned back, The New York Times on the Internet, 1–4.

Stocker, B. A., and D. T. Williams, 1991, Sediment modeling of dam removal alternative, Elwha River, Washington, *In* Hydraulic Engineering, Proceedings of the 1991 National Conference, Shane, R. M., ed., July 29–August 2, 1991, Nashville, Tennessee, American Society of Civil Engineers, New York, pp. 674–679.

Stockey, A., and R. Hunt, 1992, Fluctuating water conditions identify niches for germination in *Alisma plantago-aquatica*, Acta Oecologica **13**:227–229.

Stoecker, M. A., M. Smith, and E. D. Melton, 1995, Survival and aerenchyma development under flooded conditions of *Boltonia decurrens*, a threatened floodplains species and *Conyza canadensis*, a widely distributed competitor, American Midland Naturalist **134**:117–126.

Stott, P. A., J. G. Goldammer, and W. L. Werner, 1990, The role of fire in the tropical lowland deciduous forest of Asia, *In* Fire in the Tropical Biota, Goldammer, J. G., ed., Springer-Verlag, Berlin, pp. 32–44.

Streng, D. R., J. S. Glitzenstein, and P. A. Harcombe, 1989, Woody seedling dynamics in an East Texas floodplain forest, Ecological Monographs **59**:177–204.

Strohmeyer, D. L., and L. H. Fredrickson, 1967, An evaluation of dynamited potholes in Northwest Iowa, Journal of Wildlife Management **31**:525–532.

Stromberg, J. C., and D. T. Patten, 1989, Early recovery of an eastern Sierra Nevada riparian system after 40 years of stream diversion, *In* Proceedings of the California Riparian Systems Conference, Davis, California, USDA Forest Service General Technical Report PSW-110, Washington, D.C., pp. 399–404.

Stromberg, J. C., and D. T. Patten, 1996, Instream flow and cottonwood growth in the eastern Sierra Nevada of California, USA., Regulated Rivers: Research and Management **12**:1–12.

Stromberg, J. C., D. T. Patten, and B. D. Richter, 1991, Flood flows and dynamics of Sonoran riparian forests, Rivers **2**:221–235.

Stromberg, J. C., J. A. Tress, S. C. Wilkins, and S. D. Clark, 1992, Response of velvet mesquite to groundwater decline, Journal of Arid Environments **23**:45–58.

Stuckey, R. L., 1980, Distributional history of *Lythrum salicaria* (purple loosestrife) in North America, Bartonia **47**:3–20.

Swales, S., 1989, The use of instream habitat improvement methodology in mitigating the adverse effects of river regulation on fisheries, *In* Alternatives in Regulated River Management, Gore, J. A., and G. E. Petts, eds., CRC Press, Boca Raton, Florida, pp. 185–208.

Swanson, S., and T. Myers, 1994, Streams, geomorphology, riparian vegetation, livestock, and feedback loops: thoughts for riparian grazing by objectives, *In* Proceedings of the Annual Summer Symposium of the American Water Resources Association, Effects of Human-induced Changes on Hydrologic Systems, Marston, R. A., and V. R. Hasfurther, eds., July 26–29, 1994 Jackson Hole, Wyoming, American Water Resources Association, Minneapolis, Minnesota, pp. 255–262.

Szaro, R. C., 1990, Management of dynamic ecosystems: concluding remarks, *In* Management of Dynamic Ecosystems, Proceedings of a Symposium on the 51st Midwest Fish and Wildlife Conference, Springfield, Illinois, Sweeney, J. M., ed., December 5, 1981, The Wildlife Society, North Central Section, West Lafayette, Indiana, pp. 173–180.

Tabacchi, E., 1992, Variabilité des peuplements riverains de L'Adour. Influence de la dynamique fluviale à differentes échelles d'espace et de temps, Ph.D. thesis, Toulouse University, Toulouse, France.

Tang, Z. C., and T. T. Kozlowski, 1982, Some physiological and growth responses of *Betula papyrifera* seedlings to flooding, Physiologia Plantarum **55**:415–420.

Taylor, C. C., 1979, The drainage of Burwell Fen, Cambridgeshire, 1840–1950, *In* The Evolution of Marshland Landscapes, Oxford University Department of External Studies, Oxford, pp. 158–177.

Templeton, A. R., and D. A. Levin, 1979, Evolutionary consequences of gene pools, The American Naturalist **114**:232–249.

Theriot, R. F., 1993, Flood Tolerance of Plant Species in Bottomland Forests of the Southeastern United States, WRP-DE-6, U.S. Army Corps of Engineers, Vicksburg, Mississippi.

Thilenius, J. F., 1989, Woody plant succession on earthquake-uplifted coastal wetlands of the Copper River Delta, Alaska, Forest Ecology and Management **33/34**:439–462.

Thompson, J. R., 1992, Prairies, Forests, and Wetlands: The Restoration of Natural Landscape Communities in Iowa, University of Iowa Press, Iowa City, Iowa.

Thompson, K., and J. P. Grime, 1979, Seasonal variation in the seed banks of herbaceous species in ten contrasting habitats, Journal of Ecology **67**:893–921.

Thompson, K., and A. C. Hamilton, 1983, Peatlands and swamps of the African continent, *In* Ecosystems of the World 4B; Mires: Swamp, Bog, Fen and Moor, Gore, A. J. P., ed., Elsevier, Amsterdam, pp. 331–373.

Thompson, P. A., 1974, Effects of fluctuating temperatures on germination, Journal of Experimental Botany **25**:164–175.

Thompson, S., 1996, Involving indigenous Australians in wetland management, Wildlife Australia (Autumn):26–27.

Thoms, M. C., and K. F. Walker, 1993, Channel changes associated with two adjacent weirs on a regulated lowland alluvial river, Regulated Rivers: Research and Management **8**:271–284.

Tipton, V., and M. Schlinkmann, 1993, August 7, Flood's next stage: cleanup, St. Louis Post-Dispatch, 1–8.

Titus, J. H., 1991, Seed bank of a hardwood floodplain swamp in Florida, Castanea **56**:117–127.

Tofflemire, T. J., 1986, PCB transport in the Ft. Edward area, Northeastern Environmental Science **3**:202–208.

Tolliver, K. S., D. W. Martin, and D. R. Young, 1997, Freshwater and saltwater flooding response for woody species common to barrier island swales, Wetlands **17**:10–18.

Toole, E. R., 1965a, Water oak (*Quercus nigra* L.), *In* Silvics of Forest Trees of the United States, Fowells, H. A. ed., Agriculture Handbook 271, U.S.D.A. Forest Service, Washington, D.C., pp. 528–530.

Toole, E. R., 1965b, Willow oak (*Quercus phellos* L.), *In* Silvics of Forest Trees of the United States, Fowells, H. A. ed., Agriculture Handbook 271, U.S.D.A Forest Service, Washington, D.C., pp. 638–640.

Toth, L. A., 1990, Impacts of channelization on the Kissimmee River ecosystem, *In* Proceedings of the Kissimmee River Restoration Symposium, Loftin, M. K., L. A. Toth, and J. Obeysekera, eds., October 1988, Orlando, Florida, South Florida Water Management District, West Palm Beach, Florida, pp. 47–56.

Toth, L. A., 1991, Environmental Responses to the Kissimmee River Demonstration Project, Technical Publication 91–02, South Florida Water Management District, West Palm Beach, Florida.

Toth, L. A., 1993, The ecological basis for the Kissimmee River restoration plan, Florida Scientist **56**:25–51.

Toth, L. A., 1996, Restoring the hydrogeomorphology of the channelized Kissimmee River, *In* River Channel Restoration: Guiding Principles for Sustainable Projects, Brookes, A., and F. D. Shields, Jr., eds., John Wiley & Sons, Chichester, & United Kingdom, pp. 369–383.

Toth, L. A., D. A. Arrington, M. A. Brady, and D. A. Muszick, 1995, Conceptual evaluation of factors potentially affecting restoration of habitat structure within the channelized Kissimmee River ecosystem, Restoration Ecology **3**:160–180.

Toth, L. A., J. T. B. Obeysekera, W. A. Perkins, and M. K. Loftin, 1993, Flow regulation and restoration of Florida's Kissimmee River, Regulated Rivers: Research and Management **8**:155–166.

Townsend, C. R., 1996, Concepts in river ecology: pattern and process in the catchment hierarchy, Archiv Für Hydrobiologie, Supplement 113, Large Rivers **10**: 3–21.

Trepagnier, C. M., M. A. Kogas, and R. E. Turner, 1995, Evaluation of wetland gain and loss of abandoned agricultural impoundments in South Louisiana, 1978–1988, Restoration Ecology **3**:299–303.

Trexler, J. C., 1995, Restoration of the Kissimmee River: a conceptual model of past and present fish communities and its consequences for evaluating restoration success, Restoration Ecology **3**:195–210.

Tsuyuzaki, S., S. Urano, and T. Tsujii, 1990, Vegetation of alpine marshland and its neighboring areas, northern part of Sichuan Province, China, Vegetatio **88**: 79–86.

Turner, M. G., 1989, Landscape ecology: the effect of pattern on process, Annual Review of Ecology and Systematics **20**:171–197.

Turner, R. E., 1976, Geographic variations in salt marsh macrophyte production: a review, Contributions in Marine Science **20**:47–68.

Turner, R. E., and R. R. Lewis, III, 1997, Hydrologic restoration of coastal wetlands, Wetlands Ecology and Management **4**:65–72.

Turner, R. E., I. A. Mendelssohn, K. L. McKee, R. Costanza, C. Neill, J. P. Sikora, W. B. Sikora, and E. Swenson, 1988, Wetlands hydrology and vegetation dynamics, *In* Proceedings of the National Wetland Symposium: Mitigation of Impacts and Losses, Kusler, J. A., ed., October 8–10, 1986, New Orleans, Association of State Wetland Managers Berne, New York, pp. 135–141.

Tyurnin, B. N., 1984, Factors determining numbers of the river beaver (*Castor fiber*) in the European north, The Soviet Journal of Ecology **14**:337–344.

U.S.A.C.E., 1987, Corps of Engineers Wetlands Delineation Manual, Department of the Army, Vicksburg, Mississippi.

U.S.A.C.E., 1992, Central and South Florida, Kissimmee River, Florida. Final Feasibility Report and Environment Impact Statement: Environmental Restoration of the Kissimmee River, Florida, Jacksonville District, Jacksonville, Florida.

U.S.A.C.E., 1996, Alexander and Pulaski Counties Study: Alternative Habitat Restoration Measures, U.S. Army Corps of Engineers, St. Louis, Missouri.

U.S. Army Corps of Engineers Waterways Experiment Station, 1993, Hydraulic Structures for Wetlands, WRP Technical Note HS-EM-3.1, U.S. Army Corps of Engineers, Vicksburg, Mississippi.

U.S.D.A., 1991, Hydric Soils of the United States, U.S.D.A. Soil Conservation Service, Washington, D.C.

U.S. Department of the Interior, 1996, A Comprehensive Plan for the Restoration of the Everglades, Everglades Information Network, http://everglades.fiu.edu/.

U.S. Fish and Wildlife Service, 1988, Endangered and threatened wildlife and plants, determination of threatened status for *Boltonia decurrens* (decurrent false aster), Federal Register **53**:45858–45861.

Ungar, I. A., and T. E. Riehl, 1980, The effect of seed reserves on species composition in zonal halophyte communities, Botanical Gazette **141**:447–452.

Upper Mississippi River Summit, 1996, Upper Mississippi River Summit Semiannual Report, St. Louis.

van der Pijl, L., 1982, Principles of Dispersal in Higher Plants, Springer-Verlag, Berlin.

van der Valk, A. G., 1981, Succession in wetlands: a Gleasonian approach, Ecology **62**:688–696.

van der Valk, A. G., 1982, Succession in temperate North American wetlands, *In* Wetlands Ecology and Management, Proceedings of the First International Wetlands Conference, Gopal, B., R. E. Turner, R. G. Wetzel, and D. F. Whigham, eds., September 10–17, 1980, New Delhi, India, International Scientific Publications, Jaipur, India, pp. 169–179.

van der Valk, A. G., in press, Succession theory and wetland restoration, *In* Proceedings of INTECOL's V International Wetlands Conference, McComb, A. J., and J. A. Davis, eds., September 22–28, 1996, Perth, Australia, Gleneagles Press, Adelaide.

van der Valk, A. G., and C. B. Davis, 1978, The role of seed banks in the vegetation dynamics of prairie glacial marshes, Ecology **59**:322–335.

van der Valk, A. G., and R. L. Pederson, 1989, Seed banks and the management and restoration of natural vegetation, *In* Ecology of Soil Seed Banks, Leck,

M. A., V. T. Parker, and R. L. Simpson, eds., Academic Press, San Diego, California, pp. 329–346.

van der Valk, A. G., and T. R. Rosburg, 1997, Seed bank composition along a phosphorus gradient in the northern Florida Everglades, Wetlands **17**:228–236.

van der Valk, A. G., and J. T. A. Verhoeven, 1988, Potential role of seed banks and understory species in restoring quaking fens from floating forests, Vegetatio **76**:3–13.

van der Valk, A. G., and C. H. Welling, 1988, The development of zonation in freshwater wetlands, *In* Diversity and Pattern in Plant Communities, During, H. J., M. J. A. Werger, and J. H. Willems, eds., SPB Academic Publishing, the Hague, the Netherlands, pp. 145–158.

van der Velde, G., and L. A. van der Heijden, 1981, The floral biology and seed production of *Nymphoides peltata* (Gmel.) O. Kuntze (Menyanthaceae), Aquatic Botany **10**:261–293.

van Diggelen, R., A. Grootjans, and R. Burkunk, 1994, Assessing restoration perspectives of disturbed brook valleys: the Gorecht Area, The Netherlands, Restoration Ecology **2**:87–96.

van Dijk, G. M., E. C. L. Marteijn, and A. Schulte-Wülwer-Leidig, 1995, Ecological rehabilitation of the River Rhine: plans, progress and perspectives, Regulated Rivers: Research and Management **11**:377–388.

van Rensburg, H. J., 1972, Fire: its effects on grassland, including swamps—southern, central and eastern Africa, *In* Proceedings of the Annual Tall Timbers Fire Ecology Conference **11**:175–199.

van Urk, G., 1984, Lower-Rhine-Meuse, *In* Ecology of European Rivers, Whitton, B. A., ed., Blackwell Scientific Publications, Oxford, pp. 437–468.

van Urk, G., and H. Smit, 1989, The lower Rhine geomorphological changes, *In* Historical Change of Large Alluvial Rivers: Western Europe, Petts, G. E., H. Möller, and A. L. Roux, eds., John Wiley & Sons, Chichester, United Kingdom, pp. 167–182.

van Wieren, S. E., 1991, Management of population of large mammals, *In* The Scientific Management of Temperate Communities for Conservation, The 31st Symposium of the British Ecological Society, Southhampton, Spellerberg, I. F., F. B. Goldsmith, and M. G. Morris, eds., Blackwell, London, pp. 103–127.

van Wijk, R. J., 1983, Life-cycles and reproductive strategies of *Potamogeton pectinatus* L. in the Netherlands and the Camargue (France), *In* International Symposium on Aquatic Macrophytes, Proceedings of the International Symposium on Aquatic Macrophytes, September 18–23, 1983, Nijmegen, the Netherlands, Faculteit der Wiskunde en Natuurwetenschappen, Katholieke Universiteit, Nijmegen, The Netherlands, pp. 317–321.

van Wijk, R. J., 1986, Life cycle characteristics of *Potomogeton pectinatus* L. in relation to control, *In* Proceedings of the European Weed Research Society, 7th International Symposium on Aquatic Weeds, September 15–19, 1986, Loughborough, England, European Weed Research Society, Association of Applied Biologists, Loughborough, England, pp. 375–380.

van Wijk, R. J., 1989, Ecological studies on *Potamogeton pectinatus* L. III. Reproductive strategies and germination ecology, Aquatic Botany **33**:271–299.

van Wilgen, B. W., C. S. Everson, and W. S. W. Trollope, 1990, Fire management in southern Africa: some examples of current objectives, practices and problems, *In* Fire in the Tropical Biota. Goldammer, J. G., ed., Springer-Verlag, Berlin, pp. 179–215.

Vannote, R. L., G. W. Minshall, K. W. Cummins, J. R. Sedell, and C. E. Cushing, 1980, The river continuum concept, Canadian Journal of Fisheries and Aquatic Sciences **37**:130–137.

Vaselaar, R. T., 1997, Opening the flood gates: the 1996 Glen Canyon Dam Experiment, Restoration and Management Notes **15**:119–125.

Veldkamp, E., A. M. Weitz, I. G. Staritsky, and E. J. Huising, 1992, Deforestation trends in the Atlantic Zone of Costa Rica: a case study, Land Degradation and Rehabilitation **3**:71–84.

Vermeer, J. G., and J. H. J. Joosten, 1992, Conservation and management of bog and fen reserves in the Netherlands, *In* Fens and Bogs in the Netherlands: Vegetation, History, Nutrient Dynamics and Conservation, Verhoeven, J. T. A., ed., Kluwer, Dordrecht, the Netherlands, pp. 433–478.

Viereck, L. A., 1973, Wildfire in the taiga of Alaska, Quaternary Research **3**:465–495.

Vivian-Smith, G., and E. W. Stiles, 1994, Dispersal of salt marsh seeds on the feet and feather of waterfowl, Wetlands **14**:316–319.

Voesenek, L. A. C. J., M. C. C. De Graaf, and C. W. P. Blom, 1992, Germination and emergence of *Rumex* in river flood-plains. II. The role of perianth, temperature, light and hypoxia, Acta Botanica Neerlandica **41**:331–343.

Vogl, R. J., 1977, Fire: A destructive menace or natural process, *In* Proceedings of the Conference, Recovery and Restoration of Damaged Ecosystems, Cairns, J., Jr., K. L. Dickson, and E. E. Herricks, eds., Virginia Polytechnic Institute, March 23–25, 1975, University Press of Virginia, Charlottesville, Virginia, pp. 261–289.

Vogl, R. J., 1980, The ecological factors that produce perturbation-dependent ecosystems, *In* The Recovery Process in Damaged Ecosystems, Cairns, J., Jr., ed., Ann Arbor Science, Ann Arbor, Michigan, pp. 63–94.

Vogl, R. J., and L. T. McHargue, 1966, Vegetation of California fan palm oases on the San Andreas Fault, Ecology **47**:532–540.

Vogt, K. A., D. J. Vogt, P. Boon, A. Covich, F. N. Scatena, H. Asbjornsen, J. L. O'Hara, J. Perez, T. G. Siccama, J. Bloomfield, and J. F. Ranciato, 1996, Litter dynamics along stream, riparian and upslope areas following Hurricane Hugo, Luquillo Experimental Forest, Puerto Rico, Biotropica **28**:458–470.

Voigts, D. K., 1976, Aquatic invertebrate abundance in relation to changing marsh vegetation, The American Midland Naturalist **95**:313–322.

Volder, A., A. Bonis, and P. Grillas, 1997, Effects of drought and flooding on the reproduction of an amphibious plant, *Ranunculus peltatus*, Aquatic Botany **58**: 113–120.

Wade, D. D., J. J. Ewel, and R. H. Hofstetter, 1980, Fire in South Florida Ecosystems, Southeast Forest Experiment Station, Asheville, North Carolina.

Wade, P. M., 1990, The colonization of disturbed freshwater habitats by Characeae, Folia Geobotanica et Phytotaxonomica **25**:275–278.

Wahlenberg, W. G., 1960, Loblolly Pine, Duke University, Durham, North Carolina.

Waisel, Y., 1971, Seasonal activity and reproduction behaviour of some submerged hydrophytes in Israel, Hidrobiologia **12**:219–227.

Walker, D., 1970, Direction and rate in some British post-glacial hydroseres, *In* Studies in the Vegetational History of the British Isles, Walker, D., and R. G. West, eds., Cambridge University Press, Cambridge, pp. 117–139.

Walker, K. F., 1985, A review of the ecological effects of river regulation in Australia. Hydrobiologia **125**:111–129.

Walker, K. F., 1986, The Murray-Darling River system, *In* The Ecology of River Systems, Davies, B. R., and K. F. Walker, eds., Dr. W. Junk Publishers, Dordrecht, the Netherlands, pp. 632–659.

Walker, K. F., and M. C. Thoms, 1993, Environmental effects of flow regulation on the Lower River Murray, Australia, Regulated Rivers: Research and Management **8**:103–119.

Walker, K. F., F. Sheldon, and J. T. Puckridge, 1995, A perspective on dryland river ecosystems, Regulated Rivers: Research and Management **11**:85–104.

Walker, K. F., M. C. Thoms, and F. Sheldon, 1992, Effects of weirs on the littoral environment of the River Murray, South Australia, *In* River Conservation and Management, Boon, P. J., P. Calow, and G. E. Petts, eds., John Wiley & Sons, Chichester, United Kingdom, pp. 271–292.

Walker, L. R., J. C. Zasada, and F. S. Chapin, III, 1986, The role of life history process in primary succession on an Alaskan floodplain, Ecology **67**:1243–1253.

Wallace, J. B., and A. C. Benke, 1994, Quantification of wood habitat in subtropical Coastal Plain streams, Canadian Journal of Fisheries and Aquatic Science **41**: 1643–1652.

Warburton, D. B., W. B. Klimstra, and J. R. Nawrot, 1985, Aquatic macrophyte propagation and planting practices for wetland development. *In* Proceedings of the Conference, Wetlands and Water Management on Mined Lands, Brooks, R. P., et al., eds., Pennsylvania State University, University Park, Pennsylvania October 23–24, 1985, University Park, Pennsylvania, pp. 139–152.

Ward, D., N. Holmes, and P. José, eds., 1994, The New Rivers and Wildlife Handbook, The Royal Society for the Protection of Birds, Sandy, Bedfordshire, United Kingdom.

Ward, E., 1942, *Phragmites* management, Transactions of the North American Wildlife Conference **7**:294–298.

Ward, J. V., and J. A. Stanford, 1983, The serial discontinuity concept of lotic ecosystems, *In* Dynamics of Lotic Ecosystems, Fontaine, T. D., III., and S. M. Bartell, eds., Ann Arbor Science, Ann Arbor, Michigan, pp. 29–42.

Ward, J. V., and J. A. Stanford, 1985, The ecology of regulated streams: past accomplishments and directions for future research, *In* Regulated Streams: Advances in Ecology. Craig, J. F., and J. B. Kemper, eds., Plenum Press, New York, pp. 393–409.

Ward, J. V., and J. A. Stanford, 1995a, Ecological connectivity in alluvial river ecosystems and its disruption by flow regulation, Regulated Rivers: Research and Management **11**:105–119.

Ward, J. V., and J. A. Stanford, 1995b, The Serial Discontinuity Concept: extending the model to floodplain rivers, Regulated Rivers: Research and Management **10**: 159–168.

Ward, R., 1978, Floods: A Geographical Perspective, Macmillan, London.

Warne, E. L., 1992, Seed bank and vegetation dynamics in small reconstructed wetlands, M.S. thesis, University of Wisconsin, Milwaukee, Wisconsin.

Warren, R. S., and W. A. Niering, 1993, Vegetation change on a Northeast tidal marsh: interaction of sea-level rise and marsh accretion, Ecology **74**:96–103.

Waser, N. M., R. K. Vickery, Jr., and M. V. Price, 1982, Patterns of seed dispersal and population differentiation in *Mimulus guttatus*, Evolution **36**:753–761.

Webb, E. C., and I. A. Mendelssohn, 1996, Factors affecting vegetation dieback of an oligohaline marsh in coastal Louisiana: field manipulation of salinity and submergence, American Journal of Botany **83**:1429–1434.

Webb, J. W., M. C. Landin, and H. H. Allen, 1988, Approaches and techniques for wetlands development and restoration of dredged material disposal sites, *In* Proceedings of the National Wetland Symposium: Mitigation of Impacts and Losses, Kusler, J. A., M. L. Quammen, and G. Brooks, eds., October 8–10, 1986, New Orleans, Louisiana, Association of State Wetland Managers, Berne, New York, pp. 132–134.

Webster, J. R., and G. M. Simmons, Jr., 1978, Leaf breakdown and invertebrate colonization on a reservoir bottom, Verhandlungen Internationale Vereinigung für Theoretische und Angewandte Limnologie **20**:1587–1596.

Weinhold, C. E., and A. G. van der Valk, 1989, The impact of duration of drainage on the seed banks of northern prairie wetlands, Canadian Journal of Botany **67**: 1878–1884.

Weisner, S. E. B., and B. Ekstam, 1993, Influence of germination time on juvenile performance of *Phragmites australis* on temporarily exposed bottoms-implications for the colonization of lake beds, Aquatic Botany **45**:107–118.

Weisner, S. E. B., and W. Granéli, 1989, Influence of substrate conditions on the growth of *Phragmites australis* after a reduction in oxygen transport to below-ground parts, Aquatic Botany **35**:71–80.

Weitzman, S., and R. J. Hutnik, 1965, Silver maple (*Acer saccharinum* L.), *In* Silvics of Forest Trees of the United States, Fowells, H. A., ed., Agriculture Handbook 271, U.S.D.A. Forest Service, Washington, D.C., pp. 63–65.

Welcomme, R. L., 1974, The role of African flood plains in fisheries, Proceedings of the International Conference on the Conservation of Wetlands and Waterfowl,

Heiligenhafen, Federal Republic of Germany, December 2–4, 1974, International Waterfowl Research Bureau, Slimbridge, United Kingdom, pp. 332–344.

Welcomme, R. L., 1992, River conservation—future prospects, *In* River Conservation and Management, Boon, P. J., P. Calow, and G. E. Petts, eds., John Wiley & Sons, Chichester, United Kingdom, pp. 453–462.

Weller, J. D., 1995, Restoration of a south Florida forested wetland, Ecological Engineering **5**:143–151.

Weller, M. W., 1975, Studies of cattail in relation to management of marsh wildlife, Iowa State Journal of Research, **49**:383–412.

Weller, M. W., 1981, Freshwater Marshes, University of Minnesota Press, Minneapolis, Minnesota.

Weller, M. W., 1995, Use of two waterbird guilds as evaluation tools for the Kissimmee River restoration, Restoration Ecology **3**:211–224.

Weller, M. W., and C. S. Spatcher, 1965, Role of habitat in the distribution and abundance of marsh birds, Iowa Agriculture and Home Economics Experiment Station Special Report 43, Ames, Iowa.

Welsh, B. L., J. P. Herring, and L. M. Read, 1978, The effects of reduced wetlands and storage basins on the stability of a small Connecticut estuary, *In* Estuarine Interactions, Wiley, M. L., ed., Academic Press, New York, pp. 381–401

Wenger, K. F., 1965, Pond pine (*Pinus serotina* Michx.), *In* Silvics of Forest Trees of the United States, USDA Fowells, H. A., ed., Agriculture Handbook 271, U.S.D.A. Forest Service, Washington, D.C., pp. 411–416.

Wenzel, T. A., 1992, Minnesota Wetland Restoration Guide, Minnesota Board of Water and Soil Resources, St. Paul, Minnesota.

Westlake, D. F., 1981, The primary productivity of water plants, *In* Studies on Aquatic Vascular Plants, Symoens, J. J., S. S. Hooper, and P. Compère, eds., Royal Botanical Society of Belgium, Brussels, pp. 165–180.

Wetzel, R. G., 1983, Limnology, W. B. Saunders, Philadelphia.

Wharton, C. H., W. M. Kitchens, and T. W. Sipe, 1982, The Ecology of Bottomland Hardwood Swamps of the Southeast: A Community Profile, FWS/OBS-81/37, U.S. Fish and Wildlife Service, Washington, D.C.

Whelan, R. J., 1995, The Ecology of Fire, Cambridge University Press, Cambridge.

White, K. L., 1965, Shrub-carrs of southeastern Wisconsin. Ecology **46**:286–304.

White, P. S., 1979, Pattern, process, and natural disturbance in vegetation, Botanical Review **45**:229–299.

White, P. S., and S. T. A. Pickett, 1985, Natural disturbance and patch dynamics: an introduction, *In* The Ecology of Natural Disturbance and Patch Dynamics, Pickett, S. T. A., and P. S. White, eds., Academic Press, San Diego, California, pp. 3–13.

Whitehead, D. R., 1972, Developmental and environmental history of the Dismal Swamp, Ecological Monographs **42**:301–315.

Whitlow, T. H., and R. W. Harris, 1979, Flood Tolerance in Plants: A State-of the-art Review, Technical Report E-79-2, U.S. Army Corps of Engineers, Vicksburg, Mississippi.

Whittaker, R. H., 1953, A consideration of climax theory: the climax as a population and pattern, Ecological Monographs **23**:41–78.

Wichman, R. F., 1996, A natural history of the Cache River watershed, M.S. thesis, Southern Illinois University, Carbondale, Illinois.

Wiegers, J., 1985, Succession in Fen Woodland Ecosystems in the Dutch Haf District with Special Reference to *Betula pubescens* Ehrh., J. Cramer, Vaduz, Germany.

Wiegleb, G., and H. Brux, 1991, Comparison of life history characters of broad-leaved species of the genus *Potamogeton* L. I. General characterization of morphology and reproductive strategies, Aquatic Botany **39**:131–146.

Wilcox, D. A., 1995, Wetland and aquatic macrophytes as indicators of anthropogenic hydrologic disturbance, Natural Areas Journal **15**:240–248.

Wilkins, T. M., 1994, Restoration of degraded forested wetland ecosystems, *In* Symposium Proceedings Management of Forested Wetlands in the Central Hardwoods Region, Roberts, S. D., and R. A. Rathfon, eds., October 11–13, 1994. Evansville, Indiana, Department of Forestry and Natural Resources, Purdue University, West Lafayette, Indiana, pp. 101–104.

Willard, D. E., and A. K. Hiller, 1990, Wetland dynamics: consideration for restored and created wetlands, *In* Wetland Creation and Restoration: The Status of the Science, Kusler, J. A., and M. E. Kentula, eds., Island Press, Washington, D.C., pp. 459–466.

Williams, G. P., and M. G. Wolman, 1984, Downstream effects of dams on alluvial rivers, Geological Survey Professional Paper 1286, U.S. Geological Survey, Washington, D.C.

Williams, N., 1997, Dams drain the life out of riverbanks, Science **276**:683.

Willis, C., and W. J. Mitsch, 1995, Effects of hydrology and nutrients on seedling emergence and biomass of aquatic macrophytes from natural and artificial seed banks, Ecological Engineering **4**:65–76.

Williston, H. L., 1962, Pine Planting in a Water Impoundment Area, U.S. Forest Service Experiment Station Southern Forestry Note **137**.

Williston, H. L., F. W. Shropshire, and W. E. Balmer, 1980, Cypress management: a forgotten opportunity, Forestry Report SA-FR 8, USDA. Forest Service, Southeastern Area, Atlanta, Georgia.

Wilson, L. R., 1935, Lake development and plant succession in Vilas County, Wisconsin. Part I. The medium hard water lakes, Ecological Monographs **5**:207–247.

Windisch, A. G., 1987, The role of stream lowlands as firebreaks in the New Jersey Pine Plains region, *In* Atlantic White Cedar Wetlands, Laderman, A. D., ed., Westview Press, Boulder, Colorado, pp. 313–316.

Wind-Mulder, H. L., L. Rochefort, and D. H. Vitt, 1996, Water and peat chemistry comparisons of natural and post-harvested peatlands across Canada and their relevance to peatland restoration, Ecological Engineering **7**:161–181.

Winkley, B. R., E. J. Lesleighter, and J. R. Cooney, 1994, Instability problems of the Arial Khan River, Bangladesh, *In* The Variability of Large Alluvial Rivers,

Schumm, S. A., and B. R. Winkley, eds., American Society of Civil Engineers, New York, pp. 269–284.

Winston, R. B., 1997, Problems associated with reliably designing groundwater-dominated constructed wetlands, Wetland Journal **9**:21–25.

Wisheu, I. C., and P. A. Keddy, 1991, Seed banks of a rare wetland plant community: distrubution patterns and effects of human-induced disturbance, Journal of Vegetation Science **2**:181–188.

Woodhouse, W. W., Jr., and P. L. Knutson, 1982, Atlantic coastal marshes, *In* Creation and Restoration of Coastal Plant Communities, Lewis, R. R., III, ed., CRC Press, Boca Raton, Florida, pp. 45–70.

Woods, F. W., 1965, Live oak (*Quercus virginiana* Mill.). *In* Silvics of Forest Trees of the United States. Fowells, H. A., ed., Agriculture Handbook 271, U.S.D.A. Forest Service, Washington, D.C. pp. 584–587.

Wullschleger, J. G., S. J. Miller, and L. J. Davis, 1990, An evaluation of the effects of the restoration demonstration project on the Kissimmee River fishes, *In* Proceedings of the Kissimmee River Restoration Symposium, Loftin, M. K., L. A. Toth, and J. T. B. Obeysekera, eds., October 1988, Orlando, Florida South Florida Water Management District, West Palm Beach, Florida, pp. 67–81.

Yanosky, T. M., C. R. Hupp, and C. T. Hackney, 1995, Chloride concentration in growth rings of *Taxodium distichum* in a saltwater-intruded estuary, Ecological Applications **5**:785–792.

Yapp, G. A., 1989, Wilderness in Kakadu National Park: aboriginal and other interests, Natural Resources Journal **29**:171–184.

Yeager, L. E., 1949, Effect of permanent flooding in a river-bottom timber area, Illinois Natural History Survey Bulletin **25**:33–65.

Yeo, A. R., and T. J. Flowers, 1980, Salt tolerance in halophyte *Sueda maritima*: evaluation of the effect of salinity upon growth, Journal of Experimental Botany **31**:1171–1183.

Young, P. J., J. P. Megonigal, R. R. Sharitz, and F. P. Day, 1993, False ring formation in baldcypress (*Taxodium distichum*) saplings under two flooding regimes, Wetlands **13**:293–298.

Young, P. J., B. D. Keeland, and R. R. Sharitz, 1995, Growth response of baldcypress (*Taxodium distichum* (L.) Rich.) to an altered hydrologic regime, The American Midland Naturalist **133**:206–212.

Young, W. J., 1991, Flume study of the hydraulic effects of large woody debris in lowland rivers, Regulated Rivers: Research and Management **6**:203–211.

Youngblood, A. P., W. G. Padgett, and A. H. Winward, 1985, Riparian Community Type Classification of Eastern Idaho—Western Wyoming, USDA, Washington, D.C.

Zach, L. W., 1950, A northern climax, forest or muskeg? Ecology **31**:304–306.

Zedler, J. B., 1986, Wetland restoration: trials and errors or ecotechnology?, *In* Wetlands Functions, Rehabilitation, and Creation in the Pacific Northwest: the State of Our Understanding, Proceedings of a conference, April 30–May 2, 1986,

Port Townsend, Washington, April 30–May 2, 1986, Washington State Department of Ecology, Olympia, Washington, pp. 11–16.

Zhou, Z., and X. Pan, 1994, Lower Yellow River, *In* The Variability of Large Alluvial Rivers. Schumm, S. A., and B. R. Winkley, eds., American Society of Civil Engineers, New York, pp. 363–393.

Zimmer, D. W., and R. W. Bachman, 1976, A Study of the Effects of Stream Channelization and Bank Stabilization on Warmwater Sport Fish in Iowa; Subproject No. 4. The Effects of Long-reach Channelization on Habitat and Invertebrate Drift in Some Iowa Streams, Contract No. 14-16-0008-745, U.S. Fish and Wildlife Service, Washington, D.C.

Zinke, A., and K.-A. Gutzweiler, 1990, Possibilities for regeneration of floodplain forests within the framework on the flood-protection measures on the Upper Rhine, West Germany, Forest Ecology and Management **33/34**:13–20.

# Index